Svetlin G. Georgiev, Khaled Zennir

Multiple Integrals in Calculus

Also of Interest

Differential and Integral Calculus.
Implicit Functions, Stieltjes Integrals and Curvilinear Integrals
Svetlin G. Georgiev, Khaled Zennir, 2025
ISBN 978-3-11-914462-9, e-ISBN (PDF) 978-3-11-221808-2,
e-ISBN (EPUB) 978-3-11-221831-0

Partial Dynamic Equations. Wave, Parabolic and Elliptic Equations on Time Scales
Svetlin G. Georgiev, 2025
ISBN 978-3-11-163551-4, e-ISBN (PDF) 978-3-11-163614-6,
e-ISBN (EPUB) 978-3-11-163615-3

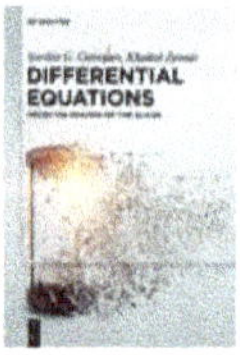

Differential Equations. Projector Analysis on Time Scales
Svetlin G. Georgiev, Khaled Zennir, 2024
ISBN 978-3-11-137509-0, e-ISBN (PDF) 978-3-11-137715-5,
e-ISBN (EPUB) 978-3-11-137771-1

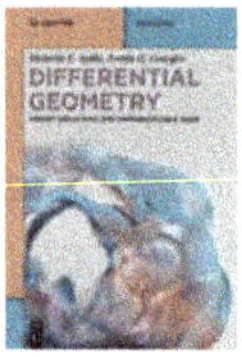

Differential Geometry. Frenet Equations and Differentiable Maps
Muhittin E. Aydin, Svetlin G. Georgiev, 2024
ISBN 978-3-11-150089-8, e-ISBN (PDF) 978-3-11-150185-7,
e-ISBN (EPUB) 978-3-11-150223-6

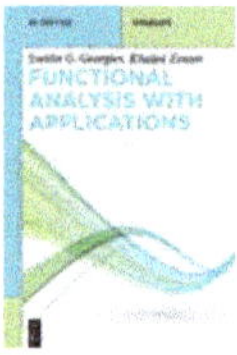

Functional Analysis with Applications
Svetlin G. Georgiev, Khaled Zennir, 2019
ISBN 978-3-11-065769-2, e-ISBN (PDF) 978-3-11-065772-2,
e-ISBN (EPUB) 978-3-11-065804-0

Svetlin G. Georgiev, Khaled Zennir

Multiple Integrals in Calculus

Improper Integrals, Line Integrals, Surface Integrals

DE GRUYTER

Mathematics Subject Classification 2020
Primary: 26-00, 26-01; Secondary: 26B15, 26B20

Authors
Dr. Svetlin G. Georgiev
Sorbonne University
1 Rue Victor Hugo
70005 Paris
France
svetlingeorgiev1@gmail.com

Dr. Khaled Zennir
Qassim University
Department of Mathematics
Ar-Rass
Qassim 51921
Saudi Arabia
khaledzennir4@gmail.com

ISBN 978-3-11-914345-5
e-ISBN (PDF) 978-3-11-221960-7
e-ISBN (EPUB) 978-3-11-221971-3

Library of Congress Control Number: 2025941718

Bibliographic information published by the Deutsche Nationalbibliothek
The Deutsche Nationalbibliothek lists this publication in the Deutsche Nationalbibliografie; detailed bibliographic data are available on the Internet at http://dnb.dnb.de.

Cover image: peepo / E+ / Getty Images
Typesetting: VTeX UAB, Lithuania

www.degruyter.com
Questions about General Product Safety Regulation:
productsafety@degruyterbrill.com

Preface

This book presents an introduction to the theory of functions of several variables. The book is primarily intended for senior undergraduate students and beginning graduate students of engineering and science courses. Students in mathematical and physical sciences will find many sections of direct relevance.

This book contains five chapters, each chapter consisting of results with their proofs, numerous examples, and exercises with solutions. Each chapter concludes with a section featuring advanced practical problems with solutions followed by a section on notes and references, explaining its context within existing literature. Many examples, exercises, and problems are included, provided with detailed explanations, solutions, or answers.

In Chapter 1, the concepts of Jordan measure and measurable sets are introduced. Some of their properties are deduced. Multiple integrals are defined and explored. Some criteria for integrability are given. Some mean value theorems are proved. In Chapter 2, some methods for computing multiple integrals are provided. The Taylor formula is deduced. Linear maps on measurable sets are investigated. Metric properties of differentiable maps are deduced. In Chapter 3, improper multiple integrals are introduced. Some criteria for their convergence are deduced. Absolute convergence of improper multiple integrals is defined and investigated. In Chapter 4, we introduce curvilinear integrals of the first and second kinds. Some of their basic properties are deduced. The Green theorem is formulated and proved. In Chapter 5, surface integrals of the first and second kinds are defined, and some of their properties are listed. The Gauss–Ostrogradsky and Stokes formulas are formulated and proved. Solutions, hints, and answers to the exercises and problems are given.

The aim of this book is to present a clear and well-organized treatment of the concepts behind the development of mathematics as well as solution techniques. The text material of this book is presented in a readable and mathematically solid format.

Paris, July 2025

Svetlin G. Georgiev
Khaled Zennir

https://doi.org/10.1515/9783112219607-202

Contents

1 Multiple integrals

In this chapter, the concepts for Jordan measure and measurable sets are introduced. Some of their properties are deduced. Multiple integrals are defined and explored. Some criteria for integrability are given. Some mean value theorems are proved.

1.1 Jordan measure. Jordan-measurable sets

1.1.1 Elementary sets

We start with some definitions needed for the construction of the so-called elementary sets.

Definition 1.1. The length of an interval I with endpoints $a \leq b$ is defined as

$$|I| = b - a.$$

Notice that, as we would expect, the length of an interval depends only on the endpoints. In particular, the intervals

$$[a,b], \quad [a,b), \quad (a,b], \quad (a,b)$$

all have the same length.

Definition 1.2. Let $B = I_1 \times I_2 \times \cdots \times I_m$, where $I_j, j \in \{1,\ldots,m\}$, are intervals. We define the volume of B as

$$|B| = \prod_{j=1}^{m} |I_j|.$$

We now move to the main object of this section, the class of sets that can be constructed using finite unions of rectangles.

Definition 1.3. A set $E \subset \mathbb{R}^n$ is said to be an elementary set if it can be written as a finite union of rectangles. By $\mathscr{E}_n$ we denote the class of elementary sets in $\mathbb{R}^n$.

When we try to combine elementary sets by taking finitely many unions and intersections, we notice that the resulting sets are still elementary. This fact is easy to visualize geometrically when $n = 2$ and can be proven for any $n \in \mathbb{N}$, as we will see in the next result. The idea of the proof is really to think as if n was 2 and to use those rectangles similarly in any dimension.

https://doi.org/10.1515/9783112219607-001

Theorem 1.1 (Boolean[1] closure of $\mathscr{E}_n$). *Let $E, R \subset \mathbb{R}^n$ be elementary sets. Then*

$$E \cup F, \quad E \cap F, \quad E \backslash F, \quad \text{and} \quad E \triangle F$$

are elementary sets. Moreover, the translation of E by $x \in \mathbb{R}^n$,

$$E + x = \{y + x : y \in E\},$$

is an elementary set.

Proof. Suppose that

$$\begin{aligned} E &= B_1 \cup \cdots \cup B_m, \\ F &= C_1 \cup \cdots \cup C_l, \end{aligned}$$

where $B_j, j \in \{1, \ldots, m\}$, and C_k, $k \in \{1, \ldots, l\}$, are rectangles. Firstly, we will prove that $E \cup F \in \mathscr{E}_n$. We have

$$E \cup F = B_1 \cup \cdots \cup B_m \cup C_1 \cup \cdots \cup C_l,$$

which is a finite union of rectangles, and so we directly get the aimed result. Now we consider $E \cap F$. We have

$$\begin{aligned} E \cap F &= (B_1 \cup \cdots \cup B_m) \cap (C_1 \cup \cdots \cup C_l) \\ &= \bigcup_{j=1}^{m} B_j \cap (C_1 \cup \cdots \cup C_l) \\ &= \bigcup_{j=m}^{l} \bigcup_{k=1}^{l} (B_j \cap C_k). \end{aligned}$$

Note that $B_j \cap C_k$ is the empty set or a rectangle, $j \in \{1, \ldots, m\}$, $k \in \{1, \ldots, l\}$. Thus $E \cap F$ is a union of a finite number of rectangles and thus is an elementary set. Now consider $E \backslash F$. Firstly, observe that the difference of two rectangles is generally not a rectangle (see Figure 1.1). We have

---

1 George Boole (November 1815–8 December 1864) was a largely self-taught English mathematician, philosopher, and logician, most of whose short career was spent as the first professor of mathematics at Queen's College, Cork in Ireland. He worked in the fields of differential equations and algebraic logic and is best known as the author of The Laws of Thought, which contains Boolean algebra. Boolean logic, essential to computer programming, and is credited with laying the foundations for the Information Age alongside the work of Claude Shannon.

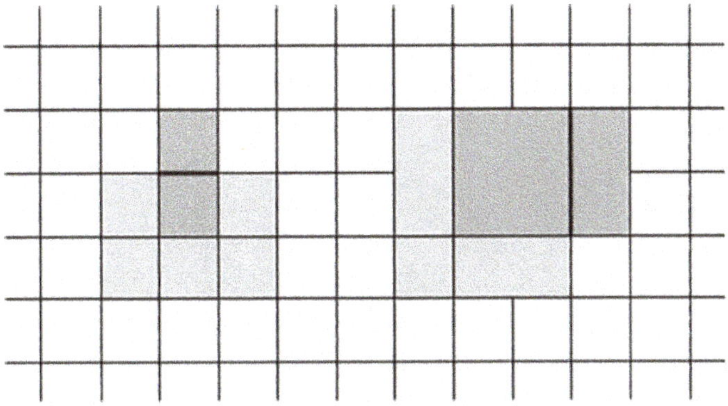

Figure 1.1: Possible shapes of $E\backslash F$.

$$\begin{aligned} E\backslash F &= \left(\bigcup_{j=1}^{m} B_j\right)\Big\backslash\left(\bigcup_{k=1}^{l} C_k\right) \\ &= \left(\bigcup_{j=1}^{m} B_j\right)\cap\left(\bigcup_{k=1}^{l} C_k\right)^c \\ &= \left(\bigcup_{j=1}^{m} B_j\right)\cap\left(\bigcup_{k=1}^{l} C_k^c\right) \\ &= \bigcup_{j=1}^{m}\bigcup_{k=1}^{l}(B_j\cap C_k^c) \\ &= \bigcup_{j=1}^{m}\bigcup_{k=1}^{l}(B_j\backslash C_k), \end{aligned}$$

i. e.,

$$E\backslash F = \bigcup_{j=1}^{m}\bigcup_{k=1}^{l}(B_j\backslash C_k). \tag{1.1}$$

Fix $j \in \{1,\dots,m\}$, $k \in \{1,\dots,l\}$. We divide $B_j \cap C_k$ into small disjoint rectangles $D_1,\dots,D_N$ so that any D_r, $r \in \{1,\dots,N\}$, is fully contained in $B_j\backslash C_k$ or fully contained in C_k (see Figure 1.1). Therefore $B_j\backslash C_k$ is a union of a finite number of rectangles and thus is an elementary set. Now using (1.1), we conclude that $E\backslash F$ is an elementary set. As above, $F\backslash E$ is an elementary set, and then

$$E \triangle F = (E\backslash F) \cup (F\backslash E)$$

is an elementary set. Now fix $x = (x_1,\dots,x_n) \in \mathbb{R}^n$ and consider $E + x$. We notice that

$$E + x = \bigcup_{j=1}^{m}(B_j + x)$$

and if

$$B_j = [a_1^j, b_1^j] \times [a_2^j, b_2^j] \times \cdots \times [a_n^j, b_n^j], \quad j \in \{1, \ldots, m\},$$

then

$$B_j + x = [a_1^j + x_1, b_1^j + x_1] \times [a_2^j + x_2, b_2^j + x_2] \times \cdots \times [a_n^j + x_n, b_n^j + x_n], \quad j \in \{1, \ldots, m\},$$

are rectangles, and $E + x$ is a union of a finite number of rectangles, i. e., $E + x$ is an elementary set. This completes the proof. □

The next result will allow us to properly define the measure of an elementary set.

Theorem 1.2. *Any elementary set $E \subset \mathbb{R}^n$ can be written as a disjoint union of rectangles.*

Proof. The main idea of the proof of this result is to divide the set E into small enough rectangles so that it is equal to the union of these rectangles. We start with the case $n = 1$. Let

$$E = I_1 \cup \cdots \cup I_m,$$

where $I_j, j \in \{1, \ldots, m\}$, have endpoints a_{2j-1} and a_{2j}, $j \in \{1, \ldots, m\}$. We write these $2m$ points $a_1, \ldots, a_{2m}$ in ascending order:

$$a_{(1)} \le a_{(2)} \le \cdots \le a_{(2m)}.$$

Let J_j have endpoints $a_{(j)}$ and $a_{(j+1)}$, $j \in \{1, \ldots, m\}$. Then each I_k, $k \in \{1, \ldots, m\}$, can be written as a disjoint union of elements of the family $J_1, \ldots, J_m$, and hence E can be represented as a disjoint union of rectangles. Note that for each $j \in \{1, \ldots, m\}$, we only need to decide correctly which boundaries of J_j must be open and which closed. Now consider the case $n \ge 1$. Suppose that

$$E = R_1 \cup \cdots \cup R_l,$$

where

$$R_j = I_{j1} \times \cdots \times I_{jd_j}, \quad j \in \{1, \ldots, m\}.$$

For each I_{jk}, $k \in \{1, \ldots, d_j\}$, $j \in \{1, \ldots, m\}$, we proceed as in the case $n = 1$, and we represent it as a disjoint union of intervals

$$I_{jk} = J_{1jk} \cup \cdots \cup J_{d_{jk}jk}, \quad k \in \{1, \ldots, d_j\}, \; j \in \{1, \ldots, m\}.$$

Thus R_j, $j \in \{1, \ldots, m\}$, is represented as a disjoint union of rectangles, and then E is represented as a disjoint union of rectangles. This completes the proof. □

Now we want to give a measure to each elementary set. The idea is simply to extend the concept of volume of rectangles to elementary sets. We have the following result.

Theorem 1.3. *Assume that $E \in \mathscr{E}_n$. can be represented as a disjoint union of rectangles, i. e.,*

$$E = B_1 \cup \cdots \cup B_r.$$

Set

$$m(E) = |B_1| + \cdots + |B_r|.$$

Then the map $m : \mathscr{E}_n \to \mathbb{R}$ is well defined in the sense that it does not depend on the choice of the rectangles $B_1, \ldots, B_m$.

Proof. Let $C_1, \ldots, C_l$ be a collection of disjoint rectangles such that

$$E = C_1 \cup \cdots \cup C_l.$$

Set

$$A_{jk} = B_j \cap C_k, \quad j \in \{1, \ldots, r\},\ k \in \{1, \ldots, l\},$$

which are rectangles. Observe that

$$\begin{aligned} B_j &= E \cap B_j \\ &= \left(\bigcup_{k=1}^{l} C_k \right) \cap B_j \\ &= \bigcup_{k=1}^{l} (C_k \cap B_j) \\ &= \bigcup_{k=1}^{l} A_{jk}, \quad j \in \{1, \ldots, r\}, \end{aligned}$$

and

$$\begin{aligned} C_k &= E \cap C_k \\ &= \left(\bigcup_{j=1}^{r} B_j \right) \cap C_k \\ &= \bigcup_{j=1}^{r} (C_k \cap B_j) \\ &= \bigcup_{j=1}^{r} A_{jk}, \quad k \in \{1, \ldots, l\}, \end{aligned}$$

and $A_{jk}, j \in \{1, \ldots, r\}, k \in \{1, \ldots, l\}$, are disjoint rectangles. Therefore

$$\left|\bigcup_{j=1}^{r} A_{jk}\right| = \sum_{j=1}^{r} |A_{jk}|,$$

and

$$\left|\bigcup_{k=1}^{l} A_{jk}\right| = \sum_{k=1}^{l} |A_{jk}|,$$

and hence

$$\begin{aligned}\left|\bigcup_{j=1}^{r} B_j\right| &= \sum_{j=1}^{r} |B_j| \\ &= \sum_{j=1}^{r}\left|\bigcup_{k=1}^{l} A_{jk}\right| \\ &= \sum_{j=1}^{r}\sum_{k=1}^{l} A_{jk} \\ &= \sum_{k=1}^{l}\sum_{j=1}^{r} A_{jk} \\ &= \sum_{k=1}^{l}\left|\bigcup_{j=1}^{r} A_{jk}\right| \\ &= \sum_{k=1}^{l} |C_k| \\ &= \left|\bigcup_{k=1}^{l} C_k\right|.\end{aligned}$$

This completes the proof. □

Definition 1.4. The map m defined in Theorem 1.3 is called the measure of a set E.

It is natural to ask about the properties of the defined mapping. In particular, we would like to understand if the defined measure of an elementary set is consistent with the previously defined volume. Some of the basic properties of the defined measure are summarized in the following result.

Theorem 1.4. *Let $E, F \subset \mathscr{E}_n$. Then the map $m : \mathscr{E}_n \to \mathbb{R}$ satisfies the following properties:*
1. $m(\emptyset) = 0$.
2. $m(B) = |B|$, *provided that B is a rectangle.*

3. *Additivity:*

$$m(E \cup F) = m(E) + m(F),$$

   provided that $E \cap F = \emptyset$.
4. *Monotonicity:*

$$m(E) \le m(F),$$

   provided that $E \subset F$.
5. *Sub-additivity*

$$m(E \cup F) \le m(E) + m(F).$$

6. *Translation invariance:*

$$m(E + x) = m(E)$$

   for all $x \in \mathbb{R}^n$.
7. *Normalization:*

$$m([0,1]^n) = 1.$$

Proof. 1. This property follows directly from the fact that the empty set can be represented as a union of disjoint rectangles of zero volumes.
2. Note that B can be represented as a union of disjoint rectangles $B_1, \dots, B_r$ by choosing $r = 1$ and $B_1 = B$, and using the independence of the choice of the rectangles, we arrive at

$$m(B) = |B|.$$

3. Let $E \cap F = \emptyset$, and let

$$E = B_1 \cup \cdots \cup B_r,$$
$$F = C_1 \cup \cdots \cup C_l$$

   be disjoint unions of rectangles. Then

$$E \cup F = B_1 \cup \cdots \cup B_r \cup C_1 \cup \cdots \cup C_l,$$

   and

$$\begin{aligned} m(E \cup F) &= |B_1| + \cdots + |B_r| + |C_1| + \cdots + |C_l| \\ &= m(E) + m(F). \end{aligned}$$

4. We have that

$$(F\backslash E) \cap (F \cap E) = \emptyset,$$
$$F \cap E = E,$$

and

$$F = (F\backslash E) \cup (F \cap E).$$

Then

$$\begin{aligned} m(F) &= m\big((F\backslash E) \cup (F \cap E)\big) \\ &= m(F\backslash E) + m(F \cap E) \\ &\geq m(F \cap E) \\ &= m(E). \end{aligned}$$

5. Note that

$$E \cup F = F \cup (E\backslash F),$$
$$m(E\backslash F) \leq m(E),$$

and

$$F \cap (E\backslash F) = \emptyset.$$

Then

$$\begin{aligned} m(E \cup F) &= m\big(F \cup (E\backslash F)\big) \\ &= m(F) + m(E\backslash F) \\ &\leq m(E) + m(F). \end{aligned}$$

6. Let

$$E = B_1 \cup \cdots \cup B_r,$$

be a disjoint union of rectangles. Then

$$E + x = (B_1 + x) \cup \cdots \cup (B_r + x).$$

Using that

$$|B_j + x| = |B_j|, \quad j \in \{1, \dots, r\},$$

we get

$$\begin{aligned} m(E + x) &= |B_1 + x| + \cdots + |B_r + x| \\ &= |B_1| + \cdots + |B_r| \\ &= m(E). \end{aligned}$$

7. By the second property we find

$$\begin{aligned} m([0,1]^n) &= \big|[0,1]^n\big| \\ &= 1. \end{aligned}$$

This completes the proof. □

Before proving the next important property of the measure, we need the following auxiliary result.

Lemma 1.1. *Let $R = \prod_{j=1}^{n}(a_j, b_j)$ and $R_\varepsilon = \prod_{j=1}^{n}(a_j + \varepsilon, b_j - \varepsilon)$ for $\varepsilon > 0$. Then*

$$m(R \backslash R_\varepsilon) \le 2\varepsilon n M^{n-1},$$

where

$$M = \max_{j \in \{1,\ldots,n\}} (b_j - a_j).$$

Moreover, if $\{B_n\}_{n\in\mathbb{N}}$ is a sequence of rectangles such that for any $\varepsilon > 0$, there is $N \in \mathbb{N}$ such that

$$R_\varepsilon \subset B_n \subset R \quad \textit{for all } n \ge N,$$

then

$$|B_n| \to |R| \quad \textit{as } n \to \infty.$$

Proof. Define

$$A_j^1 = (a_1, b_1) \times \cdots \times (a_{j-1}, b_{j-1}) \times [b_j - \varepsilon, b_j) \times (a_{j+1}, b_{j+1}) \times \cdots \times (a_n, b_n), \quad j \in \{1, \ldots, n\},$$

and

$$A_j^2 = (a_1, b_1) \times \cdots \times (a_{j-1}, b_{j-1}) \times (a_j, a_j + \varepsilon] \times (a_{j+1}, b_{j+1}) \times \cdots \times (a_n, b_n), \quad j \in \{1, \ldots, n\}.$$

Observe that

$$R \backslash R_\varepsilon \subset \bigcup_{j=1}^{n} (A_j^1 \cup A_j^2).$$

Then using the sub-additivity of the measure, we get

$$\begin{aligned} m(R \backslash R_\varepsilon) &\leq \sum_{j=1}^{n} m(A_j^1 \cup A_j^2) \\ &\leq \sum_{j=1}^{n} 2\Big(\varepsilon \prod_{k \neq j} (b_k - a_k)\Big) \\ &\leq \sum_{j=1}^{n} 2\varepsilon M^{n-1} \\ &= 2\varepsilon n M^{n-1}. \end{aligned}$$

Next, we have that

$$\begin{aligned} m(R) - m(R_\varepsilon) &= m(R \backslash R_\varepsilon) \\ &\to 0 \quad \text{as } \varepsilon \to 0. \end{aligned}$$

Then, for any n large enough, we have

$$R_\varepsilon \subset B_n \subset R,$$

and therefore by the monotonicity of the measure we get

$$\begin{aligned} m(R) &\geq \limsup_{n \to \infty} m(B_n) \\ &\geq \liminf_{n \to \infty} m(B_n) \\ &\geq m(R_\varepsilon). \end{aligned}$$

Taking the limit as $\varepsilon \to 0$, we obtain the desired result. This completes the proof. □

There is another interesting feature: some of the properties of the measure in the last result determine it in a unique way.

Theorem 1.5. *Let $m_1 : \mathscr{E}_n \to \mathbb{R}$ be a mapping satisfying finite additivity, translation invariance, and normalization. Then $m_1 = m$.*

Proof. Firstly, we will prove this for the sets $[p, q]^n$, where $p, q \in \mathbb{Q}$, and then we will extend this result by using the density of $\mathbb{Q}$ in $\mathbb{R}$. For this aim, fix $l \in \mathbb{N}$ and observe that

$$[0,1) = \bigcup_{j=1}^{l} \left[\frac{j-1}{l}, \frac{j}{l}\right)$$

and

$$[0,1]^n = \bigcup_{j_1,\ldots,j_d \in \{1,\ldots,l\}} \left[\frac{j_1 - 1}{l}, \frac{j_1}{l}\right) \times \cdots \times \left[\frac{j_d - 1}{l}, \frac{j_d}{l}\right). \tag{1.2}$$

Since m_1 satisfies the translation invariance, for any $j_1, \ldots, j_d \in \{1, \ldots, l\}$, we have

$$m_1\left(\left[0, \frac{1}{l}\right)^n\right) = m_1\left(\left[\frac{j_1 - 1}{l}, \frac{j_1}{l}\right) \times \cdots \times \left[\frac{j_d - 1}{l}, \frac{j_d}{l}\right)\right).$$

Now using the disjoint union in (1.2), we get

$$m_1([0,1]^n) = l^n m_1\left(\left[0, \frac{1}{l}\right)^n\right),$$

and for any $k \in \{0, 1, \ldots, l\}$, we have

$$\begin{aligned} m_1\left(\left[0, \frac{k}{l}\right)^n\right) &= k^n m_1\left(\left[0, \frac{1}{l}\right)^n\right) \\ &= \frac{k^n}{l^n} m_1([0,1]^n) \\ &= \frac{k^n}{l^n}, \end{aligned}$$

where we have used the normalization property of m_1. Thus, for any $q \in Q$, we have

$$m_1([0,q]^n) = q^n.$$

Moreover, since m_1 satisfies the translation invariance, for any $a_j, b_j \in \mathbb{Q}, j \in \{1, \ldots, n\}$, with $x = (-a_1, \ldots, -a_n)$, we get

$$m_1([a_1, b_1) \times \cdots \times [a_n, b_n)) = \prod_{j=1}^{n} (b_j - a_j).$$

Now, consider the rectangle

$$R = I_1 \times \cdots \times I_n,$$

where I_j are intervals with end points $a_j < b_j, j \in \{1, \ldots, n\}$. Take arbitrary $\varepsilon > 0$. By the density of Q in $\mathbb{R}$ it follows that there are $a_j^-, a_j^+, b_j^-, b_j^+ \in \mathbb{Q}$ such that

$$\begin{aligned} a_j^- < a_j < a_j^+, \\ b_j^- < b_j < b_j^+, \\ |a_j^\pm - a_j| < \varepsilon, \\ |b_j^\pm - b_j| < \varepsilon, \quad j \in \{1, \ldots, n\}. \end{aligned}$$

Then we define

$$I_j^\pm = [a_j^\mp, b_j^\pm), \quad j \in \{1, \ldots, n\},$$

and

$$R^{\pm} = I_1^{\pm} \times \cdots \times I_n^{\pm}.$$

We have that

$$R^{-} \subset R \subset R^{+},$$

and by the monotonicity of m_1 we arrive at

$$m_1(R^{-}) \leq m_1(R) \leq m_1(R^{+}).$$

By the above considerations we get

$$\begin{aligned} m_1(R^{+}) &= \prod_{j=1}^{n}(b_j^{+} - a_j^{-}) \\ &\leq \prod_{j=1}^{n}(b_j - a_j + 2\varepsilon) \end{aligned}$$

and

$$\begin{aligned} m_1(R^{-}) &= \prod_{j=1}^{n}(b_j^{-} - a_j^{+}) \\ &\leq \prod_{j=1}^{n}(b_j - a_j - 2\varepsilon). \end{aligned}$$

Let

$$M = \max_{j\in\{1,\dots,n\}}(b_j - a_j).$$

Then, applying Lemma 1.1, we get

$$\left| m_1(R^{\pm}) - \prod_{j=1}^{n}(b_j - a_j) \right| \leq 2\varepsilon M^{n-1}n.$$

Therefore

$$m_1(R) = m(R) \tag{1.3}$$

for any rectangle R. Let now E be an elementary set. Then it can be represented as a disjoint union of rectangles,

$$E = B_1 \cup \cdots \cup B_l.$$

Then, using (1.3) and the additivity of m and m_1, we get

$$\begin{aligned} m_1(E) &= m_1(B_1 \cup \cdots \cup B_l) \\ &= m_1(B_1) + \cdots + m_1(B_l) \\ &= m(B_1) + \cdots + m(B_l) \\ &= m(B_1 \cup \cdots \cup B_l) \\ &= m(E). \end{aligned}$$

This completes the proof. □

1.1.2 Area of the two-dimensional unit ball

In this section, we will see a possible way of approximating the area of the two-dimensional unit ball using elementary sets. For this aim, for $k \in \mathbb{N}$, define the set

$$Q_k = \left\{ q \in \mathbb{R} : q = \frac{j}{2^k} \text{ for some } j \in \mathbb{Z} \right\}.$$

We then define the family of cubes having vertices in Q_k and side length $\frac{1}{2^k}$, i. e.,

$$\mathscr{D}_k = \left\{ \left(p, q + \frac{1}{2^k} \right) \times \left(p, q + \frac{1}{2^k} \right) : p, q \in Q_k \right\}.$$

Note that for any $k \in \mathbb{N}$ and each $x \in Q_k + (\frac{1}{2^{k+1}}, \frac{1}{2^{k+1}})$, there exists exactly one cube $D_k(x)$ in $\mathscr{D}_k$ having center in x (see Figure 1.2). We want to approximate the area of the closed

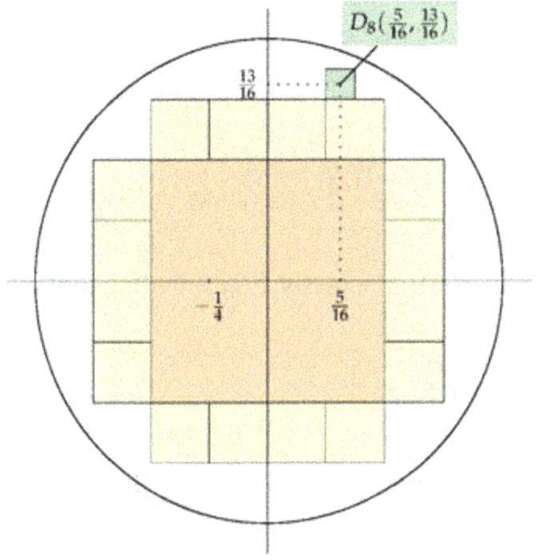

Figure 1.2: In orange, the set A_1, in yellow, the set $A_2 \setminus A_1$, and in green, an example for a cube $D_k(x)$.

unit ball $\overline{B} \subset \mathbb{R}^2$ from the inside by using the previously constructed cubes. We start with $D_1(0)$, which is not a precise approximation. We will use cubes of side length $\frac{1}{2}$. Then we take

$$A_1 = D_2\left(\frac{1}{4}, \frac{1}{4}\right) \cup D_2\left(-\frac{1}{4}, \frac{1}{4}\right) \cup D_2\left(\frac{1}{4}, -\frac{1}{4}\right) \cup D_2\left(-\frac{1}{4}, -\frac{1}{4}\right),$$

which is equal to $D_1(0)$. Things start to improve when we move to cubes of side length $\frac{1}{4}$, since now we get

$$A_2 = A_1 \cup (A_2 \backslash A_1),$$

where the part we add to the previous approximation, $A_2 \backslash A_1$, has the area

$$4 \cdot \frac{1}{4} = 1$$

(see Figure 1.2). We continue this procedure inductively, getting for each $k \geq 1$, an elementary set A_k containing A_{k-1}, which is a disjoint union of cubes in $\mathscr{D}_k$. The family $\{A_k\}_{k \in \mathbb{N}}$ is a family of elementary sets. These sets are closed. Set

$$C_k = \overline{A_k}, \quad k \in \mathbb{N}.$$

Theorem 1.6. *Let $k \geq 1$ and $r_k = 1 - \frac{\sqrt{2}}{2^k}$. Consider the closed balls*

$$\overline{B_{r_k}} = \{x \in \mathbb{R}^2 : \|x\| \leq r_k\}.$$

Then

$$\overline{B_{r_k}} \subset C_k = \overline{B_k} \subset \frac{1}{r_k} C_k, \quad k \in \mathbb{N}.$$

Proof. Take an arbitrary point $x = (x_1, x_2) \in \overline{B_{r_k}}$. Let

$$a_j = \max\left\{a \in \mathbb{Z} : \frac{a}{2^k} \leq x_j\right\}, \quad j \in \{1, 2\}.$$

Note that any cube in $\mathscr{D}_k$ has a diagonal of length $\frac{\sqrt{2}}{2^k}$, i. e., exactly $1 - r_k$. Notice that

$$D = \left(\frac{a_1}{2^k}, \frac{a_1 + 1}{2^k}\right) \times \left(\frac{a_2}{2^k}, \frac{a_2 + 1}{2^k}\right)$$

is a cube fully contained in B_{r_k} and hence one of the cubes composing A_k. Since $x \in \overline{D}$ was arbitrarily chosen, we conclude that

$$\overline{B_{r_k}} \subset C_k.$$

Moreover, we have

$$\overline{B} = \frac{1}{r_k}\overline{B_{r_k}} \subset \frac{1}{r_k}\overline{C_k} = \frac{1}{r_k}C_k,$$

which completes the proof. □

1.1.3 The Jordan measure

The example in the previous section can be used as a motivation for the following definitions, which describe in a natural way the definition of the Jordan[2] measure on $\mathbb{R}^n$.

Definition 1.5. Let $X \subset \mathbb{R}^n$ be a bounded set.

1. The Jordan inner measure $\mu_*(X)$ of X is defined as

$$\mu_*(X) = \sup\{m(E) : E \in \mathscr{E}_n,\ E \subset X\}.$$

2. The Jordan outer measure $\mu^*(X)$ of X is defined as

$$\mu^*(X) = \inf\{m(E) : E \in \mathscr{E}_n,\ X \subset E\}.$$

Definition 1.6. We say that a bounded set $X \subset \mathbb{R}^n$ is Jordan measurable if

$$\mu_*(X) = \mu^*(X),$$

and in this case, we call

$$\mu(X) = \mu_*(X) = \mu^*(X)$$

the Jordan measure of the set X. By $\mathscr{J}_n$ we will denote the set of all Jordan-measurable sets in $\mathbb{R}^n$. Then μ is the map

$$\begin{aligned} \mu : \mathscr{J}_n &\to [0,\infty), \\ X &\to \mu(X). \end{aligned}$$

Before saying anything about μ, we would like to give a few properties of μ_* and μ^*.

Theorem 1.7. *We have that μ_* and μ^* are monotone. Moreover, μ^* is sub-additive.*

2 Marie Ennemond Camille Jordan (5 January 1838–22 January 1922) was a French mathematician, known both for his foundational work in group theory and for his influential course of Analysis.

Proof. Let $X \subset Y \subset \mathbb{R}^n$. Then

$$\begin{aligned}\mu_*(X) &= \sup\{m(E) : E \in \mathscr{E}_n,\ E \subset X\}\\ &\le \sup\{m(E) : E \in \mathscr{E}_n,\ E \subset Y\}\\ &= \mu_*(Y),\end{aligned}$$

and

$$\begin{aligned}\mu^*(X) &= \inf\{m(E) : E \in \mathscr{E}_n,\ E \subset X\}\\ &\ge \inf\{m(E) : E \in \mathscr{E}_n,\ E \subset Y\}\\ &= \mu^*(Y).\end{aligned}$$

This proves the monotonicity of the inner and outer Jordan measures. Now we will prove the second part of the claim. Let $A, B \in \mathscr{E}_n$ be arbitrarily chosen. Then using the subadditivity of the measure m, we get

$$\begin{aligned}\mu^*(A \cup B) &\le m(A \cup B)\\ &\le m(A) + m(B),\end{aligned}$$

and taking infimum of both sides of the last inequality, we get

$$\mu^*(A \cup B) \le \mu^*(A) + \mu^*(B).$$

This completes the proof. □

It is not clear which sets exactly are Jordan measurable. We would like to point out that any elementary set in $\mathbb{R}^n$ is Jordan measurable.

Theorem 1.8 (Characterization of Jordan measurability). *Let $X \subset \mathbb{R}^n$ be a bounded set. Then the following are equivalent:*
1. *X is Jordan measurable.*
2. *For any $\varepsilon > 0$, there exist elementary sets Y and Z such that $Y \subset X \subset Z$ and*

$$m(Z \backslash Y) < \varepsilon.$$

3. *For any $\varepsilon > 0$, there exists an elementary set $Y \subset X$ such that*

$$\mu^*(X \backslash Y) < \varepsilon.$$

4. *For any $\varepsilon > 0$, there exists an elementary set Y such that*

$$\mu^*(X \triangle Y) < \varepsilon.$$

Proof. Suppose that $\varepsilon > 0$ is arbitrarily chosen.

$1 \Longrightarrow 2$. By the definition of $\mu_*(X)$ and $\mu^*(X)$ it follows that there exist elementary sets Y and Z such that $Y \subset X \subset Z$,

$$\mu_*(X) - m(Y) < \frac{\varepsilon}{2},$$

and

$$m(Z) - \mu^*(X) < \frac{\varepsilon}{2}.$$

Then

$$\begin{aligned} m(Z) &< \mu^*(X) + \frac{\varepsilon}{2} \\ &= \mu_*(X) + \frac{\varepsilon}{2} \\ &< m(Y) + \frac{\varepsilon}{2} + \frac{\varepsilon}{2} \\ &= m(Y) + \varepsilon, \end{aligned}$$

whereupon

$$m(Z) - m(Y) < \varepsilon.$$

Hence

$$\begin{aligned} m(Z\backslash Y) &= m(Z) - m(Y) \\ &< \varepsilon. \end{aligned}$$

$2 \Longrightarrow 3$. Let Z and Y be as in item 2. Since

$$X\backslash Y \subset Z\backslash Y,$$

by the definition of μ^* we get

$$\begin{aligned} \mu^*(X\backslash Y) &\le m(Z\backslash Y) \\ &< \varepsilon. \end{aligned}$$

$3 \Longrightarrow 4$. Let Y be as in item 3. Then

$$X \triangle Y = (X\backslash Y) \cup \emptyset,$$

and

$$\begin{aligned} \mu^*(X \triangle Y) &= \mu^*(X\backslash Y) \\ &< \varepsilon. \end{aligned}$$

$4 \Longrightarrow 1$. Let Y be as in item 4. Then

$$\mu^*(X\backslash Y) + \mu^*(Y\backslash X) < \varepsilon.$$

In particular, we have

$$\mu^*(X\backslash Y) < \varepsilon,$$
$$\mu^*(Y\backslash X) < \varepsilon.$$

By the definition of μ^* we can find elementary sets F and G such that

$$\begin{aligned} m(F) &\leq \mu^*(X\backslash Y) + \varepsilon \\ &< \varepsilon + \varepsilon \\ &= 2\varepsilon \end{aligned}$$

and

$$\begin{aligned} m(G) &\leq \mu^*(Y\backslash X) + \varepsilon \\ &< \varepsilon + \varepsilon \\ &= 2\varepsilon. \end{aligned}$$

Let

$$H = Y\backslash(F \cup G),$$

and let

$$K = F \cup G \cup H.$$

We have that H is an elementary set, and hence K is an elementary set. Note that

$$H \subset X \subset K$$

and

$$\begin{aligned} m(K) - m(H) &= m(K\backslash H) \\ &= m(F \cup G) \\ &\leq m(F) + m(G) \\ &< 2\varepsilon + 2\varepsilon \\ &= 4\varepsilon. \end{aligned}$$

Taking the limit as $\varepsilon \to 0$ concludes the proof. □

Now we will analyze in more detail Jordan-measurable sets. By the next two results we show that $\mathscr{J}_n$ satisfies properties similar to those of $\mathscr{E}_n$ and that in particular we can find statements analogous to Theorems 1.1 and 1.4.

Theorem 1.9 (Boolean closure of $\mathscr{J}_n$). *Let $X, Y \in \mathscr{J}_n$. Then*

$$X \cup Y, \quad X \cap Y, \quad X \backslash Y, \quad \text{and} \quad X \triangle Y$$

are in $\mathscr{J}_n$.

Proof. Take arbitrary $\varepsilon > 0$. Let A and B be elementary sets such that

$$A \subset X \quad \text{and} \quad B \subset Y$$

and

$$\begin{aligned} \mu^*(X \backslash A) &< \varepsilon, \\ \mu^*(Y \backslash B) &< \varepsilon. \end{aligned}$$

Then by the sub-additivity of μ^* we have

$$\begin{aligned} \mu^*((X \cup Y) \backslash (A \cup B)) &\le \mu^*(X \backslash (A \cup B)) + \mu^*(Y \backslash (A \cup B)) \\ &\le \mu^*(X \backslash A) + \mu^*(Y \backslash B) \\ &< \varepsilon + \varepsilon \\ &= 2\varepsilon \end{aligned}$$

and

$$\begin{aligned} \mu^*((X \cap Y) \backslash (A \cap B)) &\le \mu^*((X \cap Y) \backslash A) + \mu^*((X \cap Y) \backslash B) \\ &\le \mu^*(X \backslash A) + \mu^*(Y \backslash B) \\ &< \varepsilon + \varepsilon \\ &= 2\varepsilon. \end{aligned}$$

Now applying statement 3 of Theorem 1.8, we conclude that $X \cap Y$ and $X \cup Y$ are Jordan-measurable sets. Next, regarding $X \backslash Y$, we have

$$\begin{aligned} (X \backslash Y) \triangle (A \backslash B) &= ((X \backslash Y) \backslash (A \backslash B)) \cup ((A \backslash B) \backslash (X \backslash Y)) \\ &\subset ((X \backslash Y) \backslash (A \backslash Y)) \cup ((A \backslash B) \backslash (A \backslash Y)) \\ &\subset (X \backslash A) \cup (Y \backslash B). \end{aligned}$$

Applying the sub-additivity, we get

$$\begin{aligned}\mu^*((X\backslash Y) \triangle (A\backslash B)) &\le \mu^*((X\backslash A) \cup (Y\backslash B))\\ &\le \mu^*(X\backslash A) + \mu^*(Y\backslash B)\\ &< \varepsilon + \varepsilon\\ &= 2\varepsilon.\end{aligned}$$

Hence, by statement 4 of Theorem 1.8 we conclude that $X\backslash Y$ is Jordan measurable. As above, $Y\backslash X$ is Jordan measurable, and then

$$X \triangle Y = (X\backslash Y) \cup (Y\backslash X),$$

is Jordan measurable. This completes the proof. □

Theorem 1.10. *Let $X, Y \subset \mathbb{R}^n$ be Jordan-measurable sets. Then we have the following.*

1. *Additivity:*

$$\mu(X \cup Y) = \mu(X) + \mu(Y),$$

provided that X and Y are disjoint.

2. *Monotonicity:*

$$\mu(X) \le \mu(Y),$$

provided that $X \subset Y$.

3. *Sub-additivity:*

$$\mu(X \cup Y) \le \mu(X) + \mu(Y).$$

4. *Translation invariance:*

$$\mu(X + x) = \mu(X)$$

for all $x \in \mathbb{R}^n$.

Proof. 1. Suppose that X and Y are disjoint. Take arbitrary $\varepsilon > 0$. Then there are elementary sets X_+, X_-, Y_+, and Y_- such that

$$\begin{aligned}\mu(X) - \varepsilon &\le \mu(X_-)\\ &\le \mu(X)\\ &\le \mu(X_+)\\ &\le \mu(X) + \varepsilon\end{aligned}$$

and

$$\mu(Y) - \varepsilon \le \mu(Y_-)$$
$$\le \mu(Y)$$
$$\le \mu(Y_+)$$
$$\le \mu(Y) + \varepsilon.$$

Note that

$$X_- \cup Y_- \subset X \cup Y \subset X_+ \cup Y_+.$$

Then

$$\begin{aligned}\mu(X \cup Y) &\ge \mu(X_- \cup Y_-)\\ &= m(X_- \cup Y_-)\\ &= m(X_-) + m(Y_-)\\ &\ge \mu(X) - \varepsilon + \mu(Y) - \varepsilon\\ &= \mu(X) + \mu(Y) - 2\varepsilon,\end{aligned}$$

and

$$\begin{aligned}\mu(X \cup Y) &\le \mu(X_+ \cup Y_+)\\ &= m(X_+ \cup Y_+)\\ &= m(X_+) + m(Y_+)\\ &\ge \mu(X) + \varepsilon + \mu(Y) + \varepsilon\\ &= \mu(X) + \mu(Y) + 2\varepsilon.\end{aligned}$$

Taking the limit as $\varepsilon \to 0$, we obtain the desired additivity.

2. Note that

$$(Y\backslash X) \cap (Y \cap X) = \emptyset$$

and

$$Y = (Y\backslash X) \cup (Y \cap X).$$

Then

$$\begin{aligned}\mu(Y) &= \mu\big((Y\backslash X) \cup (Y \cap X)\big)\\ &= \mu(Y\backslash X) + \mu(Y \cap X)\\ &\ge \mu(Y \cap X)\\ &= \mu(X).\end{aligned}$$

3. Note that

$$X \cup Y = Y \cup (X \backslash Y)$$

and

$$Y \cap (X \backslash Y) = \emptyset.$$

Then

$$\begin{aligned}\mu(X \cup Y) &= \mu(Y \cup (X \backslash Y)) \\ &= \mu(Y) + \mu(X \backslash Y) \\ &\leq \mu(Y) + \mu(X).\end{aligned}$$

4. For $\varepsilon > 0$ and $X_\pm$ as above, we get

$$\begin{aligned}\mu(X + x) - \varepsilon &\leq m(X_+ + x) - \varepsilon \\ &= m(X_+) - \varepsilon \\ &\leq \mu(X) \\ &\leq m(X_-) + \varepsilon \\ &= m(X_- + x) + \varepsilon \\ &= \mu(X + x) + \varepsilon.\end{aligned}$$

Taking the limit as $\varepsilon \to 0$ gives both measurability and the desired equality. This completes the proof. □

1.1.4 The n-dimensional unit ball

Recall that for

$$\overline{B} = \{x \in \mathbb{R}^2 : \|x\| \leq 1\},$$

we were able to find sequences of elementary sets $\{C_k\}_{k\in\mathbb{N}}$ and $\{C_k^1\}_{k\in\mathbb{N}}$ such that

$$C_k \subset \overline{B} \subset C_k^1, \quad k \in \mathbb{N},$$

and

$$C_k^1 = \frac{1}{r_k} C_k, \quad k \in \mathbb{N}.$$

To show that $\overline{B}$ is Jordan measurable, it suffices to prove that

$$m\left(\frac{1}{r_k}C_k\right) = \frac{1}{r_k^2}m(C_k^1),$$

since then we can see that

$$\begin{aligned} r_k &= 1 - \frac{\sqrt{2}}{2^k} \\ &\to 1 \quad \text{as } k \to \infty, \end{aligned}$$

to conclude that

$$\mu_*(\overline{B}) = \mu^*(\overline{B}).$$

We can prove a more general result that the homothetic translations

$$\psi_{\lambda,a}(x) = \lambda x + a, \quad a \in \mathbb{R}^n,\ \lambda > 0,$$

preserve measurability and stretch the Jordan measure by a factor λ^n, i. e.,

$$\mu(\psi_{\lambda,a}(X)) = \lambda^n \mu(X). \tag{1.4}$$

The key observation is that the family of elementary sets contained in $X \subset \mathbb{R}^n$

$$\{E \subset X : E \text{ elementary set}\},$$

is in bijection with the family of elementary sets contained in $\lambda X + a$

$$\{F \subset \lambda X + a : F \text{ elementary set}\},$$

through the map $E \mapsto \lambda E + a$. By taking the infimum and supremum we get that for measurable X, the upper and lower measures of $\lambda X + a$ coincide. Moreover,

$$m(\lambda E) = \lambda^n m(E)$$

for all elementary sets

$$E = I_1 \times \cdots \times I_m,$$

since the endpoints of each $I_j, j \in \{1, \dots, m\}$, are stretched by a factor λ, and so, because of translation invariance of the measure m, we get the desired property (1.4). A more general result is as follows.

Theorem 1.11. *We have*

$$\mu_*(\overline{B}) = \mu^*(\overline{B}).$$

Proof. Let

$$Q_k = \left\{q \in \mathbb{R} : q = \frac{j}{2^k} \text{ for some } j \in \mathbb{Z}\right\},$$

and let

$$\mathcal{D}_k = \left\{D_k \prod_{j=1}^{n} I_j : I_j = \left(p_j, p_j + \frac{1}{2^k}\right), \, p_j \in Q_k\right\},$$
$$A_k = \{D_k \in \mathcal{D}_k : D_k \subset B\}.$$

We take $k \in \mathbb{N}$ large enough such that

$$\delta_k = \frac{\sqrt{n}}{2^k} < 1, \quad k \in \mathbb{N},$$

and fix the radius

$$\begin{aligned} r_k &= 1 - \delta_k \\ &= 1 - \frac{\sqrt{n}}{2^k}, \quad k \in \mathbb{N}. \end{aligned}$$

Let

$$D = \prod_{j=1}^{n}\left(x_j - \frac{1}{2^k}, x_j + \frac{1}{2^k}\right), \quad x = (x_1, \dots, x_n) \in \overline{B_{r_k}}.$$

Then D has the diagonal

$$\begin{aligned} \operatorname{diag}(D) &= \max_{x,y \in D} \|x - y\| \\ &= \delta_k. \end{aligned}$$

Therefore, for each $x = (x_1, \dots, x_n) \in \overline{B_{r_k}}$, the set D is fully contained in B. This allows us to find $D_k \in \mathcal{D}_k$ such that $x \in \overline{D_k}$: for each $j \in \{1, \dots, n\}$, let

$$a_j = \max\left\{a \in \mathbb{Z} : \frac{a}{2^k} \le x_j\right\}, \quad j \in \{1, \dots, n\};$$

then

$$\left(\frac{a_j}{2^k}, \frac{a_j + 1}{2^k}\right) \subset \left(x_j - \frac{1}{2^k}, x_j + \frac{1}{2^k}\right),$$

and thus

$$D_k = \prod_{j=1}\left(\frac{a_j}{2^k}, \frac{a_j + 1}{2^k}\right) \subset B.$$

Therefore

$$\overline{B_{r_k}} \subset C_k = \overline{A_k},$$

and

$$\overline{B_{r_k}} \subset C_k \subset \overline{B} \subset \frac{1}{r_k} C_k^1,$$

which can be used to get

$$\begin{aligned}\mu_*(\overline{B}) &\geq \sup_{k\in\mathbb{N}} m(C_k)\\ &= \lim_{k\to\infty} m(C_k)\\ &= \lim_{k\to\infty} \frac{1}{r_k^n} m(C_k)\\ &= \lim_{k\to\infty} m\left(\frac{1}{r_k} C_k\right)\\ &\geq \mu^*(\overline{B}).\end{aligned}$$

On the other hand, by the definition we have

$$\mu_*(\overline{B}) \leq \mu^*(\overline{B}).$$

Therefore

$$\mu_*(\overline{B}) = \mu^*(\overline{B}).$$

This completes the proof. □

1.1.5 Characterization of the Jordan measure using cubes

The above example of the n-dimensional unit ball may suggest that when checking the measurability of a set, we do not really need to work with the whole rank of elementary sets: it might be sufficient to look only at a finite disjoint union of cubes of fixed side length. The main aim of this section is to show how this can be done and what the sequences in this procedure are.

For fixed $k \in \mathbb{N}$ and $m_j \in \mathbb{N}$, $m = (m_1, \ldots, m_n)$, define

$$Q_{k,m}^n = \left\{x \in \mathbb{R}^n : \frac{m_j}{10^k} \leq x_j \leq \frac{m_j + 1}{10^k},\ m_j \in \mathbb{N},\ j \in \{1, \ldots, n\}\right\}$$

and

$$T_k = \bigcup_{m \in \mathbb{N}^n} Q^n_{k,m}.$$

Definition 1.7. The sets $Q^n_{k,m}$ are said to be cubes of rank k. The number $\frac{1}{10^{kn}}$ is said to be the n-dimensional volume of the cube $Q^n_{k,m}$ and is denoted by

$$\mu Q^n_{k,m} = 10^{-kn}.$$

Note that

$$\mathbb{R}^n = \bigcup_{Q^n_{k,m} \subset T_k} Q^n_{k,m}. \tag{1.5}$$

Exercise 1.1. Prove (1.5).

If $Q^n_{j,k,m}, Q^n_{l,k,m} \in T_k$, $Q^n_{j,k,m} \cap Q^n_{l,k,m} = \emptyset$, $j \neq l$, $j, l \in \mathbb{N}$, and

$$Y = \bigcup_j Q^n_{j,k,m},$$

then we define

$$\mu Y = \sum_j \mu Q^n_{j,k,m}. \tag{1.6}$$

We have that $\mu Y \geq 0$ or $\mu Y = \infty$. We introduce the convention $\mu \emptyset = 0$. Now for $X \subset \mathbb{R}^n$, define

$$s_k(X) = \bigcup_{Q^n_{k,m} \subset X} Q^n_{k,m}$$

and

$$S_k(X) = \bigcup_{\substack{Q^n_{k,m} \cap X \neq \emptyset \\ Q^n_{k,m} \in T_k}} Q^n_{k,m}.$$

Then

$$s_k(X) \subset X \subset S_k(X).$$

Note that

$$\partial S_k(X) \cap X = \emptyset. \tag{1.7}$$

Indeed, suppose that there is $x \in \partial S_k(X) \cap X$. Then there is $Q^n_{k,m} \in T_k$ such that $x \in \partial Q^n_{k,m}$. By the definition of $S_k(X)$, using that $x \in X$, it follows that there is a cube $\tilde{Q}^n_{k,m} \in T_k$ of rank k such that

$$\partial Q^n_{k,m} \subset \tilde{Q}^n_{k,m}.$$

Hence $x \notin \partial S_k(X)$, which is a contradiction. Thus (1.7) holds. Evidently,

$$s_0(X) \subset s_1(X) \subset s_2(X) \subset \cdots \subset s_k(X) \subset \cdots,$$

and

$$S_0(X) \supset S_1(X) \supset S_2(X) \supset \cdots \supset S_k(X) \supset \cdots.$$

Hence

$$\mu s_0 \le \mu s_1 \le \cdots \le \mu s_k \le \cdots,$$
$$\mu S_0 \ge \mu S_1 \ge \mu S_2 \ge \cdots \ge \mu S_k \ge \cdots.$$

Since $s_l(X) \subset S_l(X)$, $l \in \mathbb{N}$, we have that $\mu s_l(X) \le \mu S_l(X)$, $l \in \mathbb{N}_0$. Therefore there exist

$$\lim_{l\to\infty} \mu s_l(X) \quad \text{and} \quad \lim_{l\to\infty} \mu S_l(X),$$

and

$$\lim_{l\to\infty} \mu s_l(X) \le \lim_{l\to\infty} \mu S_l(X).$$

Definition 1.8. The limit $\lim_{l\to\infty} \mu s_l(X)$ is called the latest lower bound or inner Jordan measure of the set X. It is denoted by $\mu_* X$. The limit $\lim_{l\to\infty} \mu S_l(X)$ is called the latest upper bound or outer Jordan measure of X. It is denoted by $\mu^* X$. The set X is said to be Jordan measurable or measurable if

$$\mu X = \mu_* X = \mu^* X.$$

We have

$$0 \le \mu_* X \le \mu^* X.$$

Thus $\mu X \ge 0$. Now we give a necessary and sufficient condition for the positivity of the Jordan measure.

Theorem 1.12. *A set $X \subset \mathbb{R}^n$ has an interior point if and only if $\mu_* X > 0$.*

Proof. Suppose that $x \in X$ is an interior point of X. Then there exists $\varepsilon > 0$ such that $U(x, \varepsilon) \subset X$. Because $x \in X$, there is $l \in \mathbb{N}$ such that $\mu_* s_l(X) > 0$. Take $l_0 \in \mathbb{N}$ such that

$$10^{-l_0}\sqrt{n} < \varepsilon.$$

Hence $Q^n_{l_0,m} \subset U(x, \varepsilon)$, and thus $Q^n_{l_0,m} \in s_l(X)$. Therefore

$$\mu_* s_l(X) \geq \mu_* Q^n_{l_0,m}$$
$$> 0.$$

Let now $\mu_* X > 0$. Then there is $l \in \mathbb{N}$ such that $\mu_* s_l(X) > 0$. Thus $s_l(X)$ has an interior point, and, consequently, X has an interior point. This completes the proof. □

Corollary 1.1. *If X is an open set, then $\mu_* X > 0$.*

Proof. Since X is an open set in $\mathbb{R}^n$, it contains an interior point. Now applying Theorem 1.12, we get that $\mu_* X > 0$. This completes the proof. □

Theorem 1.13. *Let X be an open Jordan-measurable set, and let Q be a closed n-dimensional cube such that $X \cap Q \neq \emptyset$. Then*

$$\mu_*(X \cap Q) > 0.$$

Proof. For any $x \in Q \cap X$, there is a neighborhood $U(x)$ such that $U(x) \subset X$. If x is a limit point of Q, then it is a limit point for a set of interior points of Q. Thus the set $U(x)$ contains interior points of Q. Let $y \in U(x)$ be an interior point of Q. Then there is a neighborhood $V(y)$ of the point y such that $V(y) \subset Q$. Hence

$$U(x) \cap V(y) \subset U(x) \subset X,$$

and

$$U(x) \cap V(y) \subset Q.$$

Therefore

$$U(x) \cap V(y) \subset X \cap Q.$$

Note that $y \in U(x)$ and $y \in V(y)$. Therefore $U(x) \cap V(y) \neq \emptyset$. Since $U(x)$ and $V(y)$ are open sets, we conclude that $U(x) \cap V(y)$ is an open set. Because $U(x) \cap V(y)$ has an interior point, by Theorem 1.12 it follows that

$$\mu_*(U(x) \cap V(y)) > 0. \tag{1.8}$$

By the monotonicity of the Jordan measure we find

$$\mu_*(U(x) \cap V(y)) \leq \mu_*(X \cap Q).$$

Hence by (1.8) we arrive at

$$\mu_*(X \cap Q) > 0.$$

This completes the proof. □

Now we give a definition of a partition of a Jordan-measurable set, which is needed for a construction of multiple integrals.

Definition 1.9. Let X be a Jordan-measurable set, and let $m \in \mathbb{N}$. Any finite system $\tau_m\{X_j\}_{j=1}^m$ of nonempty Jordan-measurable subsets of X is said to be a partition of X if the following conditions hold:
1. $\mu(X_j \cap X_l) = 0, j \neq l, j, l \in \{1, \ldots, m\}$.
2. $\bigcup_{j=1}^m X_j = X$.

The number

$$|\tau_m| = \max_{j \in \{1,\ldots,m\}} \operatorname{diam} X_j,$$

is called the mesh of the partition τ_m.

We will prove that such partitions of a Jordan-measurable set exist. More precisely, we have the following result.

Theorem 1.14. *Let X be an arbitrary open Jordan-measurable set. Then there is a sufficiently small partition $\tau = \{X_j\}_{j=1}^{j_\tau}$ such that $\mu X_j > 0, j \in \{1, \ldots, j_\tau\}$.*

Proof. Take an arbitrary partition of $\mathbb{R}^n$ of cubes of rank k. The set τ_1 of all cubes Q of rank k such that $Q \cap X \neq \emptyset$ contains a finite number of elements or coincides with X. Let

$$\tau_1 = \{Q_j\}_{j=1}^{j_\tau}.$$

Then

$$X_j = Q_j \cap X \neq \emptyset, \quad j \in \{1, \ldots, j_\tau\},$$

are Jordan-measurable sets. Moreover,

$$X_j \subset X, \quad j \in \{1, \ldots, j_\tau\},$$

and then

$$\bigcup_{j=1}^{j_\tau} X_j \subset X. \tag{1.9}$$

Next, let $x \in X$ be arbitrarily chosen. Then there is a cube Q of rank k such that $x \in Q$, and hence $x \in X \cap Q$. Then

$$X \cap Q \in \tau$$

and

$$X \cap Q = X_l$$

for some $l \in \{1,\dots,j_\tau\}$. Therefore

$$x \in X \cap Q \subset X_l \subset \bigcap_{j=1}^{j_\tau} X_j.$$

Since $x \in X$ was arbitrarily chosen and it is an element of $\bigcup_{j=1}^{j_\tau} X_j$, we arrive at the inclusion

$$X \subset \bigcup_{j=1}^{j_\tau} X_j.$$

By this inclusion and (1.9) we get

$$X = \bigcup_{j=1}^{j_\tau} X_j.$$

Thus τ is a partition of X. Note that

$$X_j \cap X_l \subset Q_j \cap Q_l, \quad j, l \in \{1,\dots,j_\tau\},$$

and

$$\mu(Q_j \cap Q_l) = 0, \quad j, l \in \{1,\dots,j_\tau\}.$$

On the other hand, by Theorem 1.13 we get

$$\mu X_j > 0, \quad j \in \{1,\dots,j_\tau\}.$$

Take arbitrary $\delta > 0$. We choose $k \in \mathbb{N}$ such that

$$\frac{\sqrt{n}}{10^k} < \delta.$$

Hence

$$\operatorname{diam} Q = 10^{-k}\sqrt{n},$$

and

$$\begin{aligned}\operatorname{diam} X_j &= \operatorname{diam}(X \cap Q_j)\\ &\le \operatorname{diam} Q_j\\ &= 10^{-k}\sqrt{n}\\ &< \delta, \quad j \in \{1,\dots,j_\tau\}.\end{aligned}$$

Thus $|\tau| < \delta$. This completes the proof. □

An important property of the Jordan-measurable sets is as follows.

Theorem 1.15. *For any open Jordan-measurable set X, there is a sequence $\{X_m\}_{m\in\mathbb{N}}$ of Jordan-measurable sets such that*

$$X_m \subset X_{m+1}, \quad m \in \mathbb{N}, \tag{1.10}$$

$$\bigcup_{m=1}^{\infty} X_m = X, \tag{1.11}$$

and

$$X_m \subset \overline{X_m} \subset X, \quad m \in \mathbb{N}. \tag{1.12}$$

Proof. Let

$$X_m = s_m(X)_{\text{int}}, \quad m \in \mathbb{N}.$$

For these sets X_m, $m \in \mathbb{N}$, we have

$$\begin{aligned} X_m &= s_m(X)_{\text{int}} \\ &\subset s_{m+1}(X)_{\text{int}} \\ &= X_{m+1}, \quad m \in \mathbb{N}. \end{aligned}$$

Thus (1.10) holds. Note that

$$\begin{aligned} X_m &= s_m(X)_{\text{int}} \\ &\subset s_m(X) \\ &\subset X, \quad m \in \mathbb{N}, \end{aligned}$$

whereupon

$$\bigcup_{m=1}^{\infty} X_m \subset X. \tag{1.13}$$

Take arbitrary $x \in X$. Since X is an open set, there is $\varepsilon > 0$ such that $U(x, \varepsilon) \subset X$. Take $m_0 \in \mathbb{N}$ such that

$$\frac{\sqrt{n}}{10^{m_0}} < \varepsilon.$$

Then all cubes of rank m_0 that contain x lie in $U(x, \varepsilon)$. and thus they lie in X. Therefore

$$x \in s_{m_0}(X)_{\text{int}} = X_{m_0},$$

from which we have

$$x \in \bigcup_{m=1}^{\infty} X_m.$$

Because $x \in X$ was arbitrarily chosen and it is an element of $\bigcup_{m=1}^{\infty} X_m$, we arrive at the inclusion

$$X \subset \bigcup_{m=1}^{\infty} X_m.$$

By this inclusion and (1.13) we get (1.11). Moreover,

$$\begin{aligned} X_m &\subset \overline{X_m} \\ &= \overline{s_m(X)_{\text{int}}} \\ &= s_m(X)_{\text{int}} \\ &\subset s_m(X) \\ &\subset X, \end{aligned}$$

i. e., we get (1.13). This completes the proof. □

For $k \in \mathbb{N}_0$ and $X \subset \mathbb{R}^n$, by $\sigma_k(X)$ we denote the union of all cubes of rank k that belong to $S_k(X)$ and do not belong to $s_k(X)$, i. e.,

$$\sigma_k(X) = \bigcup_{\substack{Q \in T_k \\ Q \in S_k(X), Q \notin s_k(X)}} Q.$$

Then we can represent $S_k(X)$ as

$$S_k(X) = s_k(X) \cup \sigma_k(X),$$

and using (1.6), we get

$$\mu S_k(X) = \mu s_k(X) + \mu \sigma_k(X).$$

Note that for any $k \in \mathbb{N}_0$, the sets $\sigma_k(X)$, $s_k(X)$, and $S_k(X)$ are closed sets. Moreover,

$$s_k(X) = s_{k\text{int}}(X) \cup \partial s_k(X).$$

We will prove that

$$\partial s_k(X) \subset \sigma_k(X). \tag{1.14}$$

Let $x \in \partial s_k(X)$ be arbitrarily chosen. Since

$$\partial s_k(X) \subset \overline{S_k(X)} = S_k(X) \subset X,$$

we get that $x \in X$. Hence any cube of rank k containing the point x belongs to $S_k(X)$, because at least one cube Q of such cubes does not belong to $s_k(X)$. Therefore

$$x \in Q \subset S_k(X), \quad Q \notin s_k(X).$$

Hence $Q \in \sigma_k(X)$, and thus $x \in \sigma_k(X)$. Because $x \in \partial s_k(X)$ was arbitrarily chosen and it is an element of $\sigma_k(X)$, we obtain inclusion (1.14). Consequently,

$$S_k(X) = s_{k\text{int}}(X) \cup \sigma_k(X).$$

Let now $y \in s_{k\text{int}}(X)$ be arbitrarily chosen. Since there is at least one cube Q_1 such that $y \in Q_1$, $Q_1 \subset S_k(X)$, and $Q_1 \notin s_k(X)$, we conclude that $y \notin s_k(X)$. Because $y \in s_{k\text{int}}(X)$ was arbitrarily chosen, we conclude that

$$s_{k\text{int}}(X) \cap \sigma_k(X) = \emptyset.$$

The next inclusions are needed to prove the main criterion for Jordan measurability of a set.

Theorem 1.16. *For any set $X \subset \mathbb{R}^n$ and any $k \in \mathbb{N}$, we have the following inclusions:*

$$\partial X \subset \sigma_k(X) \subset S_k(\partial X).$$

Proof. Fix $k \in \mathbb{N}$. Let $x \in \partial X$ be arbitrarily chosen. Since

$$X \subset S_k(X),$$

we get

$$\begin{aligned} \overline{X} &\subset \overline{S_k}(X) \\ &= S_k(X). \end{aligned}$$

Thus

$$\begin{aligned} x \in \partial X &\subset \overline{X} \\ &\subset S_k(X) \\ &= s_{\text{int}}(X) \cup \sigma_k(X). \end{aligned}$$

By the inclusion

$$s_k(X) \subset X$$

it follows that

$$s_{k\text{int}}(X) \subset X_{\text{int}}.$$

Because $x \in \partial X$, we have that $x \notin X_{\text{int}}$, from which we conclude that $x \notin s_{k\text{int}}(X)$. Hence, using that

$$x \in s_{k\text{int}}(X) \cup \sigma_k(X),$$

we obtain $x \in \sigma_k(X)$. Because $x \in \partial X$ was arbitrarily chosen and it is an element of $\sigma_k(X)$, we obtain the inclusion

$$\partial X \subset \sigma_k(X).$$

Let now $y \in \sigma_k(X)$ be arbitrarily chosen. Then there is a cube Q of rank k such that

$$Q \subset \sigma_k(X) \quad \text{and} \quad y \in Q.$$

By the definition of σ_k we conclude that

$$Q \subset S_k(X) \quad \text{and} \quad Q \not\subset s_k(X).$$

Then

$$Q \cap X \neq \emptyset,$$

and Q contains points that do not belong to X. Therefore Q contains points of ∂X. Hence by the definition of $S_k(\partial X)$ we find that $Q \subset S_k(\partial X)$. From this, using that $y \in Q$, we obtain $y \in S_k(\partial X)$. Because $y \in \sigma_k(X)$ was arbitrarily chosen and it is an element of $S_k(\partial X)$, we arrive at the inclusion

$$\sigma_k(X) \subset S_k(\partial X).$$

This completes the proof. □

1.1.6 A measurability criterion

In this section, we give a criterion for measurability of a set. We start with the following useful result.

Theorem 1.17. *If $X \subset \mathbb{R}^n$ is bounded, then $\mu_* X < \infty$ and $\mu^* X < \infty$.*

Proof. Since X is bounded, we have that $S_0(X)$ has a finite number of cubes. Therefore $\mu S_0(X) < \infty$. Now using the relation

$$s_k(X) \subset S_k(X) \subset S_0(X), \quad k \in \mathbb{N},$$

we get

$$\mu s_k(X) \le \mu S_k(X) \le \mu S_0(X), \quad k \in \mathbb{N}.$$

Hence

$$\begin{aligned} 0 &\le \mu_* X \\ &= \lim_{k\to\infty} \mu s_k(X) \\ &\le \mu^* X \\ &= \lim_{k\to\infty} \mu S_k(X) \\ &\le \mu S_0(X) \\ &< \infty. \end{aligned}$$

This completes the proof. □

An important property of the Jordan-measurable sets reads as follows.

Theorem 1.18. *If X is Jordan measurable, then it is bounded.*

Proof. Suppose that X is unbounded. Then $S_k(X)$ contains an infinite number of cubes for any $k \in \mathbb{N}$. Therefore

$$\mu S_k(X) = \infty, \quad k \in \mathbb{N}.$$

Thus $\mu^* X = \infty$, and X is not Jordan measurable. This is a contradiction. Therefore X is bounded. This completes the proof. □

Now we are ready to prove a criterion for Jordan measurability of a set.

Theorem 1.19. *The set $X \subset \mathbb{R}^n$ is Jordan measurable if and only if it is bounded and $\mu(\partial X) = 0$.*

Proof. 1. Let X be a Jordan measurable set. By Theorem 1.18 it follows that X is bounded. By the definition of Jordan measure we have

$$\mu_* X = \mu^* X,$$

whereupon

$$\lim_{k\to\infty} \mu s_k(X) = \lim_{k\to\infty} \mu S_k(X).$$

By the definition of σ_k, $k \in \mathbb{N}$, we get

$$\mu\sigma_k(X) = \mu S_k(X) - \mu s_k(X).$$

Therefore

$$\begin{aligned}\lim_{k\to\infty} \mu\sigma_k(X) &= \lim_{k\to\infty} (\mu S_k(X) - \mu s_k(X)) \\ &= \lim_{k\to\infty} \mu S_k(X) - \lim_{k\to\infty} \mu s_k(X) \\ &= 0.\end{aligned}$$

By Theorem 1.19 we obtain

$$\partial X \subset \sigma_k(X), \quad k \in \mathbb{N}.$$

Hence, applying the monotonicity of the upper Jordan measure, we get

$$\begin{aligned}\mu^* \partial X &\le \mu^* \sigma_k(X) \\ &= \mu\sigma_k(X),\end{aligned}$$

whereupon

$$\begin{aligned}\mu_* \partial X &\le \mu^* \partial X \\ &\le \lim_{k\to\infty} \mu\sigma_k(X) \\ &= 0.\end{aligned}$$

Thus the necessity of the statement is proved.

2. Let X be bounded and $\mu\partial X = 0$. Then

$$\lim_{k\to\infty} \mu S_k(X) = 0.$$

Since

$$\sigma_k(X) \subset S_k(X), \quad k \in \mathbb{N},$$

we get

$$\mu\sigma_k(X) \le \mu S_k(X), \quad k \in \mathbb{N},$$

and thus

$$\begin{aligned}\mu S_k(X) - \mu s_k(X) &= \mu\sigma_k(X) \\ &\le \mu S_k(\partial X).\end{aligned} \tag{1.15}$$

Since X is bounded, we have

$$\mu_* X, \quad \mu^* X < \infty.$$

Therefore there exist the limits

$$\lim_{k\to\infty} \mu s_k(X), \quad \lim_{k\to\infty} \mu S_k(X), \quad k \in \mathbb{N}.$$

Now applying (1.15), we get

$$\begin{aligned}\lim_{k\to\infty} \mu S_k(X) - \lim_{k\to\infty} \mu s_k(X) &= \lim_{k\to\infty} (\mu S_k(X) - \mu s_k(X)) \\ &\le \lim_{k\to\infty} \mu S_k(\partial X) \\ &= 0,\end{aligned}$$

i. e.,

$$\mu^* X = \mu_* X,$$

and thus X is Jordan measurable. This completes the proof. □

1.2 Definition for multiple integral

Multiple integrals are a generalization of single-variable integrals to functions of several variables. These mathematical tools calculate the volumes under surfaces in multidimensional spaces, allowing for the exploration of areas and volumes within more complex geometries. Specifically, double and triple integrals are used for functions of two and three variables, respectively. Multiple integrals are useful in various scientific and engineering disciplines. By enabling the calculation of quantities like mass, volume, and energy over complex shapes and domains, they play a pivotal role in the advancement of technology and understanding of the natural world. Some applications include

- Calculating areas and volumes of irregular shapes.
- Evaluating the mass of objects with variable density.
- Quantifying the flow of liquids and gases in engineering.
- Modeling gravitational fields and electric potentials in physics.

Now, we will define the multiple Riemann[3] integral. For this aim, firstly, we will introduce a refinement of a partition.

---

3 George Friedrich Bernhard Riemann (17 September 1826–20 July 1866) was a German mathematician, who made profound contributions to analysis, number theory, and differential geometry. In the field of real analysis, he is mostly known for the first rigorous formulation of the integral, the Riemann integral, and his work on Fourier series. His contributions to complex analysis include most notably the introduction of Riemann surfaces, breaking new ground in a natural, geometric treatment of complex analysis. His 1859 paper on the prime-counting function, containing the original statement of the Riemann hy-

Let $X \subset \mathbb{R}^n$ be a Jordan-measurable set. Let $\tau_m = \{X_j\}_{j=1}^m$ be a partition of X. We will prove that

$$\mu X = \sum_{j=1}^m \mu X_j.$$

Indeed, let for any $j \in \{1, \ldots, m\}$, denote

$$X_j^* = \bigcup_{k=1,k\neq j}^m (X_k \cap X_j),$$
$$X^* = \bigcup_{j=1}^m X_j^*.$$

Then, using the definition of a partition of a set, we get

$$\mu(X_k \cap X_j) = 0, \quad k \neq j,\ k, j \in \{1, \ldots, m\},$$
$$\mu X_j^* = \sum_{k=1,k\neq j}^m \mu(X_k \cap X_j)$$
$$= 0,$$

and

$$\mu X^* \leq \sum_{j=1}^m \mu X_j^*$$
$$= 0.$$

Set

$$X_j^{**} = X_j \backslash X^*, \quad j \in \{1, \ldots, m\}.$$

Then

$$X^* \cap X_j^{**} = \emptyset, \quad j \in \{1, \ldots, m\}.$$

Hence

$$\mu X_j = \mu(X_j \backslash X^*)$$
$$= \mu X_j^{**}.$$

---

pothesis, is regarded as a foundational paper of analytic number theory. Through his pioneering contributions to differential geometry, Riemann laid the foundations of the mathematics of general relativity. He is considered by many to be one of the greatest mathematicians of all time.

Now applying the additivity of the Jordan measure, we obtain

$$\begin{aligned}\mu X &= \sum_{j=1}^{m} \mu X_j + \mu X^* \\ &= \sum_{j=1}^{m} \mu X_j.\end{aligned}$$

Definition 1.10. Let $\tau_m = \{X_j\}_{j=1}^m$ and $\tau_k = \{Y_l\}_{l=1}^k$ be two partitions of X. We say that τ_k is a refinement of τ_m and write

$$\tau_m \prec \tau_k$$

if for every $Y_l \in \tau_k$, there exists an element $X_j \in \tau_m$ such that

$$Y_l \subset X_j.$$

Suppose that τ_l, τ_m, and τ_n are three partitions of X. Then we have the following relations:

1. If $\tau_l \prec \tau_m$ and $\tau_m \prec \tau_n$, then $\tau_l \prec \tau_n$.
2. For any τ_l and τ_m, there is a partition τ of X such that

$$\tau_l \prec \tau \quad \text{and} \quad \tau_m \prec \tau.$$

Figure 1.3 illustrates the fact that if τ_l refines τ_m, then every sub-box of τ_l is contained in a sub-box of τ_m. The literal manifestation in the figure of the containment $\tau_m \prec \tau_l$ is that the set of points where a horizontal line segment and a vertical line segment meet in the right side of the figure subsumes the set of such points in the left side. An important

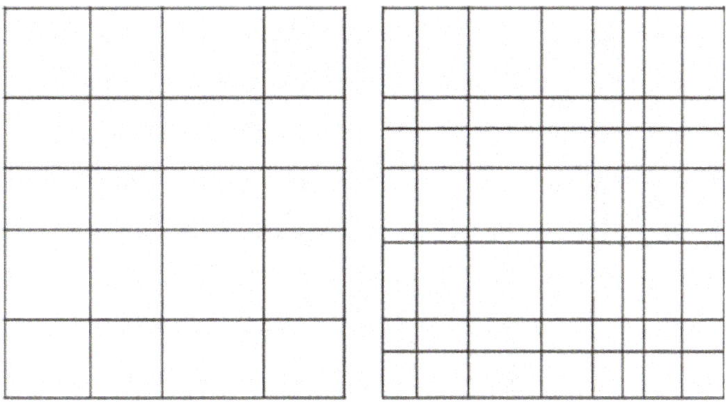

Figure 1.3: A refinement.

question is if we can find a partition of a set with arbitrarily small mesh. The answer to this question is contained in the following result.

Theorem 1.20. *For any $\varepsilon > 0$, there is a partition τ of X such that its mesh is less than ε.*

Proof. Let $k, j_k \in \mathbb{N}$ be arbitrarily chosen. For any $j \in \{1, \ldots, j_k\}$, take a cube Q_j^k of rank k such that

$$X \cap Q_j^k \neq \emptyset.$$

Denote

$$X_j^k = X \cap Q_j^k, \quad j \in \{1, \ldots, j_k\}.$$

We will prove that the system $\{X_j^k\}_{j=1}^{j_k}$ is a partition of X. Indeed, since the intersection of two Jordan-measurable sets is a Jordan-measurable set, we have that X_j^k, $j \in \{1, \ldots, j_k\}$, is a Jordan-measurable set. Moreover, we have

$$X = \bigcup_{j=1}^{j_k} X_j^k.$$

Note that

$$\begin{aligned} X_l^k \cap X_j^k &\subset Q_l^k \cap Q_j^k \\ &\subset \partial Q_l^k \cap \partial Q_k^k, \quad l \neq j, \ l, j \in \{1, \ldots, j_k\}. \end{aligned}$$

Since Q_j^k, $j \in \{1, \ldots, j_k\}$, is Jordan measurable, we get that

$$\mu \partial Q_j^k = 0, \quad j \in \{1, \ldots, j_k\}.$$

Hence

$$\begin{aligned} \mu(X_l^k \cap X_j^k) &\leq \mu(\partial Q_l^k \cap \partial Q_j^k) \\ &\leq \mu(\partial Q_j^k) \\ &= 0, \quad l \neq j, \ l, j \in \{1, \ldots, j_k\}. \end{aligned}$$

Because the volume of the n-dimensional cube of rank k is $\frac{\sqrt{n}}{10^k}$, we get

$$\begin{aligned} |\tau_k| &= \max_{j \in \{1, \ldots, j_k\}} \operatorname{diam} X_j^k \\ &\leq \operatorname{diam} Q_1^k \\ &= \frac{\sqrt{n}}{10^k} \end{aligned}$$

$$\to 0 \quad \text{as } k \to \infty.$$

This completes the proof. □

Now we give a precise definition of multiple Riemann integrals.

Definition 1.11. Let f be a function defined on a Jordan-measurable set X, and let $\tau = \{X_j\}_{j=1}^{j_\tau}$ be a partition of X. Take arbitrary points $\xi^j \in X_j, j \in \{1, \ldots, j_\tau\}$. The sum

$$\sigma_\tau = \sigma_\tau(f; \xi^1, \ldots, \xi^{j_\tau}) = \sum_{j=1}^{j_\tau} f(\xi^j)\mu X_j$$

is called the Riemann integral sum of the function f.

Definition 1.12. The number I is called the Riemann integral of f over the Jordan-measurable set X if for any sequence of partitions

$$\tau_m = \{X_j^m\}_{j=1}^{j_\tau}, \quad m \in \mathbb{N},$$

of the set X such that

$$\lim_{|\tau| \to 0} |\tau_m| = 0$$

and for any points $\xi^{j,m} \in X_j^m, j \in \{1, \ldots, j_m\}$, the sequence of Riemann integral sums

$$\sigma_{\tau_m}(f; \xi^{1,m}, \ldots, \xi^{j_m,m}), \quad m \in \mathbb{N},$$

converges to I, i. e.,

$$\lim_{m \to \infty} \sigma_{\tau_m}(f; \xi^{1,m}, \ldots, \xi^{j_m,m}) = I. \tag{1.16}$$

In such a case, we will denote the number I by

$$\int f(x)dX, \quad \int_X f(x)dx, \quad \text{or} \quad \int \cdots \int_X f(x_1, \ldots, x_n)dx_1 \ldots dx_n.$$

If the integral $\int_X f(x)dx$ exists, then we say that the function f is Riemann integrable over X. Equality (1.16) can be rewritten in the form

$$\lim_{|\tau| \to 0} \sigma_\tau = \int_X f(x)dx.$$

This definition can be reformulated in the following form.

Definition 1.13. For any $\varepsilon > 0$, there exists $\delta = \delta(\varepsilon) > 0$ such that for any partition $\tau = \{X_j\}_{j=1}^{j_\tau}$ of the set X with $|\tau| < \delta$ and for any points $\xi^j \in X_j, j \in \{1, \dots, j_\tau\}$, we have the inequality

$$\left|\sigma_\tau(f; \xi^1, \dots, \xi^{j_\tau}) - \int_X f(\tau)d\tau\right| < \varepsilon.$$

Let $X_0 \subset X$, and let $\tau = \{X_j\}_{j=1}^{j_\tau}$ be a partition of X. By $\tau(X_0)$ we denote the set of all elements of τ that do not intersect X_0, and by $\tau_0(X_0)$ we denote the set of all elements of τ that intersect X_0, i. e.,

$$\tau(X_0) = \{X_j \in \tau : X_j \cap X_0 = \emptyset\},$$
$$\tau_0(X_0) = \{X_j \in \tau : X_j \cap X_0 \neq \emptyset\}.$$

An important property of the Jordan measure of the set X_0 is as follows.

Theorem 1.21. *If* $\mu X_0 = 0$, *then*

$$\lim_{|\tau|\to 0} \sum_{X_j \in \tau_0(X_0)} \mu X_j = 0.$$

Proof. We have that

$$\mu \overline{X_0} = 0.$$

Then, for any $\varepsilon > 0$, there exists a rank k such that

$$\mu S_k(\overline{X_0}) < \varepsilon. \tag{1.17}$$

Note that

$$\overline{X_0} \subset S_k(\overline{X_0}) \quad \text{and} \quad \overline{X_0} \subset S_k(\overline{X_0})_{\text{int}}.$$

Hence, if $x \in \overline{X_0}$, then there is a neighborhood of x contained in $S_k(\overline{X_0})$, and therefore

$$\overline{X_0} \cap (\mathbb{R}^n \backslash S_k(\overline{X_0})_{\text{int}}) = \emptyset.$$

From this we conclude that

$$\delta = \operatorname{dist}(\overline{X_0}, \mathbb{R}^n \backslash S_k(\overline{X_0})) > 0$$

and any set Y with $\operatorname{diam} Y < \delta$ intersects the set X_0 and lies in $S_k(\overline{X_0})$. Now let $\tau = \{X_j\}_{j=1}^{j_\tau}$ be a partition of X such that $|\tau| < \delta$ and $X_j \in \tau_0(X_0)$. Then $X_j \cap X_0 \neq \emptyset$, and since $\operatorname{diam} X_j < \delta$, we get

$$X_j \subset S_k(\overline{X_0}).$$

Consequently,

$$\bigcup_{X_j \in \tau_0(X_0)} X_j \subset S_k(\overline{X_0}).$$

Hence, using (1.17), we find

$$\begin{aligned}\sum_{X_j \in \tau_0(X_0)} \mu X_j &= \mu \bigcup_{X_j \in \tau_0(X_0)} X_j \\ &\leq \mu S_k(\overline{X_0}) \\ &< \varepsilon.\end{aligned}$$

This completes the proof. □

So the questions are: what functions are integrable or, at least, what are some general classes of integrable functions, and how does we evaluate their integrals? We give a necessary and sufficient condition for integrability of a class of functions.

Theorem 1.22. *Let $\tau = \{X_j\}_{j=1}^{j_\tau}$ be a partition of a set X, let $X_0 \subset X$ with $\mu X_0 = 0$, and let $f : X \to \mathbb{R}$ be a bounded function. Then the integral*

$$\int_X f(x)dx$$

exists if and only if there exists the limit

$$\lim_{|\tau| \to 0} \sigma_{\tau(X_0)}(f; \xi^1, \ldots, \xi^{j_\tau})$$

for all $\xi^l \in X_l$, $l \in \{1, \ldots, j_\tau\}$. Moreover, if this limit exists, then it is equal to the integral $\int_X f(x)dx$.

Proof. Note that for any $X_j \in \tau$, $j \in \{1, \ldots, j_\tau\}$, we have

$$X_j \in \tau(X_0) \quad \text{or} \quad X_j \in \tau_0(X_0).$$

Thus, since $\tau(X_0)$ and $\tau_0(X_0)$ have no common point, we obtain

$$\tau = \tau(X_0) \cup \tau_0(X_0).$$

Moreover, if $\xi^j \in X_j$, $j \in \{1, \ldots, j_\tau\}$, then

$$\sigma_\tau(f; \xi^1, \ldots, \xi^{j_\tau}) = \sigma_{\tau(X_0)}(f; \xi^1, \ldots, \xi^{j_\tau}) + \sigma_{\tau_0(X_0)}(f; \xi^1, \ldots, \xi^{j_\tau}). \tag{1.18}$$

Since f is bounded on X, there exists a constant $M > 0$ such that

$$|f(x)| \leq M, \quad x \in X.$$

Therefore

$$\begin{aligned}|\sigma_{\tau_0(X_0)}(f;\xi^1,\ldots,\xi^{j_\tau})| &\leq \sum_{X_j\in\tau_0(X_0)} |f(\xi^j)|\mu X_j\\ &\leq M \sum_{X_j\in\tau_0(X_0)} \mu X_j.\end{aligned}$$

By Theorem 1.21 it follows that

$$\lim_{|\tau|\to 0} \sum_{X_j\in\tau_0(X_0)} \mu X_j = 0,$$

from which we have

$$\lim_{|\tau|\to 0} \sigma_{\tau_0(X_0)}(f;\xi^1,\ldots,\xi^{j_\tau}) = 0.$$

Now applying (1.18), we conclude that

$$\lim_{|\tau|\to 0} \sigma_\tau(f;\xi^1,\ldots,\xi^{j_\tau}) \quad \text{and} \quad \lim_{|\tau|\to 0} \sigma_{\tau(X_0)}(f;\xi^1,\ldots,\xi^{j_\tau})$$

exist or do not exist at the same time. This completes the proof. □

Example 1.1. Let $X \subset \mathbb{R}^n$ be a Jordan-measurable set with $\mu X = 0$, and let $f : X \to \mathbb{R}$. Then, for any partition $\tau = \{X_j\}_{j=1}^{j_\tau}$, we have

$$\mu X_j = 0, \quad j \in \{1,\ldots,j_\tau\},$$

and for $\xi_j \in X_j$, we get

$$f(\xi^j)\mu X_j = 0.$$

Hence

$$\begin{aligned}\sigma_\tau(f;\xi^1,\ldots,\xi^m) &= \sum_{j=1}^{j_\tau} f(\xi^j)\mu X_j\\ &= 0.\end{aligned}$$

Now applying the definition of multiple integrals, we arrive at

$$\int\limits_X f(x)dx = \sum_{|\tau|\to 0} \sigma_\tau(f;\xi^1,\ldots,\xi^{j_\tau})$$
$$= 0.$$

Now we give a criterion for non-integrability of a function. Before this, we need the following auxiliary result.

Theorem 1.23. *Let X be a Jordan-measurable set, let $f : X \to \mathbb{R}$, let $\tau = \{X_j\}_{j=1}^{j_\tau}$ be a partition of X, and let*

$$X_\tau^* = \bigcup_{\mu X_j > 0} X_j.$$

If f is unbounded on X_τ^*, then for any $M > 0$, there exist $\xi^j \in X_j$, $j \in \{1,\ldots,j_\tau\}$, such that

$$\left|\sum_{j=1}^{j_\tau} f(\xi^j)\mu X_j\right| > M.$$

Proof. Since any partition τ of X consists of a finite number of elements, by the definition of X_τ^* we conclude that $\mu X_\tau^* > 0$. Because f is unbounded on X, there is $l \in \{1,\ldots,j_\tau\}$ such that $\mu X_l > 0$ and f is unbounded on X_l. Without loss of generality, suppose that $l = 1$. Now since f is unbounded on X_1, we can find a sequence $\{\xi_n^1\}_{n=1}^\infty \subset X_1$ such that

$$\lim_{n\to\infty} f(\xi_n^1) = \infty.$$

Take $\xi^m \in X_m$, $m \in \{2,\ldots,j_\tau\}$. Then

$$\sum_{j=2}^{j_\tau} f(\xi^j)\mu X_j$$

is a fixed number. Hence, using that $\mu X_1 > 0$, we find

$$\lim_{n\to\infty}\left|f(\xi_n^1)\mu(X_1) + \sum_{j=2}^{j_\tau} f(\xi^j)\mu X_j\right| = \infty.$$

Therefore, for any $M > 0$, there is $n_0 \in \mathbb{N}$ such that

$$\left|f(\xi_{n_0}^1)\mu(X_1) + \sum_{j=2}^{j_\tau} f(\xi^j)\mu X_j\right| > M,$$

which completes the proof. □

A criterion for non-integrability is as follows.

Theorem 1.24. *Let X be a Jordan-measurable set in $\mathbb{R}^n$, and let $f : X \to \mathbb{R}$. If there exists a sufficiently small partition $\tau = \{X_j\}_{j=1}^{j_\tau}$ such that f is unbounded on X_τ^*, then f is not integrable on X.*

Proof. Suppose that f is integrable on X. Then

$$\lim_{|\tau|\to 0} \sum_{j=1}^{j_\tau} f(\xi^j)\mu X_j = \int_X f(x)dx, \quad \tau = \{X_j\}_{j=1}^{j_\tau},\ \xi^j \in X_j,\ j \in \{1,\dots,j_\tau\}.$$

Hence, for any $\varepsilon > 0$, there is $\delta = \delta(\varepsilon) > 0$ such that for any partition $\tau = \{X_j\}_{j=1}^{j_\tau}$ with $|\tau| < \delta$ and for any $\xi^j \in X_j, j \in \{1,\dots,j_\tau\}$, we have the inequality

$$\left|\sum_{j=1}^{j_\tau} f(\xi^j)\mu X_j - \int_X f(x)dx\right| < \varepsilon.$$

Then

$$-\varepsilon + \int_X f(x)dx < \sum_{j=1}^{j_\tau} f(\xi^j)\mu X_j < \varepsilon + \int_X f(x)dx. \tag{1.19}$$

Thus the set $\{\sigma_\tau : |\tau| < \delta\}$ is bounded. On the other hand, by Theorem 1.23 it follows that for any $M > 0$, any small enough partition $\tau = \{X_j\}_{j=1}^{j_\tau}$, and any $\xi^j \in X_j, j \in \{1,\dots,j_\tau\}$, we have

$$\left|\sum_{j=1}^{j_\tau} f(\xi^j)\mu X_j\right| > M,$$

which contradicts (1.19). Therefore f is not integrable on X. This completes the proof. □

The next result tells us that if a function is integrable on some set, then there is its subset where the function is bounded.

Theorem 1.25. *Let X be a Jordan-measurable set, and let $f : X \to \mathbb{R}$ be an integrable function on X. Then there is a set $X_0 \subset X$ such that $\mu X_0 = 0$ and f is bounded on $X \backslash X_0$.*

Proof. Assume that there is no subset X_0 of X such that f is bounded on $X\backslash X_0$. Take arbitrary $\delta > 0$ and a partition $\tau = \{X_j\}_{j=1}^{j_\tau}$ such that $|\tau| < \delta$. Then

$$X\backslash X_\tau = \{X_l \in \tau : \mu X_l = 0\}.$$

Since $X\backslash X_\tau^*$ contains a finite number of elements, we conclude that

$$\mu(X\backslash X_\tau^*) = 0.$$

By our assumption it follows that f is unbounded on X_τ^*. Now applying Theorem 1.24, we get that f is not integrable on X. This is a contradiction. This completes the proof. □

Theorem 1.26. *Let X be an open Jordan-measurable set. If f is integrable on X, then it is bounded.*

Proof. By Theorem 1.14 it follows that there is an enough small partition $\tau = \{X_j\}_{j=1}^{j_\tau}$ such that $\mu X_j > 0, j \in \{1, \dots, j_\tau\}$. Assume that f is unbounded on X. Then by Theorem 1.24 it follows that f is not integrable on X. This is a contradiction. Therefore f is bounded on X. This completes the proof. □

Remark 1.1. In Theorem 1.26 the fact that X is open is used to ensure that there is a partition $\tau = \{X_j\}_{j=1}^{j_\tau}$ of X such that $\mu X_j > 0, j \in \{1, \dots, j_\tau\}$. Thus Theorem 1.26 is valid for any Jordan-measurable set X for which there is a partition $\tau = \{X_j\}_{j=1}^{j_\tau}$ such that $\mu X_j > 0$, $j \in \{1, \dots, j_\tau\}$.

Remark 1.2. Suppose that X is an open Jordan-measurable set. Consider all cubes Q_j of rank k such that $Q_j \cap X \neq \emptyset$. Then $Q_j \cap \overline{X} \neq \emptyset$. Since $S_k(X)$ is a closed set and

$$X \subset S_k(X),$$

we get

$$\overline{X} \subset S_k(X).$$

Set

$$P_j = Q_j \cap \overline{X}, \quad Q_j \cap X \neq \emptyset.$$

Then $\tau = \{P_j\}$ is a cover of the set $\overline{X}$, and $S_k(X)$ consists of the cubes Q_j for which $Q_j \cap X \neq \emptyset$. Therefore for $\overline{X}$, there is a partition $\tau_1 = \{X_j\}_{j=1}^{j_{\tau_1}}$ such that $\mu X_j > 0, j \in \{1, \dots, j_{\tau_1}\}$.

As in the one-dimensional case, if the function f is bounded on a Jordan-measurable set, then it is possible to define the lower and upper Darboux[4] sums for this function. This can be done as follows.

Definition 1.14. Let X be a Jordan-measurable set, let $f : X \to \mathbb{R}$ be a bounded function, let $\tau = \{X_j\}_{j=1}^{j_\tau}$ be a partition of X, and let

$$m_j = \inf_{x \in X_j} f(x),$$
$$M_j = \sup_{x \in X_j} f(x), \quad j \in \{1, \dots, j_\tau\}.$$

4 Jean-Gaston Darboux(14 August 1842–23 February 1917) was a French mathematician. Darboux made several important contributions to geometry and mathematical analysis. He was a biographer of Henri Poincaré, and he edited the Selected Works of Joseph Fourier.

The sums

$$s_\tau = \sum_{j=1}^{j_\tau} m_j \mu X_j,$$

$$S_\tau = \sum_{j=1}^{j_\tau} M_j \mu X_j$$

are called the lower and upper Darboux sums of f, respectively. Since f is a bounded function on X, we have that the lower and upper Darboux sums exist.

By the definition of lower and upper Darboux sums, we have

$$s_\tau \leq \sigma_\tau \leq S_\tau$$

for any partition τ of X.

Graphing f over X in the usual fashion when $n = 2$ and interpreting the lower and upper sums as sums of box-volumes shows that they are respectively too small and too big to be the volume under the graph (see Figure 1.4). Alternatively, if $n = 2$ or $n = 3$, then thinking of f as the density of a plate or a block occupying the box B shows that the lower and upper sums are too small and too big to be the object mass. Again, the hope is that as the partitions become finer, the lower and upper sums will converge to a common value that they are trapping from either side.

Figure 1.4: Too small and too big.

Refining a partition brings the lower and upper sums nearer to each other.

Theorem 1.27. *Let X be a Jordan-measurable set, and let $f : X \to \mathbb{R}$ be a bounded function. If $\tau_1 \prec \tau_2$, then*

$$s_{\tau_1} \leq s_{\tau_2} \quad \text{and} \quad S_{\tau_2} \leq S_{\tau_1}.$$

See Figure 1.5 for a picture-proof for lower sums when $n = 1$, thinking of the sums in terms of area. The formal proof is just a symbolic rendition of the figure features.

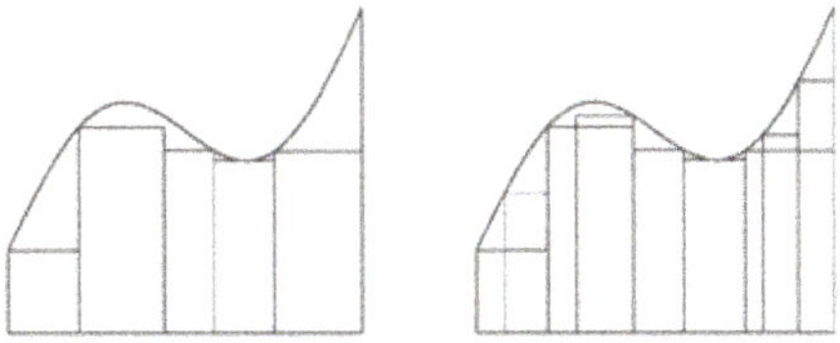

Figure 1.5: Lower sum increasing under refinement.

Proof. Let $\tau_1 = \{X_j^1\}_{j=1}^{j_{\tau_1}}$, $\tau_2 = \{X_j^2\}_{j=1}^{j_{\tau_2}}$, and

$$\begin{aligned} m_j^1 &= \inf_{x\in X_j^1} f(x), \\ M_j^1 &= \sup_{x\in X_j^1} f(x), \quad j \in \{1,\dots,j_{\tau_1}\}, \\ m_l^2 &= \inf_{x\in X_l^2} f(x), \\ M_l^2 &= \sup_{x\in X_l^2} f(x), \quad l \in \{1,\dots,j_{\tau_2}\}. \end{aligned}$$

If $X_l^2 \subset X_l^1$, then we have

$$\begin{aligned} m_l^1 &\le m_l^2, \\ M_l^2 &\le M_l^1. \end{aligned}$$

Because $\tau_1 \prec \tau_2$, we have that any $X \in \tau_1$ is a union of elements of τ_2 denoted by

$$X_j^1 = \bigcup_{l_j} X_{l_j}^2.$$

Then

$$\mu X_j^1 = \sum_{l_j} \mu X_{l_j}^2.$$

Hence

$$\begin{aligned} s_{\tau_1} &= \sum_{j=1}^{j_{\tau_1}} m_j \mu X_j^1 \\ &= \sum_{j=1}^{j_{\tau_1}} m_j \sum_{l_j} \mu X_{l_j}^2 \\ &= \sum_{j=1}^{j_{\tau_1}} \sum_{l_j} m_j \mu X_{l_j}^2 \end{aligned}$$

$$
\begin{aligned}
&\le \sum_{j=1}^{j_{\tau_1}} \sum_{l_j} m_{l_j} \mu X_{l_j}^2 \\
&= \sum_{l=1}^{j_{\tau_2}} m_l \mu X_l^2 \\
&= s_{\tau_2},
\end{aligned}
$$

and

$$
\begin{aligned}
S_{\tau_1} &= \sum_{j=1}^{j_{\tau_1}} M_j \mu X_j^1 \\
&= \sum_{j=1}^{j_{\tau_1}} M_j \sum_{l_j} \mu X_{l_j}^2 \\
&= \sum_{j=1}^{j_{\tau_1}} \sum_{l_j} M_j \mu X_{l_j}^2 \\
&\ge \sum_{j=1}^{j_{\tau_1}} \sum_{l_j} M_{l_j} \mu X_{l_j}^2 \\
&= \sum_{l=1}^{j_{\tau_2}} M_l \mu X_l^2 \\
&= S_{\tau_2}.
\end{aligned}
$$

This completes the proof. □

A relation between Riemann sums and lower and upper Darboux sums is as follows.

Theorem 1.28. *Let X be a Jordan-measurable set, let $f : X \to \mathbb{R}$ be a bounded function, and let $\sigma_\tau = \sigma_\tau(f, \xi^1, \ldots, \xi^{j_\tau})$ be any Riemannian sum of f with respect to a partition $\tau = \{X_j\}_{j=1}^{j_\tau}$ of X. Then*

$$
\begin{aligned}
s_\tau &= \inf_{\xi^1,\ldots,\xi^{j_\tau}} \sigma_\tau(f, \xi^1, \ldots, \xi^{j_\tau}), \\
S_\tau &= \sup_{\xi^1,\ldots,\xi^{j_\tau}} \sigma_\tau(f, \xi^1, \ldots, \xi^{j_\tau}).
\end{aligned}
$$

Proof. We have

$$
\begin{aligned}
s_\tau &= \sum_{j=1}^{j_\tau} m_j \mu X_j \\
&= \sum_{j=1}^{j_\tau} \inf_{x \in X_j} f(x) \mu X_j
\end{aligned}
$$

$$= \inf_{\xi^1,\ldots,\xi^{j_\tau}} \sum_{j=1}^{j_\tau} f(\xi^j)\mu X_j$$
$$= \inf_{\xi^1,\ldots,\xi^{j_\tau}} \sigma_\tau(f,\xi^1,\ldots,\xi^{j_\tau})$$

and

$$\begin{aligned} S_\tau &= \sum_{j=1}^{j_\tau} M_j \mu X_j \\ &= \sum_{j=1}^{j_\tau} \sup_{x\in X_j} f(x)\mu X_j \\ &= \sup_{\xi^1,\ldots,\xi^{j_\tau}} \sum_{j=1}^{j_\tau} f(\xi^j)\mu X_j \\ &= \sup_{\xi^1,\ldots,\xi^{j_\tau}} \sigma_\tau(f,\xi^1,\ldots,\xi^{j_\tau}). \end{aligned}$$

This completes the proof. □

The difference of the upper Darboux sum and lower Darboux sum can be represented in terms of the modulus of continuity.

Theorem 1.29. *Let X be a Jordan-measurable set, let $f : X \to \mathbb{R}$ be a bounded function, and let $\tau = \{X_j\}_{j=1}^{j_\tau}$ be a partition of X. Then*

$$S_\tau - s_\tau = \sum_{j=1}^{j_\tau} \omega(f, X_j).$$

Proof. We have

$$\begin{aligned} M_j - m_j &= \sum_{x\in X_j} - \inf_{x\in X_j} f(x) \\ &= \sup_{x,y\in X_j} (f(x) - f(y)) \\ &= \omega(f, X_j). \end{aligned}$$

Thus

$$\begin{aligned} S_\tau - s_\tau &= \sum_{j=1}^{j_\tau} M_j\mu X_j - \sum_{j=1}^{j_\tau} m_j \mu X_j \\ &= \sum_{j=1}^{j_\tau} (M_j - m_j)\mu X_j \end{aligned}$$

$$= \sum_{j=1}^{j_\tau} \omega(f, X_j)\mu X_j.$$

This completes the proof. □

Definition 1.15. Let X be a Jordan-measurable set, and let $f : X \to \mathbb{R}$ be a bounded function. The numbers

$$I_* = \sup_\tau s_\tau \quad \text{and} \quad I^* = \inf_\tau S_\tau$$

are said to be the lower and upper Darboux integrals of f over X. We have

$$s_\tau \leq I_* \leq I^* \leq S_\tau.$$

The next result will help us to see that any continuous function is Riemann integrable.

Theorem 1.30. *Let X be a Jordan-measurable set, and let $f : X \to \mathbb{R}$ be a bounded function. Then f is Riemann integrable over X if and only if*

$$\lim_{|\tau|\to 0} (S_\tau - s_\tau) = 0. \tag{1.20}$$

Proof. 1. Let f be a Riemann-integrable function over X, and let

$$I = \int_X f(x)dx.$$

Take arbitrary $\varepsilon > 0$. Then there is $\delta = \delta(\varepsilon) > 0$ such that for any partition τ of X for which $|\tau| < \delta$, we have

$$|\sigma_\tau - I| < \frac{\varepsilon}{2}$$

or

$$I - \frac{\varepsilon}{2} < \sigma_\tau < I + \frac{\varepsilon}{2},$$

and

$$I - \frac{\varepsilon}{2} < s_\tau \leq \sigma_\tau \leq S_\tau < I + \frac{\varepsilon}{2}.$$

Hence

$$\begin{aligned} S_\tau - s_\tau &< I + \frac{\varepsilon}{2} - I + \frac{\varepsilon}{2} \\ &= \varepsilon. \end{aligned}$$

So (1.20) holds.

2. Assume that (1.20) holds. Then

$$\begin{aligned} 0 &\le I^* - I_* \\ &\le S_\tau - s_\tau, \end{aligned}$$

and

$$\begin{aligned} 0 &\le \lim_{|\tau|\to 0} (I^* - I_*) \\ &\le \lim_{|\tau|\to 0} (S_\tau - s_\tau) \\ &= 0. \end{aligned}$$

Therefore

$$I_* = I^* = I.$$

We have

$$s_\tau \le I \le S_\tau$$

and

$$\begin{aligned} 0 \le I - s_\tau \le S_\tau - s_\tau, \\ 0 \le S_\tau - I \le S_\tau - s_\tau. \end{aligned}$$

Now applying (1.20), we get

$$\begin{aligned} 0 &\le \lim_{|\tau|\to 0} (I - s_\tau) \\ &= \lim_{|\tau|\to 0} (S_\tau - I) \\ &\le \lim_{|\tau|\to 0} (S_\tau - s_\tau) \\ &= 0, \end{aligned}$$

i. e.,

$$\lim_{|\tau|\to 0} S_\tau = \lim_{|\tau|\to 0} s_\tau = I.$$

Note that

$$s_\tau \le \sigma_\tau \le S_\tau,$$

and then

$$\begin{aligned} I &= \lim_{|\tau|\to 0} s_\tau \\ &\le \lim_{|\tau|\to 0} \sigma_\tau \\ &\le \lim_{|\tau|\to 0} S_\tau \\ &= I, \end{aligned}$$

i. e.,

$$\lim_{|\tau|\to 0} \sigma_\tau = I.$$

So f is Riemann integrable over X. This completes the proof. □

Although the integrability criterion gives a test for the integrability of any specific function f, it is cumbersome to apply case by case. But handily, it will provide the punchline of the proof of the next theorem, which says that a natural class of functions is integrable.

Theorem 1.31. *Let X be a Jordan-measurable compact set, and let $f : X \to \mathbb{R}$ be a continuous function on X. Then f is integrable over X.*

Proof. Since X is a compact set in $\mathbb{R}^n$ and f is a continuous function on X, we have that f is bounded and uniformly continuous on X. Let $\varepsilon > 0$ be arbitrarily chosen. Using that f is uniformly continuous on X, there is $\delta = \delta(\varepsilon) > 0$ such that if $|x - y| < \delta$, then

$$|f(x) - f(y)| < \frac{\varepsilon}{\mu X}.$$

Take a partition $\tau = \{X_j\}_{j=1}^{j_\tau}$ of X such that $|\tau| < \delta$. Let

$$\begin{aligned} m_j &= \inf_{x\in X_j} f(x), \\ M_j &= \sup_{x\in X_j} f(x), \quad j \in \{1, \dots, n\}. \end{aligned}$$

Then there are $\xi^j \in X_j$ and $\eta^j \in X_j, j \in \{1, \dots, j_\tau\}$, such that

$$\begin{aligned} M_j &= f(\xi^j), \\ m_j &= f(\eta^j), \quad j \in \{1, \dots, j_\tau\}. \end{aligned}$$

Because $|\tau| < \delta$, we have that

$$|\xi^j - \eta^j| < \delta, \quad j \in \{1, \dots, j_\tau\},$$

and thus

$$\begin{aligned} M_j - m_j &= f(\xi^j) - f(\eta^j) \\ &< \frac{\varepsilon}{\mu X}, \quad j \in \{1, \dots, j_\tau\}. \end{aligned}$$

Therefore

$$\begin{aligned} 0 &\le S_\tau - s_\tau \\ &= \sum_{j=1}^{j_\tau} (M_j - m_j)\mu X_j \\ &< \sum_{j=1}^{j_\tau} \frac{\varepsilon}{\mu X} \mu X_j \\ &= \frac{\varepsilon}{\mu X} \sum_{j=1}^{j_\tau} \mu X_j \\ &= \frac{\varepsilon}{\mu X} \mu X \\ &= \varepsilon. \end{aligned}$$

Now applying Theorem 1.30, we conclude that f is integrable over X. This completes the proof. □

The continuity of a function is not a necessary condition for its integrability. There are discontinuous integrable functions. A wide class of discontinuous integrable functions is contained in the next result.

Theorem 1.32. *Let X be a Jordan-measurable compact set, and let $f : X \to \mathbb{R}$ be a bounded function. If the Jordan measure of the set of all points of discontinuity of the function f is zero, then f is Riemann integrable over X.*

Proof. Since f is bounded over X, there is a constant $M > 0$ such that

$$|f(x)| \le M, \quad x \in X.$$

Let X_0 be the set of all points of discontinuity of the function f. Then $\mu X_0 = 0$, and

$$\lim_{m \to \infty} \mu S_m(X_0) = 0.$$

Fix arbitrary $\varepsilon > 0$. Then there is a rank k such that

$$\mu S_k(X_0) < \frac{\varepsilon}{3^n 4M}. \tag{1.21}$$

Let

$$S_k(X_0) = \bigcup_{j=1}^{l} Q_j.$$

By P_j, $j \in \{1, \dots, l\}$, we will denote the cube obtained by the cube Q_j, $j \in \{1, \dots, l\}$, via the similarity transformation centered at the center of the cube Q_j, $j \in \{1, \dots, l\}$, and the similarity coefficient 3. Then

$$\mu P_j = 3^n \mu Q_j, \quad j \in \{1, \dots, l\}.$$

Set

$$P = \bigcup_{j=1}^{l} P_j.$$

Then applying inequality (1.21), we get

$$\begin{aligned} \mu P &= \mu\left(\bigcup_{j=1}^{l} P_j\right) \\ &\leq \sum_{j=1}^{l} \mu P_j \\ &= \sum_{j=1}^{l} 3^n \mu Q_j \\ &= 3^n \mu S_k(X_0) \\ &< 3^n \frac{\varepsilon}{3^n 4M} \\ &= \frac{\varepsilon}{4M}. \end{aligned}$$

Note that for any set A with $\operatorname{diam} A < 10^{-k}$ and

$$A \cap S_k(X_0) \neq \emptyset,$$

we have $A \subset P$. Let G be the set of all interior points of $S_k(X_0)$. We have that G is an open set. Since X is a compact set, it is a bounded and closed set. Therefore the set $F = X \backslash G$ is a closed and bounded set, i. e., F is a compact set. Note that $X_0 \subset S_k(X_0)$ and thus $X_0 \subset G$. Therefore f is a continuous function on F. Moreover, since F is a difference of two Jordan-measurable sets, it is a Jordan-measurable set. Now applying Theorem 1.31, we conclude that f is Riemann integrable over F. Then there is $\delta = \delta(\varepsilon) > 0$ such that for any partition τ_F of F with $|\tau_F| < \delta$, we have

$$S_{\tau_F} - s_{\tau_F} < \frac{\varepsilon}{2}. \tag{1.22}$$

Let

$$\delta_0 = \min\{10^{-k}, \delta\},$$

and let $\tau = \{X_j\}_{j=1}^{j_\tau}$ be any partition of X with $|\tau| < \delta$. Observe that

$$\tau_F = \{X_j \cap F\}_{j=1}^{j_\tau}, \quad X_j \cap F \neq \emptyset,\ j \in \{1, \dots, j_\tau\},$$

is a partition of F with

$$\begin{aligned} |\tau_F| &\le |\tau| \\ &< \delta_0, \end{aligned}$$

and inequality (1.22) holds. Set

$$\begin{aligned} M_j &= \sup_{x\in X_j} f(x), \\ m_j &= \inf_{x\in X_j} f(x), \\ S_\tau &= \sum_{j=1}^{j_\tau} M_j \mu X_j, \\ s_\tau &= \sum_{j=1}^{j_\tau} m_j \mu X_j, \\ M_j' &= \sup_{x\in X_j\cap F} f(x), \\ m_j' &= \inf_{x\in X_j\cap F} f(x), \\ S_{\tau_F} &= \sum_{X_j\cap F\neq\emptyset} M_j' \mu X_j, \\ s_\tau &= \sum_{X_j\cap F\neq\emptyset} m_j' \mu X_j. \end{aligned}$$

If $X_j \cap G = \emptyset, j \in \{1, \dots, j_\tau\}$, then $X_j \subset F, j \in \{1, \dots, j_\tau\}$, and in this case, we have

$$\begin{aligned} M_j &= M_j', \\ X_j \cap F &= X_j, \quad j \in \{1, \dots, j_\tau\}, \end{aligned}$$

and

$$\begin{aligned}\sum_{X_j\cap G=\emptyset}(M_j-m_j)\mu X_j &= \sum_{X_j\cap G=\emptyset}(M_j'-m_j')\mu X_j\\ &= \sum_{X_j\cap G=\emptyset}(M_j'-m_j')\mu(X_j\cap F)\\ &= S_{\tau_F}-s_{\tau_F}\\ &< \frac{\varepsilon}{2}.\end{aligned} \tag{1.23}$$

If $X_j\cap G\neq\emptyset, j\in\{1,\dots,j_\tau\}$, then $X_j\subset P, j\in\{1,\dots,j_\tau\}$, and

$$\begin{aligned}\sum_{X_j\cap G\neq\emptyset}\mu X_j &= \mu\Big(\bigcup_{X_j\cap G\neq\emptyset}X_j\Big)\\ &\le \mu P\\ &< \frac{\varepsilon}{4M}.\end{aligned}$$

Note that

$$|m_j|,\ |M_j|\le M,\quad j\in\{1,\dots,j_\tau\}.$$

Then

$$\begin{aligned}\sum_{X_j\cap G\neq\emptyset}(M_j-m_j)\mu X_j &\le \sum_{X_j\cap G\neq\emptyset}(|M_j|+|m_j|)\mu X_j\\ &\le 2M\sum_{X_j\cap G\neq\emptyset}\mu X_j\\ &< 2m\frac{\varepsilon}{4M}\\ &< \frac{\varepsilon}{2}.\end{aligned}$$

Now applying the last inequality and inequality (1.23), we obtain

$$\begin{aligned}S_\tau - s_\tau &= \sum_{j=1}^{j_\tau}(M_j-m_j)\mu X_j\\ &= \sum_{X_j\cap G\neq\emptyset}(M_j-m_j)\mu X_j+\sum_{X_j\cap G=\emptyset}(M_j-m_j)\mu X_j\\ &< \frac{\varepsilon}{2}+\frac{\varepsilon}{2}\\ &= \varepsilon.\end{aligned}$$

Because $\varepsilon > 0$ was arbitrarily chosen, we get

$$\lim_{|\tau|\to 0}(S_\tau - s_\tau) = 0.$$

Hence by Theorem 1.30 it follows that f is Riemann integrable over X. This completes the proof. □

1.3 Criteria for integrability

In the previous section, we have studied some criteria for the integrability of a function. In this section, we give another characterization of integrable functions. For this aim, we need the following quantities. Let $X \subset \mathbb{R}^n$ be a Jordan-measurable set, let $f : X \to \mathbb{R}$ be a bounded function, let $\tau = \{X_j\}_{j=1}^{j_\tau}$ be a partition of X, and let

$$\begin{aligned} m_j &= \inf_{x \in X_j} f(x), \\ M_j &= \sup_{x \in X_j} f(x), \quad j \in \{1, \ldots, j_\tau\}, \\ s_\tau &= \sum_{j=1}^{j_\tau} m_j \mu X_j, \\ S_\tau &= \sum_{j=1}^{j_\tau} M_j \mu X_j, \\ I_* &= \sup_\tau s_\tau, \\ I^* &= \inf_\tau S_\tau. \end{aligned}$$

Relations for the defined quantities are given in the following result.

Theorem 1.33. *We have*

$$\begin{aligned} I_* &= \lim_{|\tau| \to 0} s_\tau, \\ I^* &= \lim_{|\tau| \to 0} S_\tau. \end{aligned}$$

Proof. Since f is a bounded function on X, there is a positive constant M such that

$$|f(x)| \le M, \quad x \in X.$$

Take arbitrary $\varepsilon > 0$. By the definition of I_* it follows that there is a partition $\tau_1 = \{X_{1j}\}_{j=1}^{j_{\tau_1}}$ of the set X such that

$$s_{\tau_1} > I_* - \frac{\varepsilon}{3}, \tag{1.24}$$

where

$$s_{\tau_1} = \sum_{j=1}^{j_{\tau_1}} m_{1j} \mu X_{1j}$$

and

$$m_{1j} = \inf_{x \in X_{1j}} f(x), \quad x \in \{1, \dots, j_{\tau_1}\}.$$

Set

$$X_0 = \bigcup_{j=1}^{j_{\tau_1}} \partial X_{1j}.$$

Since $X_{1j}, j \in \{1, \dots, j_{\tau_1}\}$, are measurable sets, we have that

$$\mu(\partial X_{1j}) = 0, \quad j \in \{1, \dots, j_{\tau_1}\}.$$

Therefore

$$\mu X_0 = 0.$$

Hence there is a rank $k = k(\varepsilon)$ such that

$$\mu S_k(X_0) < \frac{\varepsilon}{3^{n+1} M}.$$

We have that

$$s_\tau \le I_*.$$

We will prove that

$$I_* - \varepsilon < s_\tau.$$

Suppose that

$$S_k(X_0) = \bigcup_{j=1}^{m} Q_j.$$

Let $P_j, j \in \{1, \dots, m\}$, be the cubes obtained by the cubes $Q_j, j \in \{1, \dots, m\}$, by the similarity transformations centered at the centers of $Q_j, j \in \{1, \dots, m\}$, and the coefficients equal to 3. Set

$$P = \bigcup_{j=1}^{m} P_j,$$
$$G = X \backslash P.$$

By the definition of P we have

$$
\begin{aligned}
\mu P &= \mu\left(\bigcup_{j=1}^{m} P_j\right) \\
&\leq \sum_{j=1}^{m} \mu P_j \\
&= 3^n \sum_{j=1}^{m} \mu Q_j \\
&= 3^n \mu S_k(X_0) \\
&< 3^n \frac{\varepsilon}{3^{n+1} M} \\
&= \frac{\varepsilon}{3M}.
\end{aligned}
$$

Suppose that $A \subset X$ with $\operatorname{diam} A < 10^{-k}$ and $A \cap G \neq \emptyset$. Let $x \in A \cap G$ be arbitrarily chosen. Since $A \subset X$, we have that $x \in X$, and then $x \in X_{1j}$ for some $j \in \{1, \ldots, j_{\tau_1}\}$. Assume that $A \not\subset X_{1j}$. Then there is $y \in A \backslash X_{1j}$. Because $x \in A$, $y \in A$, and $\operatorname{diam} A < 10^{-k}$, we get

$$d(x, y) < 10^{-k}.$$

Hence the segment with endpoints x and y has the length 10^{-k} and does not intersect the set $S_k(X_0)$. Because $x \in X_{1j}$, we have that there is $z \in \partial X_{1j} \cap [x, y]$. By the definition of X_0 we get

$$\partial X_{1j} \subset X_0 \subset S_k(X_0).$$

Therefore $z \in S_k(X_0)$, and the segment with endpoints x and y intersects $S_k(X_0)$. This is a contradiction. Consequently,

$$A \subset X_{1j}. \tag{1.25}$$

Now assume that there is $X_{1k} \in \tau_1$ such that

$$A \subset X_{1k}, \quad k \neq j.$$

Then

$$A \subset X_{ij} \cap X_{1k}.$$

If X_{1j} and X_{1k} have a common interior point, then this interior point will be an interior point for their intersection $X_{1j} \cap X_{1k}$, and then

$$\mu(X_{1j} \cap X_{1k}) > 0,$$

which is a contradiction because $\tau_1 = \{X_{1j}\}_{j=1}^{j_{\tau_1}}$, and by the definition of a partition of a set we have

$$\mu(X_{1j} \cap X_{1k}) = 0.$$

Thus any point of $X_{1j} \cap X_{1k}$ is a boundary point of $X_{1j} \cap X_{1k}$. Then any point of A is a boundary point for X_{1j} or X_{1k}. Hence

$$\begin{aligned} A &\subset \bigcup_{j=1}^{j_{\tau_1}} \partial X_{1j} \\ &= X_0 \\ &\subset S_k(X_0), \end{aligned}$$

which is impossible because $A \cap G \neq \emptyset$ and $G \cap S_k(X_0) = \emptyset$. Thus there is a unique $X_{1j} \in \tau_1$ such that (1.25) holds. Now suppose that $\tau = \{X_j\}_{j=1}^{j_\tau}$ is a partition of X such that $|\tau| < 10^{-k}$. If

$$m_j = \inf_{x \in X_j} f(x), \quad j \in \{1, \dots, j_\tau\},$$

then

$$\begin{aligned} s_\tau &= \sum_{j=1}^{j_\tau} m_j \mu X_j \\ &= \sum_{X_j \cap G \neq \emptyset} m_j \mu X_j + \sum_{X_j \subset P} m_j \mu X_j. \end{aligned}$$

Note that

$$\begin{aligned} \left| \sum_{X_j \subset P} m_j \mu X_j \right| &\leq \sum_{X_j \subset P} |m_j| \mu X_j \\ &\leq M \sum_{X_j \subset P} \mu X_j \\ &= M \mu P \\ &< M \frac{\varepsilon}{3M} \\ &= \frac{\varepsilon}{3}. \end{aligned}$$

In particular,

$$\sum_{X_j \subset P} m_j \mu X_j > -\frac{\varepsilon}{3}.$$

Consequently,

$$s_\tau > \sum_{X_j \cap G \neq \emptyset} m_j \mu X_j - \frac{\varepsilon}{3}. \tag{1.26}$$

Note that

$$\begin{aligned} \operatorname{diam} X_j &\le |\tau| \\ &< 10^{-k}. \end{aligned}$$

Therefore, for any $X_j \in \tau$ with $X_j \cap G \neq \emptyset$, there is a unique $X_{1k} \in \tau_1$ such that

$$X_j \subset X_{1k}.$$

Let G_k be the union of all X_τ such that $X_j \cap G \neq \emptyset$ and $X_j \subset X_{1k}$, i. e.,

$$G_k = \bigcup_{\substack{X_j \subset X_{1k} \\ X_j \cap G \neq \emptyset}} X_\tau.$$

Then

$$\sum_{X_j \cap G \neq \emptyset} m_j \mu X_j = \sum_{j=1}^{j_\tau} \sum_{X_j \subset G_k} m_{1j} \mu X_j. \tag{1.27}$$

Observe that

$$\begin{aligned} X_{1j} &= (X_{1j} \cap G) \cup (X_{1j} \backslash G) \\ &= G_j \cup (X_{1j} \backslash G_j). \end{aligned}$$

Then

$$\begin{aligned} m_{1j} \mu X_{1j} &= m_{1j} \mu G_j + m_{1j} \mu (X_{1j} \backslash G_j) \\ &= m_{1j} \mu \Big(\bigcup_{X_k \subset G_j} X_k \Big) + m_{1j} \mu (X_k \backslash G_j) \\ &= m_{1j} \sum_{X_k \subset G_j} \mu X_k + m_{1j} \mu (X_{1j} \backslash G_j). \end{aligned} \tag{1.28}$$

Note that for any $x \in X_{1j} \backslash G_j$, there is $X_k \in \tau$ such that $x \in X_k$. Assume that $X_k \cap G \neq \emptyset$. Since $x \in X_{1j}$ and $x \in X_k$, we have that $x \in X_k \cap X_{1j}$ and $X_k \cap X_{1j} \neq \emptyset$. Moreover, $X_k \cap X_{1j} = X_k$, and by the definition of G_j we get $X_k \subset G_j$, and hence $x \in G_j$, which is a contradiction because $x \in X_{1j} \backslash G_j$. Therefore $X_k \cap G = \emptyset$. Then $X_k \subset P$ and $x \in P$. Because $x \in X_{1j} \backslash G_j$ was arbitrarily chosen and it is an element of P, we get the inclusions

$$X_{1j}\backslash G_j \subset P$$

and

$$X_{1j}\backslash G_j \subset X_{1j} \cap P.$$

Consequently,

$$m_{1j}\mu(X_{1j}\backslash G_j) \le M\mu(X_{1j} \cap P).$$

Now applying (1.28), we obtain

$$\begin{aligned} m_{1j}\mu X_{1j} &\le m_{1j} \sum_{X_k \subset G_j} \mu X_k + M\mu(X_{1j} \cap P) \\ &\le \sum_{X_k \subset G_j} m_k \mu X_k + M\mu(X_{1j} \cap P), \end{aligned}$$

whereupon

$$\sum_{X_k \subset G_j} m_k \mu X_k \ge m_{1j}\mu X_{1j} - M\mu(X_{1j} \cap P)$$

and

$$\begin{aligned} \sum_{X_k \subset G_j} m_k \mu X_k &\ge \sum_{j=1}^{j_{\tau_1}} m_{1j}\mu X_{1j} - M \sum_{j=1}^{j_{\tau_1}} \mu(X_{1j} \cap P) \\ &= s_{\tau_1} - M \sum_{j=1}^{j_{\tau_1}} \mu(X_{1j} \cap P). \end{aligned}$$

On the other hand,

$$\begin{aligned} \sum_{j=1}^{j_{\tau_1}} \mu(X_{1j} \cap P) &= \mu\left(\bigcup_{j=1}^{j_{\tau_1}} (X_{ij} \cap P)\right) \\ &\le \mu P \\ &< \frac{\varepsilon}{3M}, \end{aligned}$$

from which we get

$$\begin{aligned} \sum_{X_k \cap G \neq \emptyset} m_k \mu X_k &\ge s_{\tau_1} - M\frac{\varepsilon}{3M} \\ &= s_{\tau_1} - \frac{\varepsilon}{3}. \end{aligned}$$

Then applying (1.26), we get

$$\begin{aligned} s_\tau &> \sum_{X_k \cap G \neq \emptyset} m_k \mu X_k - \frac{\varepsilon}{3} \\ &> s_{\tau_1} - \frac{\varepsilon}{3} - \frac{\varepsilon}{3} \\ &= s_{\tau_1} - \frac{2\varepsilon}{3} \\ &> I_* - \frac{\varepsilon}{3} - \frac{2\varepsilon}{3} \\ &= I_* - \varepsilon. \end{aligned}$$

Thus

$$I_* - \varepsilon < s_\tau \leq I_*.$$

Because $\varepsilon > 0$ was arbitrarily chosen, we get

$$\lim_{|\tau| \to 0} s_\tau = I_*.$$

We leave the equation

$$\lim_{|\tau| \to 0} S_\tau = I^*$$

to the reader for an exercise. This completes the proof. □

Now we relate the upper and lower Riemann integrals to Riemann integrability.

Theorem 1.34 (The Darboux criterion). *The function f is Riemann integrable over X if and only if*

$$I_* = I^*.$$

Proof. Firstly, note that for any partition τ of X, we have

$$s_\tau \leq I_* \leq I^* \leq S_\tau.$$

1. Let $I_* = I^*$. Then

$$\begin{aligned} \lim_{|\tau| \to 0} (S_\tau - s_\tau) &= \lim_{|\tau| \to 0} S_\tau - \lim_{|\tau| \to 0} s_\tau \\ &= I^* - I_* \\ &= 0. \end{aligned}$$

Now applying Theorem 1.30, we conclude that f is Riemann integrable over X.

2. Let F be Riemann integrable over X. Then applying Theorems 1.30 and 1.33, we get

$$\begin{aligned} 0 &= \lim_{|\tau|\to 0}(S_\tau - s_\tau) \\ &= \lim_{|\tau|\to 0} S_\tau - \lim_{|\tau|\to 0} s_\tau \\ &= I^* - I_*, \end{aligned}$$

whereupon $I_* = I^*$. This completes the proof. □

Using this concept, we can reformulate our first criterion into our second integrability criterion. Its proof is immediate from the first one.

Theorem 1.35 (The Riemann criterion). *The function f is Riemann integrable over X if and only if for any $\varepsilon > 0$, there is a partition τ of X such that*

$$S_\tau - s_\tau < \varepsilon.$$

Proof. 1. Let f be Riemann integrable over X. Take arbitrary $\varepsilon > 0$. Then, applying Theorem 1.30, we get

$$\lim_{|\tau|\to 0}(S_\tau - s_\tau) = 0.$$

Hence there is $\delta = \delta(\varepsilon) > 0$ such that for any partition τ of X with $|\tau| < \delta$, we have

$$S_\tau - s_\tau < \varepsilon.$$

2. Let for any $\varepsilon > 0$, τ be a partition of X such that

$$S_\tau - s_\tau < \varepsilon.$$

Hence

$$I^* - I_* < \varepsilon$$

for any $\varepsilon > 0$. Therefore $I_* = I^*$. Now applying Theorem 1.34, we conclude that f is Riemann integrable over X. This completes the proof. □

Corollary 1.2. *Let X be a Jordan-measurable set. A function $f : X \to \mathbb{R}$ is Riemann integrable over X if and only if*

$$\lim_{|\tau|\to 0} \sum_{j=1}^{j_\tau} \omega(f, X_j)\mu X_j = 0.$$

Here $\tau = \{X_j\}_{j=1}^{j_\tau}$ is a partition of X.

1.4 Properties of multiple integrals

Multiple integrals have many properties common to those of integrals of functions of one variable, such as linearity, commutativity, and monotonicity. We start with the following property.

Theorem 1.36 (Volume of a set). *Let X be a Jordan-measurable set. Then*

$$\int_X dx = \mu X.$$

Proof. Let $\tau = \{X_j\}_{j=1}^{j_\tau}$ be any partition of X. Then

$$\begin{aligned}\int_X dx &= \lim_{|\tau|\to 0} \sum_{j=1}^{j_\tau} \mu X_j \\ &= \mu X.\end{aligned}$$

This completes the proof. □

If a function f is integrable over a given set, then f is integrable over any of its subsets.

Theorem 1.37. *Let X and Y be Jordan-measurable sets, and let $Y \subset X$. If $f : X \to \mathbb{R}$ is bounded and Riemann integrable over X, then f is Riemann integrable over Y.*

Proof. Let $Z = X\backslash Y$. Let also, $\tau_Y = \{Y_j\}_{j=1}^{j_{\tau_Y}}$ and $\tau_Z = \{Z_j\}_{j=1}^{j_{\tau_Z}}$ be partitions of Y and Z, respectively. Then

$$\tau = \tau_Y \cup \tau_Z = \{X_j\}_{j=1}^{j_\tau}$$

is a partition of X. We have

$$\sum_{j=1}^{j_\tau} \omega(f, X_j)\mu X_j = \sum_{j=1}^{j_{\tau_Y}} \omega(f, Y_j)\mu Y_j + \sum_{j=1}^{j_{\tau_Z}} \omega(f, Z_j)\mu Z_j,$$

whereupon

$$\sum_{j=1}^{j_{\tau_Y}} \omega(f, Y_j)\mu Y_j \le \sum_{j=1}^{j_\tau} \omega(f, X_j)\mu X_j,$$

and

$$\begin{aligned}\lim_{|\tau|\to 0} \sum_{j=1}^{j_{\tau_Y}} \omega(f, Y_j)\mu Y_j &\le \lim_{|\tau|\to 0} \sum_{j=1}^{j_\tau} \omega(f, X_j)\mu X_j \\ &= 0.\end{aligned}$$

Now applying Theorems 1.29 and 1.30, we conclude that f is integrable over Y and

$$\int\limits_Y f(y)dy \le \int\limits_X f(x)dx.$$

This completes the proof. □

The integral over the whole is the sum of the integrals over the pieces.

Theorem 1.38. *Let X, Y, and Z be Jordan-measurable sets such that*

$$X = Y \cup Z, \quad Y \cap Z = \emptyset,$$

and let $f : X \to \mathbb{R}$ *be a bounded and Riemann-integrable function over X. Then f is Riemann integrable over Y and Z, and*

$$\int\limits_X f(x)dx = \int\limits_Y f(y)dy + \int\limits_Z f(z)dz.$$

Proof. Let $\tau_Y = \{Y_j\}_{j=1}^{j_{\tau_Y}}$ and $\tau_Z = \{Z_j\}_{j=1}^{j_{\tau_Z}}$ be partitions of Y and Z, respectively. Then

$$\tau = \tau_Y \cup \tau_Z = \{X_j\}_{j=1}^{j_\tau}$$

is a partition of X with

$$|\tau| = \max\{|\tau_Y|, |\tau_Z|\}.$$

Take $\xi^j \in X_j, j \in \{1, \ldots, j_{\tau_Y}\}, \eta^l \in \{1, \ldots, j_{\tau_Z}\}$, and

$$\sigma_{\tau_Y} = \sum_{j=1}^{j_{\tau_Y}} f(\xi^j)\mu Y_j,$$

$$\sigma_{\tau_Z} = \sum_{l=1}^{j_{\tau_Z}} f(\eta^l)' \mu Y_l,$$

$$\sigma_\tau = \sigma_{\tau_Y} + \sigma_{\tau_Z}.$$

By Theorem 1.37 it follows that f is Riemann integrable over Y and Z, and

$$\lim_{|\tau|\to 0} \sigma_\tau = \int\limits_X f(x)dx,$$

$$\lim_{|\tau|\to 0} \sigma_{\tau_Y} = \int\limits_Y f(y)dy,$$

$$\lim_{|\tau|\to 0} \sigma_{\tau_Z} = \int\limits_Z f(z)dz.$$

Hence

$$\begin{aligned}\int_X f(x)dx &= \lim_{|\tau|\to 0} \sigma_\tau \\ &= \lim_{|\tau|\to 0} (\sigma_{\tau_Y} + \sigma_{\tau_Z}) \\ &= \lim_{|\tau|\to 0} \sigma_{\tau_Y} + \lim_{|\tau|\to 0} \sigma_{\tau_Z} \\ &= \int_Y f(y)dy + \int_Z f(z)dz.\end{aligned}$$

This completes the proof. □

The next result tells us that the Riemann integral does not depend on the boundaries of the set.

Theorem 1.39. *Let X be a Jordan-measurable set, and let $f : X \to \mathbb{R}$ be bounded and integrable over X. Then f is integrable over X_{int}, and*

$$\int_X f(x)dx = \int_{X_{\text{int}}} f(x)dx.$$

Proof. We have

$$X = X_{\text{int}} \cup (X \backslash X_{\text{int}}), \quad X_{\text{int}} \cap (X \backslash X_{\text{int}}) = \emptyset,$$

and

$$\mu(X \backslash X_{\text{int}}) = 0.$$

Thus applying Theorem 1.38, we get

$$\begin{aligned}\int_X f(x)dx &= \int_{X_{\text{int}}} f(x)dx + \int_{X\backslash X_{\text{int}}} f(x)dx \\ &= \int_{X_{\text{int}}} f(x)dx.\end{aligned}$$

This completes the proof. □

Remark 1.3. There are cases where X, Y, and Z are Jordan measurable sets such that

$$X = Y \cup Z, \quad Y \cap Z = \emptyset,$$

the function $f : X \to \mathbb{R}$ is Riemann integrable over Y and Z, but it is not Riemann integrable over X. We will see this in the next example.

Example 1.2. Let (r, ϕ) denote the polar coordinates of a point in $\mathbb{R}^2$. Consider the function

$$f(r,\phi) = \begin{cases} 0 & \text{if } r < 1, \\ \frac{1}{\phi} & \text{if } 0 < \phi \leq 2\pi. \end{cases}$$

Define the sets

$$\begin{aligned} Y &= \{(r,\phi) : r < 1\}, \\ Z &= \{(r,\phi) : r = 1\}, \\ X &= Y \cup Z. \end{aligned}$$

We have

$$\mu Z = 0$$

and

$$\begin{aligned} &\int_Y f(y)dy = 0, \\ &\int_Z f(z)dz = 0. \end{aligned}$$

Assume that f is integrable over X. Note that $X = \overline{Y}$. Then by Remark 1.2 it follows that there is a partition $\tau = \{X_j\}_{j=1}^{j_\tau}$ of X such that $\mu X_j > 0, j \in \{1, \ldots, j_\tau\}$. Now applying Theorem 1.26 and Remark 1.1, we conclude that f is bounded on X. This is a contradiction because f is unbounded on X. Therefore f is not integrable over X.

In the next theorem, we show that the Riemann integral satisfies the linearity and homogeneity properties.

Theorem 1.40. *Let X be a Jordan-measurable set, and let $f_1, f_2 : X \to \mathbb{R}$ be Riemann integrable over X. Then, for any $\lambda_1, \lambda_2 \in \mathbb{C}$, the function $\lambda_1 f_1 + \lambda_2 f_2$ is Riemann integrable over X, and*

$$\int_X (\lambda_1 f_1(x) + \lambda_2 f_2(x))dx = \lambda_1 \int_X f_1(x)dx + \lambda_2 \int_X f_2(x)dx.$$

Proof. Let $\tau = \{X_j\}_{j=1}^{j_\tau}$ be a partition of X. Take $\xi^j \in X_j, j \in \{1, \ldots, j_\tau\}$. Note that

$$\int_X (\lambda_1 f_1(x) + \lambda_2 f_2(x))dx = \lim_{|\tau| \to 0} \sigma_\tau(\lambda_1 f_1 + \lambda_2 f_2; \xi^1, \ldots, \xi^{j_\tau}),$$

$$\int_X f_1(x)dx = \lim_{|\tau|\to 0} \sigma_\tau(f_1; \xi^1, \dots, \xi^{j_\tau}),$$
$$\int_X f_2(x)dx = \lim_{|\tau|\to 0} \sigma_\tau(f_2; \xi^1, \dots, \xi^{j_\tau})$$

and

$$\begin{aligned}\sigma_\tau(\lambda_1 f_1 + \lambda_2 f_2; \xi^1, \dots, \xi^{j_\tau}) &= \sum_{j=1}^{j_\tau} (\lambda_1 f_1 + \lambda_2 f_2)(\xi^j)\mu X_j \\ &= \sum_{j=1}^{j_\tau} (\lambda_1 f_1(\xi^j) + \lambda_2 f_2(\xi^j))\mu X_j \\ &= \lambda_1 \sum_{j=1}^{j_\tau} f_1(\xi^j)\mu X_j + \lambda_2 \sum_{j=1}^{j_\tau} f_2(\xi^j)\mu X_j \\ &= \lambda_1 \sigma_\tau(f_1; \xi^1, \dots, \xi^{j_\tau}) + \lambda_2 \sigma_\tau(f_2; \xi^1, \dots, \xi^{j_\tau}).\end{aligned}$$

Hence

$$\begin{aligned}\int_X (\lambda_1 f_1(x) + \lambda_2 f_2(x))dx &= \lim_{|\tau|\to 0} \sigma_\tau(\lambda_1 f_1 + \lambda_2 f_2; \xi^1, \dots, \xi^{j_\tau}) \\ &= \lambda_1 \lim_{|\tau|\to 0} \sigma_\tau(f_1; \xi^1, \dots, \xi^{j_\tau}) + \lambda_2 \lim_{|\tau|\to 0} \sigma_\tau(f_2; \xi^1, \dots, \xi^{j_\tau}) \\ &= \lambda_1 \int_X f_1(x)dx + \lambda_2 \int_X f_2(x)dx.\end{aligned}$$

This completes the proof. □

The product of two integrable functions is also an integrable function.

Theorem 1.41. *Let X be a Jordan-measurable set, and let $f, g : X \to \mathbb{R}$ be Riemann integrable over X and bounded on X. Then fg is Riemann integrable over X.*

Proof. Since $f, g : X \to \mathbb{R}$ are bounded on X, there is a constant $M > 0$ such that

$$|f(x)|,\ |g(x)| \le M, \quad x \in X.$$

Let $\tau = \{X_j\}_{j=1}^{j_\tau}$ be a partition of X. Note that for any $x, y \in X$, we have

$$\begin{aligned}|f(x)g(x) - f(y)g(y)| &= |f(x)(g(x) - g(y)) + g(y)(f(x) - f(y))| \\ &\le |f(x)||g(x) - g(y)| + |g(y)||f(x) - f(y)| \\ &\le M(|f(x) - f(y)| + |g(x) - g(y)|).\end{aligned}$$

Therefore

$$\omega(fg, X_j) \le M(\omega(f, X_j) + \omega(g, X_j)), \quad j \in \{1, \dots, j_\tau\}. \tag{1.29}$$

Because $f, g : X \to \mathbb{R}$ are Riemann integrable over X, applying Corollary 1.2, we get

$$\lim_{|\tau|\to 0} \sum_{j=1}^{j_\tau} \omega(f, X_j)\mu X_j = 0,$$

$$\lim_{|\tau|\to 0} \sum_{j=1}^{j_\tau} \omega(g, X_j)\mu X_j = 0.$$

Hence by (1.29) we arrive at

$$\begin{aligned}\lim_{|\tau|\to 0} \sum_{j=1}^{j_\tau} \omega(fg, X_j)\mu X_j &\le M\left(\lim_{|\tau|\to 0} \sum_{j=1}^{j_\tau} \omega(f, X_j)\mu X_j + \lim_{|\tau|\to 0} \sum_{j=1}^{j_\tau} \omega(g, X_j)\mu X_j\right)\\ &= 0.\end{aligned}$$

Now we conclude that fg is Riemann integrable over X. This completes the proof. □

We will give a criterion when the division of two integrable functions is an integrable function.

Theorem 1.42. *Let X be a Jordan-measurable set, and let $f, g : X \to \mathbb{R}$ be Riemann integrable over X and bounded on X, and suppose $\inf_{x\in X} |g(x)| = m > 0$. Then $\frac{f}{g}$ is Riemann integrable over X.*

Proof. Because $f, g : X \to \mathbb{R}$ are bounded on X, there is a constant $M > 0$ for which

$$|f(x)|,\ |g(x)| \le M, \quad x \in X.$$

Let $\tau = \{X_j\}_{j=1}^{j_\tau}$ be a partition of X. Observe that for any $x, y \in X$, we have

$$\begin{aligned}\left|\frac{f(x)}{g(x)} - \frac{f(y)}{g(y)}\right| &= \frac{1}{|g(x)||g(y)|}|f(x)g(y) - f(y)g(x)|\\ &\le \frac{1}{m^2}|f(x)(g(y) - g(x)) - g(x)(f(y) - f(x))|\\ &\le \frac{1}{m^2}(|f(x)||g(x) - g(y)| + |g(x)||f(x) - f(y)|)\\ &\le \frac{M}{m^2}(|f(x) - f(y)| + |g(x) - g(y)|).\end{aligned}$$

Thus

$$\omega\left(\frac{f}{g}, X_j\right) \le \frac{M}{m^2}(\omega(f, X_j) + \omega(g, X_j)), \quad j \in \{1, \dots, j_\tau\}. \tag{1.30}$$

Since $f, g : X \to \mathbb{R}$ are Riemann integrable over X, applying Corollary 1.2, we get

$$\lim_{|\tau|\to 0}\sum_{j=1}^{j_\tau}\omega(f,X_j)\mu X_j = 0,$$

$$\lim_{|\tau|\to 0}\sum_{j=1}^{j_\tau}\omega(g,X_j)\mu X_j = 0.$$

Hence by (1.30) we arrive at

$$\begin{aligned}\lim_{|\tau|\to 0}\sum_{j=1}^{j_\tau}\omega\left(\frac{f}{g},X_j\right)\mu X_j &\le \frac{M}{m^2}\left(\lim_{|\tau|\to 0}\sum_{j=1}^{j_\tau}\omega(f,X_j)\mu X_j + \lim_{|\tau|\to 0}\sum_{j=1}^{j_\tau}\omega(g,X_j)\mu X_j\right)\\ &= 0.\end{aligned}$$

From this we conclude that $\frac{f}{g}$ is Riemann integrable over X. This completes the proof. □

Next, we consider how multiple integration behaves with respect to the order relation on functions.

Theorem 1.43. *Let X be a Jordan-measurable set, let $f, g : X \to \mathbb{R}$ be Riemann integrable over X, and suppose*

$$f(x) \le g(x), \quad x \in X.$$

Then we have the following:

1.
$$\int_X f(x)dx \le \int_X g(x)dx.$$

2. *$|f|$ is Riemann integrable over X, and*
$$\left|\int_X f(x)dx\right| \le \int_X |f(x)|dx.$$

Proof. 1. Let $\tau = \{X_j\}_{j=1}^{j_\tau}$ be a partition of X, and let $\xi^j \in X_j$, $j \in \{1,\dots,j_\tau\}$. Then

$$f(\xi^j) \le g(\xi^j), \quad j \in \{1,\dots,j_\tau\},$$

and

$$\begin{aligned}\sigma_\tau(f;\xi^1,\dots,\xi^n) &= \sum_{j=1}^{j_\tau} f(\xi^j)\mu X_j\\ &\le \sum_{j=1}^{j_\tau} g(\xi^j)\mu X_j\\ &= \sigma_\tau(g;\xi^1,\dots,\xi^n).\end{aligned}$$

Hence

$$\begin{aligned}\int_X f(x)dx &= \lim_{|\tau|\to 0} \sigma_\tau(f;\xi^1,\dots,\xi^n)\\ &\le \lim_{|\tau|\to 0} \sigma_\tau(g;\xi^1,\dots,\xi^n)\\ &= \int_X g(x)dx.\end{aligned}$$

2. Let $\tau = \{X_j\}_{j=1}^{j_\tau}$ be a partition of X, and let $\xi^j \in X_j, j \in \{1,\dots,j_\tau\}$. Note that for any $x, y \in X$, we have

$$\big||f(x)| - |f(y)|\big| \le |f(x) - f(y)|.$$

Then

$$\omega_\tau(|f|, X_j) \le \omega_\tau(f, X_j), \quad j \in \{1,\dots,j_\tau\}.$$

Since f is Riemann integrable over X, applying Corollary 1.2, we have that

$$\lim_{|\tau|\to 0} \sum_{j=1}^{j_\tau} \omega_\tau(f, X_j)\mu X_j = 0.$$

Moreover,

$$\int_X f(x)dx = \lim_{|\tau|\to 0} \sigma_\tau(f;\xi^1,\dots,\xi^{j_\tau}).$$

Then

$$\begin{aligned}0 &\le \lim_{|\tau|\to 0} \sum_{j=1}^{j_\tau} \omega_\tau(|f|, X_j)\mu X_j\\ &\le \lim_{|\tau|\to 0} \sum_{j=1}^{j_\tau} \omega_\tau(f, X_j)\mu X_j\\ &= 0,\end{aligned}$$

whereupon

$$\lim_{|\tau|\to 0} \sum_{j=1}^{j_\tau} \omega_\tau(|f|, X_j)\mu X_j = 0.$$

Now applying Corollary 1.2, we conclude that $|f|$ is Riemann integrable over X. Next,

$$\left|\int_X f(x)dx\right| = \left|\lim_{|\tau|\to 0} \sigma_\tau(f;\xi^1,\dots,\xi^n)\right|$$
$$= \left|\lim_{|\tau|\to 0} \sum_{j=1}^{j_\tau} f(\xi^j)\mu X_j\right|$$
$$\leq \lim_{|\tau|\to 0} \sum_{j=1}^{j_\tau} |f(\xi^j)|\mu X_j$$
$$= \int_X |f(x)|dx.$$

This completes the proof. □

With the next result, we will show the monotonicity of the Riemann integral.

Theorem 1.44. *Let X and Y be Jordan-measurable sets such that $Y \subset X$. If $f : X \to \mathbb{R}$ is nonnegative, bounded, and Riemann integrable over X, then*

$$\int_Y f(x)dx \leq \int_X f(x)dx.$$

Proof. By Theorem 1.38 it follows that f is Riemann integrable over Y and $X\backslash Y$ and that

$$\int_X f(x)dx = \int_Y f(x)dx + \int_{X\backslash Y} f(x)dx. \tag{1.31}$$

Since f is nonnegative on $X\backslash Y$, applying Theorem 1.43, we get

$$\int_{X\backslash Y} f(x)dx \geq 0.$$

Hence by (1.31) we have

$$\int_X f(x)dx \geq \int_Y f(x)dx.$$

This completes the proof. □

Now we will give a criterion for non-negativity of the Riemann integral.

Theorem 1.45. *Let X be an open Jordan-measurable set, let $f : X \to \mathbb{R}$ be a non-negative Riemann integrable function over X, and let $x^0 \in X$ be such that $f(x^0) > 0$ and f is continuous at x^0. Then*

$$\int_X f(x)dx > 0.$$

Proof. Since X is open, $x^0 \in X$, and f is continuous at x^0, there is a neighborhood $U(x^0)$ of the point x^0 such that $U(x^0) \subset X$ and

$$f(x) > \frac{f(x^0)}{2}, \quad x \in U(x^0).$$

Note that $\mu(U(x^0)) > 0$ because $U(x^0)$ is an open set in X. Now applying Theorem 1.44 and then Theorems 1.43 and 1.36, we get

$$\begin{aligned}\int_X f(x)dx &\geq \int_{U(x^0)} f(x)dx \\ &\geq \int_{U(x^0)} \frac{f(x^0)}{2}dx \\ &= \frac{f(x^0)}{2} \int_{U(x^0)} dx \\ &= \frac{f(x^0)}{2}\mu(U(x^0)) \\ &> 0.\end{aligned}$$

This completes the proof. □

Corollary 1.3. *Let X be an open Jordan-measurable set, and let $f : X \to \mathbb{R}$ be continuous Riemann integrable over X and bounded on X such that $f \not\equiv 0$ on X. Then*

$$\int_X f(x)dx > 0. \tag{1.32}$$

Proof. Since $f \not\equiv 0$ on X, there is a point $x^0 \in X$ such that $f(x^0) > 0$. Now applying Theorem 1.45, we get (1.32). This completes the proof. □

The next result shows the full additivity of the multiple integrals over open sets.

Theorem 1.46. *Let X, X_m, $m \in \mathbb{N}$, be open Jordan-measurable sets such that*

$$X_m \subset X_{m+1}, \quad m \in \mathbb{N}, \tag{1.33}$$

and

$$\bigcup_{m=1}^{\infty} X_m = X, \tag{1.34}$$

and let $f : X \to \mathbb{R}$ be Riemann integrable. Then

$$\lim_{m\to\infty}\int\limits_{X_m} f(x)dx = \int\limits_X f(x)dx.$$

Proof. Firstly, it follows that there is a sequence $\{X_m\}_{m\in\mathbb{N}}$ of Jordan-measurable sets that satisfy (1.33) and (1.34). Take arbitrary $\varepsilon > 0$. Then there is a rank k such that

$$\mu X - \varepsilon < \mu s_k(X). \tag{1.35}$$

Since X is Jordan measurable, by Theorem 1.18 it follows that X is bounded. Because $s_k(X)$ is a union of a finite number of closed cubes, we conclude that $s_k(X)$ is a compact set. By the union (1.34) it follows that the set $s_k(X)$ is covered by the system $\{X_m\}_{m\in\mathbb{N}}$. Then by the Heine–Borel theorem there is a finite number of elements of the system $\{X_m\}_{m\in\mathbb{N}}$ that covers the set $s_k(X)$. Let m_0 be the largest number such that X_{m_0} is in this cover. Using inclusion (1.33), we get that

$$s_k(X) \subset X_{m_0}$$

and

$$s_k(X) \subset X_{m_0} \subset X_m \subset X, \quad m > m_0.$$

From this and from (1.35) we get

$$\begin{aligned}\mu X - \varepsilon &< \mu s_k(X)\\ &\le \mu X_m\\ &\le \mu X, \quad m > m_0.\end{aligned}$$

Because $\varepsilon > 0$ was arbitrarily chosen, by the last chain of inequalities we obtain

$$\lim_{m\to\infty} \mu X_m = \mu X.$$

Since X is open and f is Riemann integrable over X, by Theorem 1.26 it follows that f is bounded on X. Therefore there is a constant $M > 0$ such that

$$|f(x)| \le M, \quad x \in X.$$

Hence

$$\begin{aligned}\left|\int\limits_X f(x)dx - \int\limits_{X_m} f(x)dx\right| &= \left|\int\limits_{X\setminus X_m} f(x)dx\right|\\ &\le \int\limits_{X\setminus X_m} |f(x)|dx\\ &\le M\mu(X\setminus X_m)\end{aligned}$$

$$\begin{aligned} &= M(\mu X - \mu X_m) \\ &\to 0 \quad \text{as } m \to \infty. \end{aligned}$$

This completes the proof. □

1.5 Mean value theorems

In calculus, the mean value theorems for integrals join the fundamental concepts of integration and continuity. These theorems, instrumental cornerstones of integral calculus, furnish a powerful tool for deciphering the intricate interplay between areas under curves and average values of continuous functions. The mean value theorems for integrals underpin many mathematical and real-world applications, such as proving inequalities, estimating the errors in numerical integration, and solving differential equations. In fields like physics and engineering, they are instrumental in understanding phenomena described by continuous functions over a set. The first mean value theorem for integrals reads as follows.

Theorem 1.47. *Let X be a Jordan-measurable set, and let $f, g : X \to \mathbb{R}$ be Riemann-integrable functions over X such that*

$$m \leq f(x) \leq M, \quad x \in X,$$

and the function g has a constant sign on X. Then there is a constant $\mu \in [m, M]$ such that

$$\int_X f(x)g(x)dx = \nu \int_X g(x)dx. \tag{1.36}$$

Proof. Without loss of generality, assume that $g(x) \geq 0$, $x \in X$. Then

$$mg(x) \leq f(x)g(x) \leq Mg(x), \quad x \in X.$$

Hence by Theorem 1.43 we get

$$m \int_X g(x)dx \leq \int_X f(x)g(x)dx \leq M \int_X g(x)dx. \tag{1.37}$$

First, suppose that

$$\int_X g(x)dx = 0.$$

Then applying inequalities (1.37), we get

$$\int_X f(x)g(x)dx = 0.$$

Hence in this case, equality (1.36) holds for all $v \in [m, M]$. Let now

$$\int_X g(x)dx \neq 0.$$

Since $g(x) \geq 0$, $x \in X$, by the last relation we get

$$\int_X g(x)dx > 0.$$

From this and from (1.37) we arrive at

$$m \leq \frac{\int_X f(x)g(x)dx}{\int_X g(x)dx} \leq M.$$

Let

$$v = \frac{\int_X f(x)g(x)dx}{\int_X g(x)dx}.$$

Then $v \in [m, M]$, and (1.36) holds. This completes the proof. □

There are various slightly different theorems, called the second mean value theorem for integrals. A commonly found version is as follows.

Theorem 1.48. *Let X be a domain, let $f, g : X \to \mathbb{R}$ be Riemann integrable over X such that $f \in \mathscr{C}(X)$ and g has a constant sign on X. Then there is $\xi \in X$ such that*

$$\int_X f(x)g(x)dx = f(\xi) \int_X g(x)dx. \tag{1.38}$$

Proof. Without loss of generality, assume that $g(x) \geq 0$, $x \in X$. Because X is a domain, it is an open set. Since f is Riemann integrable over X, by Theorem 1.26 we conclude that f is bounded on X. Therefore there exist

$$m = \inf_{x \in X} f(x),$$
$$M = \sup_{x \in X} f(x)$$

and

$$m \leq f(x) \leq M, \quad x \in X.$$

Let $\nu \in (m, M)$. Then there are $x^1, x^2 \in X$ such that

$$f(x^1) < \nu < f(x^2).$$

Because X is a domain, there is $\xi \in X$ such that

$$\nu = f(\xi).$$

Then applying (1.36), we obtain (1.38). Let now $\nu = M$. Then, applying (1.36), we get

$$\int_X (M - f(x))g(x)dx = 0. \tag{1.39}$$

If

$$\int_X g(x)dx = 0,$$

then using inequalities (1.37), we get

$$\int_X f(x)g(x)dx = 0,$$

and (1.38) holds for all $\xi \in X$. Let

$$\int_X g(x)dx \neq 0.$$

Then

$$\int_X g(x)dx > 0.$$

Now by Theorems 1.46 and 1.15 and by the last inequality it follows that there is an open Jordan-measurable set X_0 such that

$$X_0 \subset \overline{X_0} \subset X$$

and

$$\int_{X_0} g(x)dx > 0.$$

By Theorem 1.18 we have that X is bounded. Hence $\overline{X_0}$ is a compact set. Let there be $\xi_0 \in X_0$ such that $f(\xi) = M$. Then by (1.36), we get (1.38). Let there be no $\xi \in X_0$ such that $f(\xi) = M$. Then

$$M - f(x) > 0, \quad x \in \overline{X_0}.$$

Because $M - f$ is a continuous function on $\overline{X_0}$, there is $m_0 > 0$ such that

$$m_0 = \min_{x \in \overline{X_0}} (M - f(x)).$$

Hence

$$\begin{aligned} \int_X (M - f(x))g(x)dx &\geq \int_{\overline{X_0}} (M - f(x))g(x)dx \\ &\geq m_0 \int_{\overline{X_0}} g(x)dx \\ &> 0, \end{aligned}$$

which contradicts with (1.39). We leave the case $v = m$ to the reader as an exercise. This completes the proof. □

1.6 Advanced practical problems

Problem 1.1. Find the Jordan measure of the following sets:
1. Open rectangle with sides a and b.
2. Closed rectangle with sides a and b.

Problem 1.2. Prove that the following sets are Jordan measurable:
1.

$$\{(x, y) \in \mathbb{R}^2 : x \geq 0,\ y \geq 0,\ x + y \leq 1\}.$$

2.

$$\{(x, y) \in \mathbb{R}^2 : -1 \leq x \leq 1,\ 0 \leq y \leq \sqrt{1 - y^2}\}.$$

3.

$$\{(x, y) \in \mathbb{R}^2 : 0 \leq x \leq 1,\ 0 \leq y \leq e^x\}.$$

4.

$$\{(x, y) \in \mathbb{R}^2 : 1 \leq x \leq e,\ 0 \leq y \leq \log x\}.$$

5.

$$\{(x, y) \in \mathbb{R}^2 : 0 \leq x \leq \pi,\ 0 \leq y \leq \sin x\}.$$

Problem 1.3. Prove that any finite set has Jordan measure 0.

Problem 1.4. Prove that the set of all rational numbers in $[0,1]$ is not Jordan measurable.

Problem 1.5. Prove that $\mathbb{R}^3$ is not a Jordan measurable set.

Problem 1.6. Let $X \subset \mathbb{R}^n$ be a non-Jordan-measurable set. Prove that $\mathbb{R}^n \backslash X$ is a Jordan-measurable set.

Problem 1.7. Let

$$X = \{(x,y) \in \mathbb{R}^2 : (x-1)^2 + y^2 \le 1\},$$
$$X_n = \left\{(x,y) \in \mathbb{R}^2 : \left(x - \frac{1}{n}\right)^2 + y^2 \le \frac{1}{4^{2n}}\right\}, \quad n \in \mathbb{N}.$$

Prove that $X \backslash \bigcup_{n=1}^{\infty} X_n$ is Jordan measurable.

Problem 1.8. Prove that the set

$$\left\{(x,y) \in \mathbb{R}^2 : 0 < x \le \frac{1}{\pi},\ 0 \le y \le \left|\sin\left(\frac{1}{x}\right)\right|\right\}$$

is Jordan measurable.

Problem 1.9. Prove that the circle in $\mathbb{R}^2$ is Jordan measurable.

Problem 1.10. Prove that the ellipsoid in $\mathbb{R}^3$ is Jordan measurable.

Problem 1.11. Let X be a Jordan-measurable set in $\mathbb{R}^n$, $a \in \mathbb{R}^n$, and

$$X_a = \{x = x^1 + a,\ x^1 \in X\}.$$

Prove that X_a is Jordan measurable and

$$\mu X_a = \mu X.$$

Problem 1.12. Let X be a Jordan-measurable set in $\mathbb{R}^n$, let A be an $n \times n$ orthogonal matrix, and let

$$X_A = \{x = Ax^1 : x^1 \in X\}.$$

Prove that X_A is Jordan measurable and

$$\mu X = \mu X_A.$$

Problem 1.13. Let $X = [-2,2] \times [-1,1]$ and

$$X_{ij} = \left[\frac{2(i-1)}{n}, \frac{2i}{n}\right] \times \left[\frac{j-1}{n}, \frac{j}{n}\right], \quad i,j \in \{-n+1, -n, \ldots, n\},$$

$\tau_n = \{X_{ij}\}_{i,j=-n+1}^{n}$. Find s_{τ_n}, S_{τ_n} and

$$\lim_{n\to\infty} s_{\tau_n}, \quad \lim_{n\to\infty} S_{\tau_n}$$

for the function f, where

1. $f(x,y) = 2x - y$, $(x,y) \in X$.
2. $f(x,y) = e^{x-y}$, $(x,y) \in X$.
3. $f(x,y) = xy$, $(x,y) \in X$.
4. $f(x,y) = x^2 + y^2$, $(x,y) \in X$.

Problem 1.14. Let

$$X = \{(x,y) \in \mathbb{R}^2 : x^2 + y^2 \le 1\},$$

let $\tau = \{X_j\}_{j=1}^{j_\tau}$ be a partition of X, let Θ_τ be a system of points $\xi^j \in X_j$, $j \in \{1,\dots,j_\tau\}$, and let $\varepsilon > 0$. Find $\delta = \delta(\varepsilon) > 0$ such that for any partition τ with $|\tau| < \delta$ and for any system Θ_τ, we have the inequality

$$\left|\sigma_\tau(f;\Theta_\tau) - \int_X f(x)dx\right| < \varepsilon,$$

where

1.
$$f(x,y) = \sin(px + qy), \quad (x,y) \in X,\ p,q \in \mathbb{R}.$$

2.
$$f(x,y) = e^{px+qy}, \quad (x,y) \in X,\ p,q \in \mathbb{R}.$$

3.
$$f(x,y) = \log(2 + px + qy), \quad (x,y) \in X,\ p,q \in \mathbb{R},\ \sqrt{p^2+q^2} < 2.$$

4.
$$f(x,y) = e^{x^2-y^2}, \quad (x,y) \in X.$$

5.
$$f(x,y) = \frac{1}{1+3x^2+y^2}, \quad (x,y) \in X.$$

Problem 1.15. Let $X = [a,b] \times [c,d]$ and $g(x,y) = f(x)$, $(x,y) \in X$, where f is a Riemann-integrable function over $[a,b]$. Prove that g is Riemann integrable over X and

$$\int\limits_X g(x,y)dxdy = (d-c)\int\limits_a^b f(x)dx.$$

Problem 1.16. Let X^1 be Jordan measurable in $\mathbb{R}^{n-1}$,

$$\begin{aligned} x &= (x_1,\ldots,x_n) \in \mathbb{R}^n,\\ x^1 &= (x_1,\ldots,x_{n-1}) \in \mathbb{R}^{n-1}, \end{aligned}$$

$X = X^1 \times [c,d]$, and

$$f(x) = f(x^1), \quad x \in \mathbb{R}^n,$$

where f is bounded Riemann-integrable function over X^1. Prove that g is Riemann integrable over X and

$$\int\limits_X g(x)dx = (d-c)\int\limits_{X^1} f(x^1)dx^1.$$

Problem 1.17. Let X^1 be a Jordan-measurable set in $\mathbb{R}^k$, let X^2 be a Jordan-measurable set in $\mathbb{R}^l$, and let

$$\begin{aligned} x^1 &= (x_1,\ldots,x_k),\\ x^2 &= (x_1,\ldots,x_k,x_{k+1},\ldots,x_{k+l}),\\ X &= X^1 \times X^2\\ &= \{(x^1,x^2) \in \mathbb{R}^{k+l},\ x^1 \in X^1,\ x^2 \in X^2\}. \end{aligned}$$

Let also, f be bounded and Riemann integrable over X^1 and

$$g(x^1,x^2) = f(x^1), \quad x \in X.$$

Prove that g is Riemann integrable over X and

$$\int\limits_X g(x^1)dx^1dx^2 = \mu_l(X^2)\int\limits_{X^1} f(x^1)dx^1.$$

Problem 1.18. Let f be a Riemann-integrable function over $[a,b]$, let g be a Riemann-integrable function over $[c,d]$, and let

$$h(x,y) = f(x)g(y), \quad (x,y) \in X = [a,b] \times [c,d].$$

Prove that h is Riemann integrable over X and

$$\int_X h(x,y)dxdy = \int_a^b f(x)dx \int_c^d g(y)dy.$$

Problem 1.19. Prove that $I_* < I^*$, where $X = [a,b] \times [c,d]$ and

1.
$$f(x,y) = \begin{cases} 1 & \text{if } x, y \in \mathbb{Q}, \\ 0 & \text{otherwise.} \end{cases}$$

2.
$$f(x,y) = \begin{cases} 1 & \text{if } x + y \in \mathbb{Q}, \\ 0 & \text{otherwise.} \end{cases}$$

Problem 1.20. Prove that f is Riemann integrable over X, where

1. $X = [0,1] \times [0,1]$, and
$$f(x,y) = \begin{cases} (-1)^n & \text{if } \frac{1}{n+1} < x < \frac{1}{n},\ 0 \le y \le 1,\ n \in \mathbb{N}, \\ 1 & x = 0,\ 0 \le y \le 1. \end{cases}$$

2.
$$X = \{(x,y) \in \mathbb{R}^2 : |x| + |y| \le 1\},$$
and
$$f(x,y) = \begin{cases} (-1)^n, & 1 - \frac{1}{n} \le |x| + |y| < 1 - \frac{1}{n+1},\ n \in \mathbb{N}, \\ 0, & |x| + |y| = 1. \end{cases}$$

3.
$$X = \{(x,y) : x^2 + y^2 \le 1\},$$
and
$$f(x,y) = \begin{cases} \frac{1}{n}, & \frac{1}{n+1} < \sqrt{x^2 + y^2} \le \frac{1}{n},\ n \in \mathbb{N}, \\ 0, & (x,y) = (0,0). \end{cases}$$

4. $X = [-1,1] \times [-1,1]$, and
$$f(x,y) = \operatorname{sign}(xy), \quad (x,y) \in X.$$

5.
$$X = \{(x,y) : 0 \le x \le 1,\ 0 \le y \le 1 - x\},$$
and

$$f(x,y) = \begin{cases} (-1)^{n-1}, & \frac{1}{2^n} \le x+y \le \frac{1}{2^{n-1}},\ n \in \mathbb{N}, \\ 1, & (x,y) = (0,0). \end{cases}$$

6. $X = [0,1] \times [0,1]$, and

$$f(x,y) = \begin{cases} \frac{1}{q_1 q_2} & \text{if } (x,y) = (\frac{p_1}{q_1}, \frac{p_2}{q_2}),\ \frac{p_j}{q_j} \in \mathbb{Q}\backslash\mathbb{N},\ p_j, q_j \in \mathbb{N},\ j \in \{1,2\}, \\ 0 & \text{otherwise.} \end{cases}$$

7. $X = [0,1] \times [0,1]$, and

$$f(x,y) = \begin{cases} \frac{1}{q_1 + q_2} & \text{if } (x,y) = (\frac{p_1}{q_1}, \frac{p_2}{q_2}),\ p_j, q_j \in \mathbb{N},\ \frac{p_j}{q_j} \in \mathbb{Q}\backslash\mathbb{N},\ j \in \{1,2\}, \\ 0 & \text{otherwise.} \end{cases}$$

Problem 1.21. Let $X = [0,1]\times[0,1]$. Prove that f is not Riemann integrable over X, where

1.

$$f(x,y) = \begin{cases} 1 & \text{if } x, y \in \mathbb{Q}, \\ 0 & \text{otherwise.} \end{cases}$$

2.

$$f(x,y) = \begin{cases} 1 & \text{if } x \in \mathbb{Q}, \\ 0 & \text{otherwise.} \end{cases}$$

3.

$$f(x,y) = \begin{cases} 1 & \text{if } y \in \mathbb{R}\backslash\mathbb{Q}, \\ x & \text{otherwise.} \end{cases}$$

4.

$$f(x,y) = \begin{cases} 1 & \text{if } (x,y) = (\frac{p_1}{q}, \frac{p_2}{q}),\ \frac{p_j}{q} \in \mathbb{Q}\backslash\mathbb{N},\ p_j, q \in \mathbb{N},\ j \in \{1,2\}, \\ 0 & \text{otherwise.} \end{cases}$$

5.

$$f(x,y) = \begin{cases} \frac{1}{q_1 q_2} & \text{if } (x,y) = (\frac{p_1}{q_1}, \frac{p_2}{q_2}),\ \frac{p_j}{q_j} \in \mathbb{Q}\backslash\mathbb{N},\ p_j, q_j \in \mathbb{N},\ j \in \{1,2\}, \\ 1 & \text{otherwise.} \end{cases}$$

Problem 1.22. Let X be an open Jordan-measurable set, and let $f \in \mathscr{C}(X)$. Suppose that for any open set $X_1 \subset X$, we have

$$\int_{X_1} f(x)dx = 0.$$

Prove that $f \equiv 0$ on X.

Problem 1.23. Let X be a Jordan-measurable set, and let $f : X \to \mathbb{R}$ be uniformly continuous on X. Prove that f is Riemann integrable over X.

Problem 1.24. Let X be a Jordan-measurable set, $x_0 \in X$, and let $f, g : X \to \mathbb{R}$ be Riemann integrable over X, continuous at x_0, and such that $f(x_0) < g(x_0)$ and $f(x) \le g(x)$, $x \in X$. Prove that

$$\int_X f(x)dx < \int_X g(x)dx.$$

Problem 1.25. Prove the following inequalities:

1.

$$1.96 < \int_X \left(100 + (\cos x)^2 + (\cos y)^2\right)^{-1} dxdy \le 1,$$
$$X = \{(x,y) \in \mathbb{R}^2 : |x| + |y| \le 10\}.$$

2.

$$\frac{\sqrt{2}}{3}\pi < \int_X \frac{\sqrt{x^4 + y^4}}{4 + x^4 + y^4} dxdy < \frac{3}{4}\pi,$$
$$X = \{(x,y) \in \mathbb{R}^2 : 1 \le x^2 + y^2 \le 4\}.$$

3.

$$\frac{4 - \sqrt{2}}{7 \log 2} < \int_X \frac{2^{x-y}}{1 + 2^{x+y}} dxdy < \frac{16 - 2\sqrt{2}}{31 \log 2},$$
$$X = [2.5, 3] \times [-1, -0.5].$$

Problem 1.26. Let $X_n = [0, n] \times [0, n]$ and $Y_n = X_{n+1} \backslash X_n$, $n \in \mathbb{N}$. Prove that

$$\lim_{n \to \infty} \int_{Y_n} f(x,y)dxdy = a,$$

where:

1. $a = 0$, and

$$f(x,y) = (x^2 + y^2)^\alpha, \quad \alpha < -\frac{1}{2}, \; (x,y) \in Y_n, \; n \in \mathbb{N}.$$

2. $a = 0$, and

$$f(x,y) = (1 + xy)^{-2}, \quad (x,y) \in Y_n,\ n \in \mathbb{N}.$$

3. $a = 0$, and

$$f(x,y) = \frac{xy}{(x+y)^4}, \quad (x,y) \in Y_n,\ n \in \mathbb{N}.$$

4. $a = 0$, and

$$f(x,y) = e^{-|x^2-y^2|}, \quad (x,y) \in Y_n,\ n \in \mathbb{N}.$$

5. $a = \infty$, and

$$f(x,y) = (2x + y)^{-\frac{1}{2}}, \quad (x,y) \in Y_n,\ n \in \mathbb{N}.$$

6. $a = \infty$, and

$$f(x,y) = \frac{1}{1 + |x-y|}, \quad (x,y) \in Y_n,\ n \in \mathbb{N}.$$

7. $a = 1$, and

$$f(x,y) = \frac{1}{1 + y \log|x+y|}, \quad (x,y) \in Y_n,\ n \in \mathbb{N}.$$

2 Computing of multiple integrals

In this chapter, we provide some methods for computing multiple integrals are investigated The Taylor formula is deduced. Linear maps on measurable sets are investigated. Metric properties of differentiable maps are deduced.

2.1 Reducing multiple integrals to iterated integrals

In this section, we show that the computation of multiple integrals can be reduced to integration of functions of one real variable.

2.1.1 Double integrals

The evaluation of double integrals by using the definition can be a very lengthy process. Therefore we need a practical and convenient technique for computing double integrals. We need to learn how to compute double integrals without employing the definition that uses limits and double sums. The basic idea is that the evaluation becomes easier if we can break a double integral into single integrals by iterating first with respect to one variable and then with respect to the other.

In this section, we consider double integrals of functions defined on a bounded region D on the plane. We consider two types of planar bounded regions.

Definition 2.1. A region D in the (x,y)-plane is said to be of type I if it lies between two vertical lines and the graphs of two continuous functions g_1 and g_2, that is (see Figure 2.1),

$$D = \{(x,y) \in \mathbb{R}^2 : a \le x \le b,\ g_1(x) \le y \le g_2(x)\}. \tag{2.1}$$

A region D in the (x,y)-plane is said to be a region of type II if it lies between two horizontal lines and the graphs of two continuous functions h_1 and h_2, that is (see Figure 2.2),

$$D = \{(x,y) \in \mathbb{R}^2 : c \le y \le d,\ h_1(y) \le x \le h_2(y)\}. \tag{2.2}$$

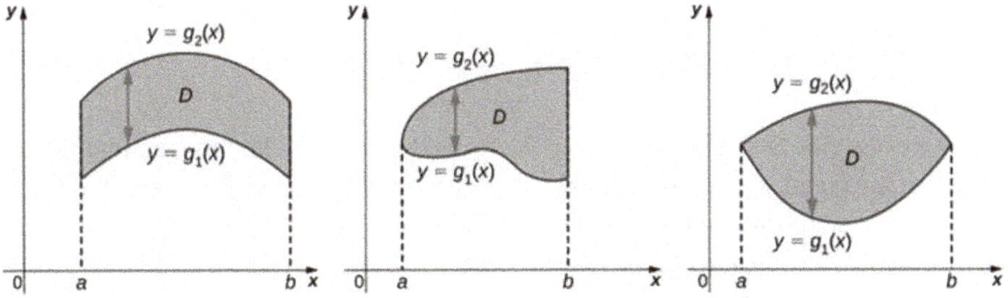

Figure 2.1: A region type I.

https://doi.org/10.1515/9783112219607-002

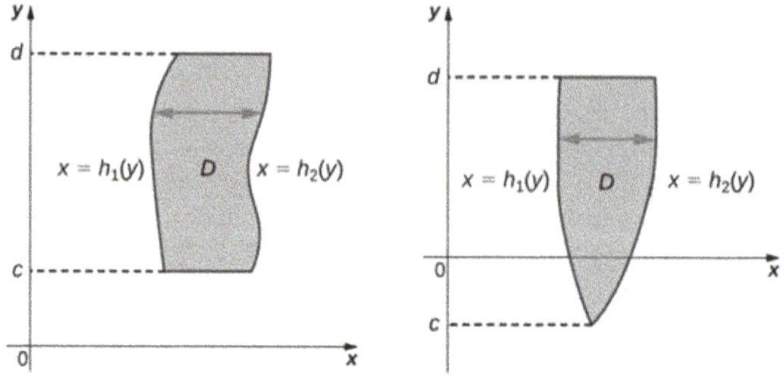

Figure 2.2: A region of type II.

Example 2.1. Consider the region in the first quadrant between the functions

$$y = \sqrt{x} \quad \text{and} \quad y = x^3.$$

We will describe it as a region of type I and as a region of type II. Note that both curves have two points of intersection, $(0,0)$ and $(1,1)$. When describe this region as a region of type I, we need to identify the function that lies above the region and the function that lies below the region. Here the region is bounded above by $y = \sqrt{x}$ and below by $y = x^3$ when $x \in [0,1]$. Hence we can represent the region D as a region of type I as follows Figure 2.3):

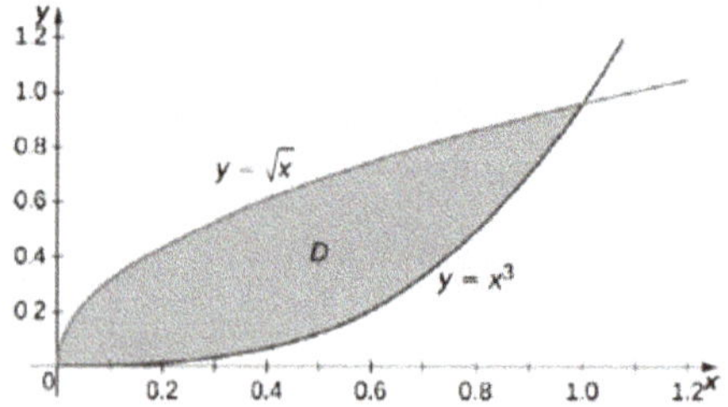

Figure 2.3: The region D as a region of type I and as a region of type II.

$$D = \{(x,y) \in \mathbb{R}^2 : 0 \le x \le 1,\ x^3 \le y \le \sqrt{x}\}.$$

When we describe D as a region of type II, we need to identify the function that lies on the left of the region and the function that lies on the right of the region. Here the region D is bounded on the left by $x = y^2$ and on the right by $x = \sqrt[3]{y}$ when $y \in [0,1]$. Thus D can be represented as a region of type II as follows:

$$D = \{(x,y) \in \mathbb{R}^2 : 0 \le y \le 1,\ y^2 \le x \le \sqrt[3]{y}\}.$$

Note that regions (2.1) and (2.2) are Jordan-measurable sets. We can see this in the following result.

Theorem 2.1. *The set D defined by* (2.1) *is a Jordan-measurable compact set.*

Proof. Since $g_1, g_2 \in \mathscr{C}([a,b])$, we have that g_1 and g_2 are bounded functions on $[a,b]$. Thus D is a bounded set. Note that

$$\partial D = \{x = a\} \cup \{x = b\} \cup \{g_1(x) : a \le x \le b\} \cup \{g_2(x) : a \le x \le b\}$$

and each of the sets $\{x = a\}$, $\{x = b\}$, $\{g_1(x) : a \le x \le b\}$, and $\{g_2(x) : a \le x \le b\}$ have Jordan measure zero. Then ∂D has Jordan measure zero. Now applying Theorem 1.19, we conclude that D is a Jordan-measurable set. Let $\{(x_n, y_n)\}_{n\in\mathbb{N}}$ be an arbitrary sequence of elements of D that converges to (x,y). Then

$$\begin{aligned} \lim_{n\to\infty} x_n &= x, \\ \lim_{n\to\infty} y_n &= y, \end{aligned}$$

and

$$\begin{aligned} a \le x_n &\le b, \\ g_1(x_n) \le y_n &\le g_2(x_n), \quad n \in \mathbb{N}, \end{aligned}$$

and thus

$$\begin{aligned} a &\le \lim_{n\to\infty} x_n \\ &= x \\ &\le b \end{aligned}$$

and

$$\begin{aligned} g_1(x) &= \lim_{n\to\infty} g_1(x_n) \\ &\le \lim_{n\to\infty} y_n \\ &= y \\ &\le \lim_{n\to\infty} g_2(x_n) \\ &= g_2(x). \end{aligned}$$

Therefore $(x,y) \in D$, and D is a closed set. Consequently, D is a Jordan-measurable compact set. This completes the proof. □

As above, we can prove that the set D defined by (2.2) is a Jordan-measurable compact set.

For evaluations of double integrals over a bounded region D, we need to represent it as a region of type I, or a region of type II, or a mix of both. Without understanding the region D, we will not be able to decide the limits of integration in double integrals. As a matter of fact, if the region D is bounded by smooth curves on a plane and we are able to describe it as a region of type I, or a region of type II, or a mix of both, then we can use the following theorem to evaluate double integrals.

Theorem 2.2. *Let* $f \in \mathscr{C}(D)$. *Then*

$$\iint_D f(x,y)dxdy = \int_a^b \left(\int_{g_1(x)}^{g_2(x)} f(x,y)dy \right) dx. \tag{2.3}$$

Proof. Define the function

$$F(x) = \int_{g_1(x)}^{g_2(x)} f(x,y)dy, \quad x \in [a,b]. \tag{2.4}$$

Firstly, we will prove that $F \in \mathscr{C}([a,b])$. Since $f \in \mathscr{C}(D)$, for any $x \in [a,b]$, the function $f(x,\cdot)$ is a continuous function on $[g_1(x), g_2(x)]$. Therefore, for any $x \in [a,b]$, the integral (2.4) exists. For $x \in [a,b]$, set

$$y = g_1(x) + (g_2(x) - g_1(x))t, \quad t \in [0,1].$$

Then (2.4) takes the form

$$F(x) = \int_0^1 f(x, g_1(x) + (g_2(x) - g_1(x))t)(g_2(x) - g_1(x))dt.$$

Let

$$l(x,t) = f(x, g_1(x) + (g_2(x) - g_1(x))t)(g_2(x) - g_1(x)), \quad x \in [a,b],\ t \in [0,1].$$

Because $f \in \mathscr{C}(D)$, $g_1, g_2 \in \mathscr{C}([a,b])$, and D is a compact set, we have that l is a uniformly continuous function on D. Take arbitrary $\varepsilon > 0$. Then, for any $x \in [a,b]$, there is $\delta = \delta(\varepsilon) > 0$ such that if $|r| < \delta$, then

$$\left|l(x+r,t) - l(x,t)\right| < \varepsilon, \quad t \in [0,1],$$

and therefore

$$\begin{aligned}|F(x+r)-F(x)| &= \left|\int_0^1 l(x+r,t)dt - \int_0^1 l(x,t)dt\right| \\ &= \left|\int_0^1 (l(x+r,t)-l(x,t))dt\right| \\ &\le \int_0^1 |l(x+r,t)-l(x,t)|dt \\ &< \varepsilon.\end{aligned}$$

Consequently, $F \in \mathscr{C}([a,b])$. Therefore the integral in the right-hand side of (2.3) exists.
Fix $k \in \mathbb{N}$. Consider

$$x^j = a + \frac{b-a}{k}j, \quad j \in \{0,1,\dots,k\}.$$

Define

$$\begin{aligned}\phi_0(x) &= g_1(x), \\ \phi_1(x) &= g_1(x) + \frac{g_2(x)-g_1(x)}{k}, \\ &\vdots \\ \phi_j(x) &= g_1(x) + \frac{g_2(x)-g_1(x)}{k}j, \\ &\vdots \\ \phi_k(x) &= g_1(x) + \frac{g_2(x)-g_1(x)}{k}k \\ &= g_2(x), \quad a \le x \le b.\end{aligned}$$

By the above definition it follows that

$$\begin{aligned}\phi_j(x) &= g_1(x) + \frac{g_2(x)-g_1(x)}{k}j \\ &= g_1(x) + \frac{g_2(x)-g_1(x)}{k}(j-1+1) \\ &= g_1(x) + \frac{g_2(x)-g_1(x)}{k}(j-1) + \frac{g_2(x)-g_1(x)}{k} \\ &= \phi_{j-1}(x) + \frac{g_2(x)-g_1(x)}{k}, \quad j \in \{1,\dots,k\},\ x \in [a,b].\end{aligned}$$

Define the sets

$$X_{ij}^k = \{(x,y) : x^{i-1} \le x \le x^i,\ \phi_{j-1}(x) \le y \le \phi_j(x)\}, \quad i,j \in \{1,\dots,k\}.$$

Then $\tau_k = \{X_{ij}^k\}_{i,j=1}^k$ is a partition of X. Since $\phi, \psi \in \mathscr{C}([a,b])$, there is a constant $c > 0$ such that

$$|g_1(x)|, \ |g_2(x)| \le c, \quad x \in [a,b].$$

Hence

$$\begin{aligned} |\phi_j(x) - \phi_{j-1}(x)| &= \left|\frac{g_2(x) - g_1(x)}{k}\right| \\ &\le \frac{1}{k}(|g_2(x)| + |g_1(x)|) \\ &\le \frac{2c}{k}, \quad x \in [a,b]. \end{aligned}$$

Now for $(x_1, x_2), (y_1, y_2) \in X_{ij}^k$, we have

$$\begin{aligned} |y_1 - x_1| &\le x^i - x^{i-1} \\ &= a + \frac{b-a}{k} i - \left(a + \frac{b-a}{k}(i-1)\right) \\ &= \frac{b-a}{k} \end{aligned}$$

and

$$\begin{aligned} |y_2 - x_2| &= |y_2 - \phi_j(y_1) + \phi_j(y_1) - \phi_j(x_1) + \phi_j(x_1) - x_2| \\ &\le |y_2 - \phi_j(y_1)| + |\phi_j(y_1) - \phi_j(x_1)| + |\phi_j(x_1) - x_2| \\ &\le |\phi_j(y_1) - \phi_{j-1}(y_1)| + |\phi_j(y_1) - \phi_j(x_1)| + |\phi_j(x_1) - \phi_{j-1}(x_1)| \\ &\le \frac{2c}{k} + |\phi_j(y_1) - \phi_j(x_1)| + \frac{2c}{k} \\ &\le \frac{4c}{k} + \omega_i(\phi_j), \end{aligned}$$

where

$$\begin{aligned} \omega_i(\phi_j) &= \sup_{x_1, y_1 \in [x_{i-1}, x_i]} |\phi_j(y_1) - \phi_j(x_1)| \\ &= \sup_{x_1, y_1 \in [x_{i-1}, x_i]} \left| g_1(y_1) + \frac{g_2(y_1) - g_1(y_1)}{k} j - g_1(x_1) - \frac{g_2(x_1) - g_1(x_1)}{k} j \right| \\ &= \sup_{x_1, y_1 \in [x_{i-1}, x_i]} \left| (g_1(y_1) - g_1(x_1)) + \frac{g_2(y_1) - g_2(x_1)}{k} j + \frac{g_1(x_1) - g_1(y_1)}{k} j \right| \\ &\le \sup_{x_1, y_1 \in [x_{i-1}, x_i]} |g_1(y_1) - g_1(x_1)| \\ &\quad + \frac{j}{k}\Big(\sup_{x_1, y_1 \in [x_{i-1}, x_i]} |g_2(y_1) - g_2(x_1)| + \sup_{x_1, y_1 \in [x_{i-1}, x_i]} |g_1(x_1) - g_1(y_1)| \Big) \end{aligned}$$

$$\leq 2\omega_i(g_1) + \omega_i(g_1)$$
$$\leq 2\omega\left(\frac{b-a}{k}, g_1\right) + \omega\left(\frac{b-a}{k}, g_2\right).$$

Therefore

$$|y_2 - x_2| \leq \frac{4c}{k} + 2\omega\left(\frac{b-a}{k}, g_1\right) + \omega\left(\frac{b-a}{k}, g_2\right),$$

and

$$\begin{aligned} d((x_1,x_2),(y_1,y_2)) &= \sqrt{(y_1-x_1)^2 + (y_2-x_2)^2} \\ &\leq |y_1 - x_1| + |y_2 - x_2| \\ &\leq \frac{b-a}{k} + \frac{4c}{k} + 2\omega\left(\frac{b-a}{k}, g_1\right) + \omega\left(\frac{b-a}{k}, g_2\right), \end{aligned}$$

from which we get

$$\begin{aligned} \operatorname{diam} X_{ij}^k &= \sup_{(x_1,x_2),(y_1,y_2)\in X_{ij}^k} d((x_1,x_2),(y_1,y_2)) \\ &\leq \frac{b-a}{k} + \frac{4c}{k} + 2\omega\left(\frac{b-a}{k}, g_1\right) + \omega\left(\frac{b-a}{k}, g_2\right). \end{aligned}$$

Since $g_1, g_2 \in \mathscr{C}([a,b])$, we have that

$$\lim_{k\to\infty} \omega\left(\frac{b-a}{k}, g_1\right) = 0,$$
$$\lim_{k\to\infty} \omega\left(\frac{b-a}{k}, g_2\right) = 0.$$

Take arbitrary $\varepsilon > 0$. Then there is $k_0 \in \mathbb{N}$ such that for any $k \geq k_0$, we have

$$\begin{aligned} \frac{b-a}{k} &< \frac{\varepsilon}{3}, \\ \frac{4c}{k} &< \frac{\varepsilon}{3}, \\ 2\omega\left(\frac{b-a}{k}, g_1\right) + \omega\left(\frac{b-a}{k}, g_2\right) &< \frac{\varepsilon}{3}. \end{aligned}$$

Then, for all $k \geq k_0$ and $i, j \in \{1, \ldots, k\}$, we get

$$\begin{aligned} \operatorname{diam} X_{ij}^k &< \frac{\varepsilon}{3} + \frac{\varepsilon}{3} + \frac{\varepsilon}{3} \\ &= \varepsilon \end{aligned}$$

and

$$
\begin{aligned}
|\tau_k| &= \max_{i,j\in\{1,\dots,k\}} \operatorname{diam} X_{ij}^k \\
&< \varepsilon.
\end{aligned}
$$

Because $\varepsilon > 0$ was arbitrarily chosen, we get

$$\lim_{k\to\infty} |\tau_k| = 0.$$

Let

$$
\begin{aligned}
m_{ij}^k &= \inf_{(x,y)\in X_{ij}^k} f(x,y), \\
M_{ij}^k &= \sup_{(x,y)\in X_{ij}^k} f(x,y).
\end{aligned}
$$

Then

$$
\begin{aligned}
\int_a^b \left(\int_{g_1(x)}^{g_2(x)} f(x,y)dy \right) dx &= \sum_{i=1}^k \int_{x^{i-1}}^{x^i} \left(\int_{g_1(x)}^{g_2(x)} f(x,y)dy \right) dx \\
&= \sum_{i=1}^k \int_{x^{i-1}}^{x^i} \left(\sum_{j=1}^k \int_{\phi_{j-1}(x)}^{\phi_j(x)} f(x,y)dy \right) dx \\
&= \sum_{i,j=1}^k \int_{x^{i-1}}^{x^i} \left(\int_{\phi_{j-1}(x)}^{\phi_j(x)} f(x,y)dy \right) dx \\
&\le \sum_{i,j=1}^k M_{ij}^k \int_{x^{i-1}}^{x^i} (\phi_j(x) - \phi_{j-1}(x))dx \\
&= \sum_{i,j=1}^k M_{ij}^k \mu X_{ij}^k,
\end{aligned}
$$

and

$$
\begin{aligned}
\int_a^b \left(\int_{g_1(x)}^{g_2(x)} f(x,y)dy \right) dx &= \sum_{i=1}^k \int_{x^{i-1}}^{x^i} \left(\int_{g_1(x)}^{g_2(x)} f(x,y)dy \right) dx \\
&= \sum_{i,j=1}^k \int_{x^{i-1}}^{x^i} \left(\int_{\phi_{j-1}(x)}^{\phi_j(x)} f(x,y)dy \right) dx
\end{aligned}
$$

$$\geq \sum_{i,j=1}^{k} m_{ij}^{k} \int_{x^{i-1}}^{x^{i}} (\phi_j(x) - \phi_{j-1}(x))dx$$
$$= \sum_{i,j=1}^{k} m_{ij}^{k} \mu X_{ij}^{k}.$$

Therefore

$$\begin{aligned} s_{\tau_k} &= \sum_{i,j=1}^{k} m_{ij}^{k} \mu X_{ij}^{k} \\ &\leq \int_a^b \left(\int_{g_1(x)}^{g_2(x)} f(x,y)dy \right) dx \\ &\leq \sum_{i,j=1}^{k} M_{ij}^{k} \mu X_{ij}^{k} \\ &= S_{\tau_k}. \end{aligned}$$

Hence

$$\begin{aligned} \lim_{|\tau_k|\to 0} s_{\tau_k} &= \lim_{|\tau_k|\to 0} S_{\tau_k} \\ &= \int_D f(x,y)dydx \\ &= \int_a^b \left(\int_{g_1(x)}^{g_2(x)} f(x,y)dy \right) dx. \end{aligned}$$

This completes the proof. □

As we have proved Theorem 2.2, we can prove the following result.

Theorem 2.3. *Let $f \in \mathscr{C}(D)$. Then*

$$\int\int_D f(x,y)dydx = \int_c^d \left(\int_{h_1(y)}^{h_2(y)} f(x,y)dx \right) dy.$$

Example 2.2. We will evaluate the integral

$$I = \int\int_D x^2 e^{xy} dxdy,$$

where D is the region in Figure 2.4. Firstly, we will construct the region as a region of type I. We have

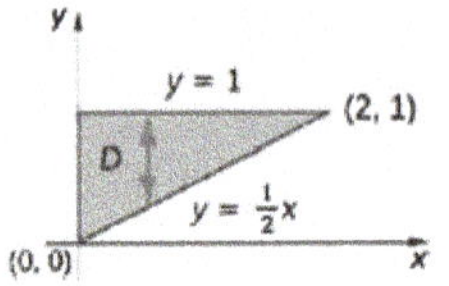

Figure 2.4: A region of type I.

$$D = \left\{(x,y) \in \mathbb{R}^2 : 0 \le x \le 2,\ \frac{1}{2}x \le y \le 1\right\}.$$

Then applying Theorem 2.2, we get

$$\begin{aligned}
I &= \int_0^2 x^2\left(\int_{\frac{1}{2}x}^{1} e^{xy}\,dy\right)dx \\
&= \int_0^2 x\left(\int_{\frac{1}{2}x}^{1} e^{xy}\,d(xy)\right)dx \\
&= \int_0^2 x\left(e^{xy}\Big|_{y=\frac{1}{2}x}^{y=1}\right)dx \\
&= \int_0^2 x\left(e^x - e^{\frac{1}{2}x^2}\right)dx \\
&= \int_0^2 xe^x\,dx - \int_0^2 xe^{\frac{1}{2}x^2}\,dx \\
&= xe^x\Big|_{x=0}^{x=2} - \int_0^2 e^x\,dx - \int_0^2 e^{\frac{1}{2}x^2}\,d\left(\frac{1}{2}x^2\right) \\
&= 2e^2 - e^x\Big|_{x=0}^{x=2} - e^{\frac{1}{2}x^2}\Big|_{x=0}^{x=2} \\
&= 2e^2 - e^2 + 1 - e^2 + 1 \\
&= 2.
\end{aligned}$$

We can consider the region D as a region of type II. In this case, we have

$$D = \{(x,y) \in \mathbb{R}^2 : 0 \le y \le 1,\ 0 \le x \le 2y\}.$$

If we first integrate with respect to x, this integral is lengthy to compute because we have to use integration by parts twice.

Example 2.3. Now we will evaluate the integral

$$I = \iint_D (3x^2 + y^2)\,dxdy,$$

where D is the region

$$D = \{(x,y) \in \mathbb{R}^2 : -2 \le y \le 3,\ y^2 - 3 \le x \le y + 3\}.$$

Notice that the region D can be considered as a region of type I and as a region of type II. The case where D is a region of type I is more complicated than the case where D is a region of type II. Thus we will consider D as a region of type I (see Figure 2.5). Applying Theorem 2.3, we find

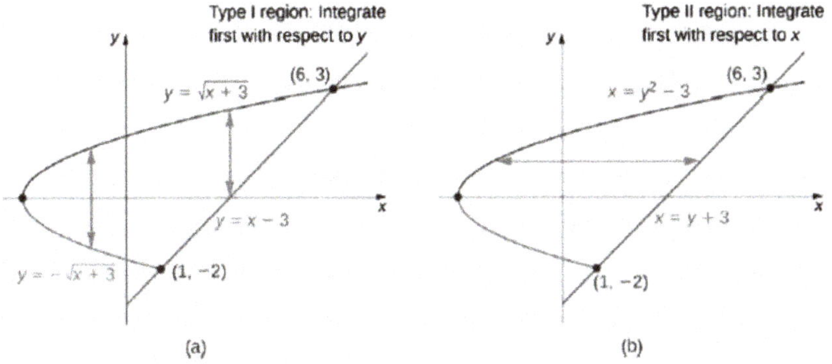

Figure 2.5: A region of type I (a) and a region of type II (b).

$$\begin{aligned}
I &= \int_{-2}^{3}\left(\int_{y^2-3}^{y+3}(3x^2 + y^2)dx\right)dy \\
&= \int_{-2}^{3}\left(3\int_{y^2-3}^{y+3}x^2dx + y^2\int_{y^2-3}^{y+3}dx\right)dy \\
&= \int_{-2}^{3}\left(x^3\Big|_{x=y^2-3}^{x=y+3} + y^2(y + 3 - y^2 + 3)\right)dy \\
&= \int_{-2}^{3}((y+3)^3 - (y^2-3)^3 + y^3 - y^4 + 6y^2)dy \\
&= \int_{-2}^{3}(y^3 + 9y^2 + 27y + 27 - y^6 + 9y^4 - 27y^2 + 27 + y^3 - y^4 + 6y^2)dy
\end{aligned}$$

$$
\begin{aligned}
&= \int_{-2}^{3} (-y^6 + 8y^4 + 2y^3 - 12y^2 + 27y + 54)dy \\
&= -\int_{-2}^{3} y^6 dy + 8\int_{-2}^{3} y^4 dy + 2\int_{-2}^{3} y^3 dy - 12\int_{-2}^{3} y^2 dy + 27\int_{-2}^{3} y dy \\
&\quad + 54\int_{-2}^{3} dy \\
&= -\frac{1}{7}y^7\Big|_{y=-2}^{y=3} + \frac{8}{5}y^5\Big|_{y=-2}^{y=3} + \frac{1}{2}y^4\Big|_{y=-2}^{y=3} - 4y^3\Big|_{y=-2}^{y=3} \\
&\quad + \frac{27}{2}y^2\Big|_{y=-2}^{y=3} + 54y\Big|_{y=-2}^{y=3} \\
&= -\frac{1}{7}(3^7 - (-2)^7) + \frac{8}{5}(3^5 - (-2)^5) + \frac{1}{2}(3^4 - (-2)^4) \\
&\quad - 4(3^3 - (-2)^3) + \frac{27}{2}(3^2 - (-2)^2) + 54(3 + 2) \\
&= -\frac{1}{7}(2187 + 128) + \frac{8}{5}(243 + 32) + \frac{1}{2}(81 - 16) \\
&\quad - 4(27 + 8) + \frac{27}{2}(9 - 4) + 270 \\
&= -\frac{2315}{7} + 440 + \frac{65}{2} - 140 + \frac{135}{2} + 270 \\
&= -\frac{2315}{7} + 670 \\
&= \frac{-2315 + 4690}{7} \\
&= \frac{2375}{7}.
\end{aligned}
$$

Example 2.4. We will find the area I of the region D given by Figure 2.6. The region D is not easily decomposed into any one type. We will represent it as the union of the regions

$$
\begin{aligned}
D_1 &= \{(x,y) \in \mathbb{R}^2 : -2 \le x \le 0,\ 0 \le y \le (x+2)^2\}, \\
D_2 &= \left\{(x,y) \in \mathbb{R}^2 : 0 \le y \le 4,\ 0 \le x \le y - \frac{1}{16}y^3\right\}, \\
D_3 &= \left\{(x,y) \in \mathbb{R}^2 : -4 \le y \le 0,\ -2 \le x \le y - \frac{1}{16}y^3\right\}.
\end{aligned}
$$

These regions are illustrated in Figure 2.7. Then

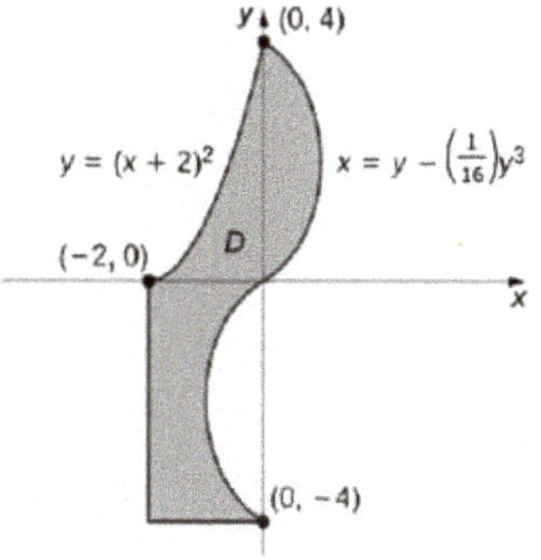

Figure 2.6: Region D.

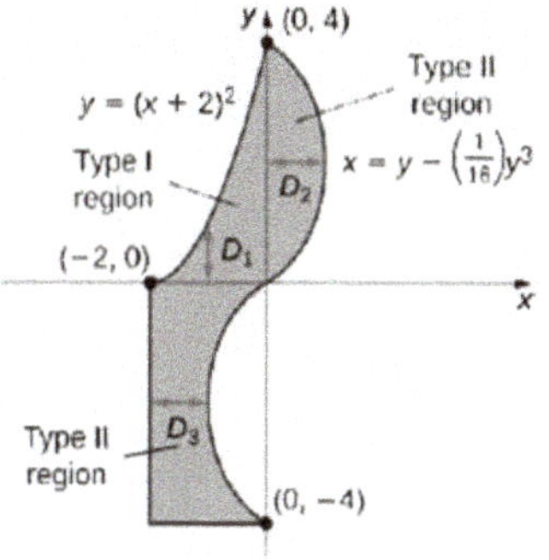

Figure 2.7: The region D represented as the union of three regions D_1, D_2, and D_3.

$$
\begin{aligned}
I &= \iint\limits_D dxdy \\
&= \int_{-2}^{0}\left(\int_{0}^{(x+2)^2} dy\right)dx + \int_{0}^{4}\left(\int_{0}^{y-\frac{1}{16}y^3} dx\right)dy \\
&\quad + \int_{-4}^{0}\left(\int_{-2}^{y-\frac{1}{16}y^3} dx\right)dy \\
&= \int_{-2}^{0}(x+2)^2dx + \int_{0}^{4}\left(y - \frac{1}{16}y^3\right)dy \\
&\quad + \int_{-4}^{0}\left(y - \frac{1}{16}y^3 + 2\right)dy \\
&= \frac{(x+2)^3}{3}\Bigg|_{x=-2}^{x=0} + \frac{1}{2}y^2\Bigg|_{y=0}^{y=4} - \frac{1}{64}y^4\Bigg|_{y=0}^{y=4}
\end{aligned}
$$

$$
\begin{aligned}
&+\frac{1}{2}y^2\Big|_{y=-4}^{y=0}-\frac{1}{64}y^4\Big|_{y=-4}^{y=0}+2\cdot 4\\
&=\frac{8}{3}+8-4-8+4+8\\
&=\frac{8}{3}+8\\
&=\frac{32}{3}.
\end{aligned}
$$

Exercise 2.1. Evaluate

1.
$$\int_0^{\frac{\pi}{2}}\left(\int_0^x \cos(x+y)dy\right)dx.$$

2.
$$\int_{-1}^{1}\left(\int_{2y}^{y}(x-y)e^y dy\right)dx.$$

Exercise 2.2. Evaluate

1.
$$\int_X x^2y^2dxdy,$$

   where X is bounded by
$$x=y^2,\quad x=1.$$

2.
$$\int_X xy^2dxdy,$$

   where
$$X=\{(x,y)\in\mathbb{R}^2: x^2+y^2\le a^2,\ x\ge 0\}.$$

3.
$$\int_X (x^3+y^3)dxdy,$$

   where

$$X = \{(x,y) \in \mathbb{R}^2 : x^2 + y^2 \le R^2,\ y \ge 0\}.$$

4.

$$\int_X (x + 2y)\,dxdy,$$

where X is bounded by the lines

$$y = x, \quad y = 2x, \quad x = 2, \quad x = 3.$$

5.

$$\int_X (x^2 + y^2)\,dxdy,$$

where X is bounded by the lines

$$y = x, \quad y = x + a, \quad y = a, \quad y = 3a.$$

6.

$$\int_X \sqrt{x - y}\,dxdy,$$

where

$$X = \left\{\frac{4}{5}x \le y \le x,\ 1 \le y \le 4\right\}.$$

7.

$$\int_X \sin(\pi(x - y))\,dxdy,$$

where X is the triangle with vertices

$$(-4, 1), \quad \left(-1, -\frac{1}{2}\right), \quad \left(\frac{7}{2}, \frac{17}{2}\right).$$

8.

$$\int_X e^{x-y}\,dxdy,$$

where X is bounded by the lines

$$x = -1, \quad x = 1, \quad y = x, \quad y = 2x.$$

9.

$$\int_X (x+y)dxdy,$$

where

$$X = \{(x,y) \in \mathbb{R}^2 : x^2 + y^2 \le R^2,\ y \ge x\}.$$

10.

$$\int_X xydxdy,$$

where

$$X = \{(x,y) \in \mathbb{R}^2 : x^2 + y^2 \le 25,\ 3x + y \ge 5\}.$$

2.1.2 Triple integrals

In this section, we consider triple integrals of continuous functions over a bounded region E in $\mathbb{R}^3$. Let D be the bounded region that is a projection of E into the (x,y)-plane. Suppose that the region E has the form

$$E = \{(x,y,z) : (x,y) \in D,\ u_1(x,y) \le z \le u_2(x,y)\},$$

where $u_1, u_2 \in \mathscr{C}(D)$ (see Figure 2.8). Then the triple integral of a continuous function f over E is represented in the form

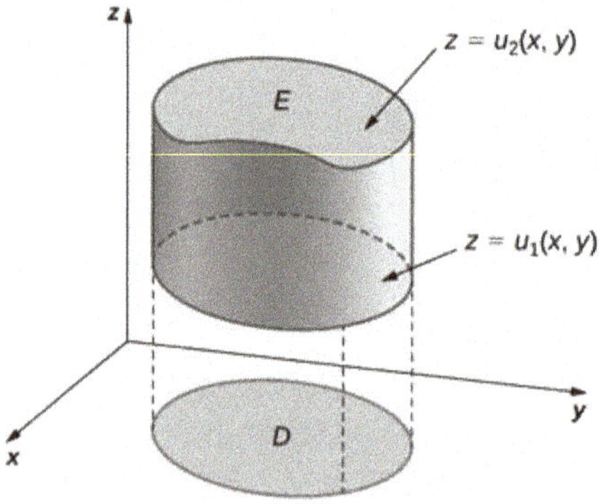

Figure 2.8: The region E.

$$\iiint_E f(x,y,z)dxdydz = \iint_D \left(\int_{u_1(x,y)}^{u_2(x,y)} f(x,y,z)dz \right) dxdy.$$

Let now D be the bounded region that is a projection of E into the (y,z)-plane. Then the region E has the form

$$E = \{(x,y,z) : (y,z) \in D,\ u_1(y,z) \le x \le u_2(y,z)\},$$

where $u_1, u_2 \in \mathscr{C}(D)$. Then the triple integral of a continuous function f over E is represented in the form

$$\iiint_E f(x,y,z)dxdydz = \iint_D \left(\int_{u_1(y,z)}^{u_2(y,z)} f(x,y,z)dx \right) dydz.$$

If D is the bounded region that is a projection of E into the (x,z)-plane, then the region E has the form

$$E = \{(x,y,z) : (x,z) \in D,\ u_1(x,z) \le y \le u_2(x,z)\}, \tag{2.5}$$

where $u_1, u_2 \in \mathscr{C}(D)$. In this case, the triple integral of a continuous function f over E is represented in the following form

$$\iiint_E f(x,y,z)dxdydz = \iint_D \left(\int_{u_1(x,z)}^{u_2(x,z)} f(x,y,z)dy \right) dxdz. \tag{2.6}$$

Consider (2.5) and (2.6). The region D may be of type I or of type II. If D in the (x,y)-plane is of type I (see Figure 2.9), then

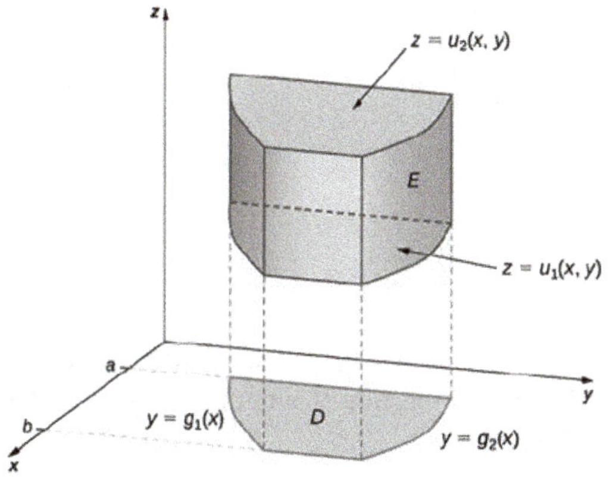

Figure 2.9: The region E.

$$E = \{(x,y,z) : a \le x \le b,\ g_1(x) \le y \le g_2(x),\ u_1(x,y) \le z \le u_2(x,y)\},$$

where $g_1, g_2 \in \mathscr{C}([a,b])$. Then the triple integral becomes

$$\iiint_D f(x,y,z)dxdydz = \int_a^b \int_{g_1(x)}^{g_2(x)} \int_{u_1(x,y)}^{u_2(x,y)} f(x,y,z)dzdydx.$$

Now suppose that D in the (x,y)-plane is of type II (see Figure 2.10). Then

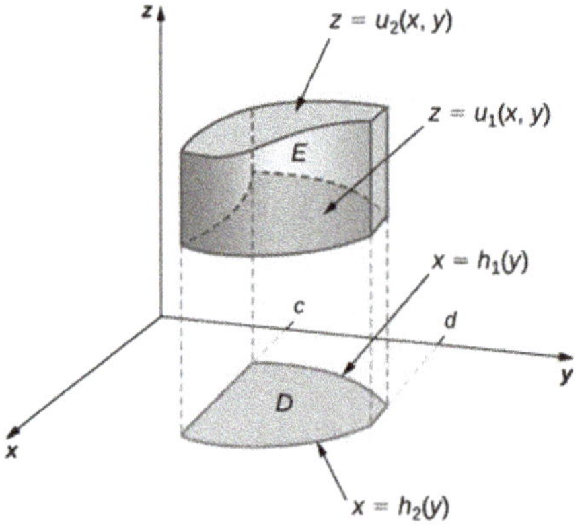

Figure 2.10: The region E.

$$E = \{(x,y,z) : c \le y \le d,\ h_1(y) \le x \le h_2(y),\ u_1(x,y) \le z \le u_2(x,y)\},$$

where $h_1, h_2 \in \mathscr{C}([c,d])$. In this case, the triple integral becomes

$$\iiint_D f(x,y,z)dxdydz = \int_c^d \int_{h_1(y)}^{h_2(y)} \int_{u_1(x,y)}^{u_2(x,y)} f(x,y,z)dzdxdy.$$

Example 2.5. We will evaluate the integral

$$I = \iiint_E (x-y)dxdydz,$$

where E is the solid tetrahedron bounded by the planes $x = 0, y = 0, z = 0$, and $x+y+z = 1$. Figure 2.11 shows the solid tetrahedron E and its projection D onto the (x,y)-plane. We describe the solid region tetrahedron as follows:

$$E = \{(x,y,z) : 0 \le x \le 1,\ 0 \le y \le 1-x,\ 0 \le z \le 1-x-y\}.$$

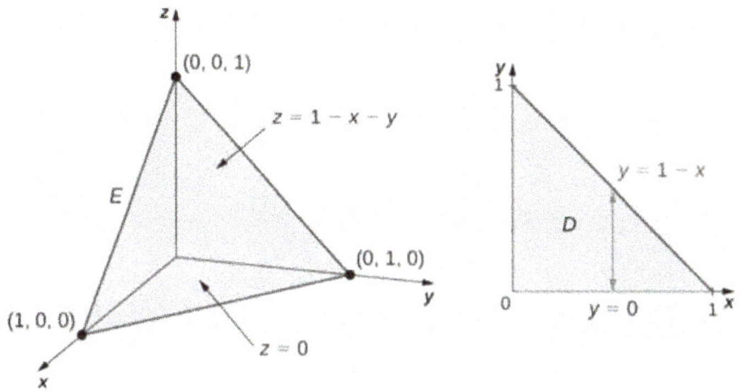

Figure 2.11: The solid tetrahedron E and its projection D onto the (x, y)-plane.

Then

$$\begin{aligned}
I &= \int_0^1 \left(\int_0^{1-x} \left(\int_0^{1-x-y} (x-y)dz \right) dy \right) dx \\
&= \int_0^1 \left(\int_0^{1-x} (x-y)(1-x-y)dy \right) dx \\
&= \int_0^1 \left(\int_0^{1-x} (x - x^2 - xy - y + xy + y^2)dy \right) dx \\
&= \int_0^1 \left(\int_0^{1-x} (x - x^2 - y + y^2)dy \right) dx \\
&= \int_0^1 \left((x - x^2)(1-x) - \frac{1}{2}(1-x)^2 + \frac{1}{3}(1-x)^3 \right) dx \\
&= \int_0^1 \left(x - x^2 - x^2 + x^3 - \frac{1}{2} + x - \frac{x^2}{2} + \frac{1}{3} - x + x^2 - \frac{1}{3}x^3 \right) dx \\
&= \int_0^1 \left(\frac{2}{3}x^3 - \frac{3}{2}x^2 + x - \frac{1}{6} \right) dx \\
&= \frac{1}{6} - \frac{1}{2} + \frac{1}{2} - \frac{1}{6} \\
&= 0.
\end{aligned}$$

Example 2.6. We will find the volume I of a right pyramid that has the square base in the (x, y)-plane $[-1, 2] \times [-1, 1]$ and vertex at the point $(0, 0, 1)$, as it is shown in Figure 2.12.

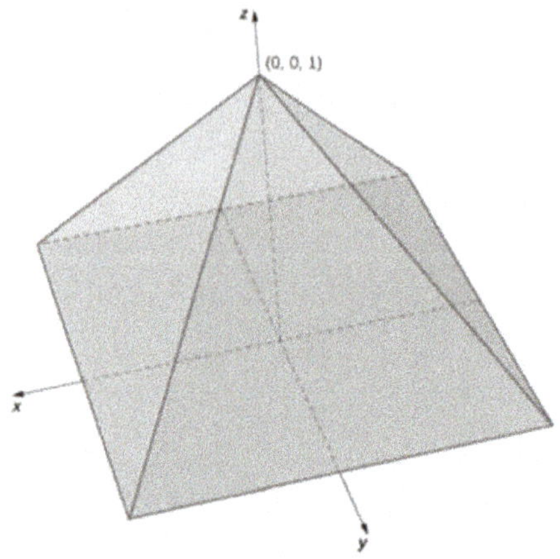

Figure 2.12: A pyramid with a square base.

The value of z changes from 0 to 1, and at each height z, the cross-section of the pyramid for any value of z is the square

$$[-1+z, 1-z] \times [-1+z, 1-z].$$

Then we can describe this pyramid as follows:

$$E = \{(x,y,z) \in \mathbb{R}^3 : 0 \le z \le 1,\ -1+z \le y \le 1-z,\ -1+z \le x \le 1-z\}.$$

Hence

$$\begin{aligned} I &= \iiint_E dxdydz \\ &= \int_0^1 \left(\int_{-1+z}^{1-z} \left(\int_{-1+z}^{1-z} dx \right) dy \right) dz \\ &= \int_0^1 \left(\int_{-1+z}^{1-z} 2(1-z) dy \right) dz \\ &= 4 \int_0^1 (1-z)^2 dz \\ &= -\frac{4}{3}(1-z)^3 \Big|_{z=0}^{z=1} \\ &= \frac{4}{3}. \end{aligned}$$

Example 2.7. We will evaluate the integral

$$\iiint\limits_{E} z dx dy dz,$$

where E is the region bounded by the paraboloid $y = x^2 + z^2$ and the plane $y = 4$ (see Figure 2.13). The projection of the region E onto the (x, y)-plane is the region bounded above by $y = 4$ and below by the parabola $y = x^2$, as it is shown in Figure 2.14. Then the region E can be described as follows:

$$E = \{(x, y, z) \in \mathbb{R}^3 : -2 \le x \le 2,\ x^2 \le y \le 4,\ -\sqrt{y - x^2} \le z \le \sqrt{y - x^2}\}.$$

Hence

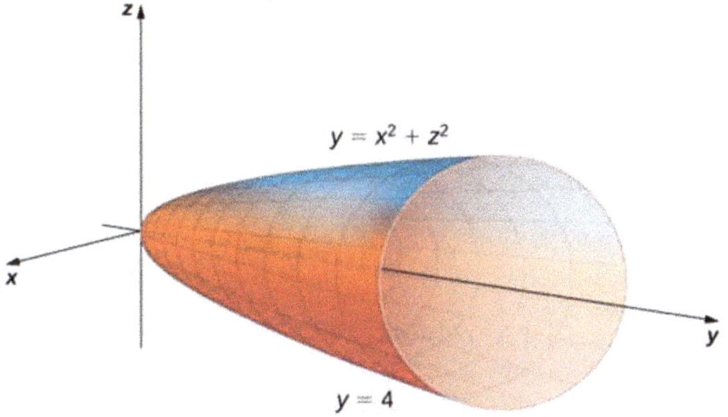

Figure 2.13: A paraboloid.

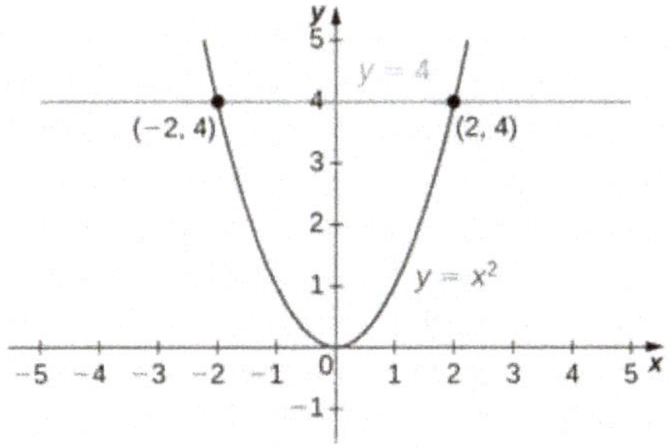

Figure 2.14: A projection of a paraboloid onto (x, y)-plane.

$$I = \int_{-2}^{2}\left(\int_{x^2}^{4}\left(\int_{-\sqrt{y-x^2}}^{\sqrt{y-x^2}} zdz\right)dy\right)dx$$
$$= \frac{1}{2}\int_{-2}^{2}\left(\int_{x^2}^{4} z^2\Big|_{z=-\sqrt{y-x^2}}^{z=\sqrt{y-x^2}} dy\right)dx$$
$$= 0.$$

Exercise 2.3. Evaluate

1.
$$\int_0^a\int_0^b\int_0^c ze^x \sin y dxdydz.$$

2.
$$\int_0^a\int_0^b\int_0^c (x+y)e^{x-y} dxdydz.$$

3.
$$\int_0^1\int_{-1}^1\int_0^1 xyzdxdydz.$$

4.
$$\int_0^1\int_0^1\int_0^1 (xy+yz+xz)dxdydz.$$

5.
$$\int_0^\pi\int_0^\pi\int_0^\pi \sin(x+y+z)dxdydz.$$

6.
$$\int_X (x+2y+3z)dxdydz,$$

where X is bounded by

$$y=0, \quad z=0, \quad z=2, \quad x+y=2, \quad 2x-y+2=0.$$

7.

$$\int_X (xy)^2 dxdydz,$$

where

$$X = \{(x, y, z) \in \mathbb{R}^3 : 0 \le x \le y \le z \le 1\}.$$

2.1.3 Multiple integrals

In this section, we generalize the technique proposed in the previous sections for evaluation of double and triple integrals. Consider $\mathbb{R}^n$, $n > 3$, and $\mathbb{R}^{n-1}$, the hyperplane $x_n = 0$. Let $X \subset \mathbb{R}^n$ be a Jordan-measurable closed set, let $f : X \to \mathbb{R}$ be a continuous function, and let $X_{x_1 \cdots x_{n-1}}$ be the projection of X over $\mathbb{R}^{n-1}$. Suppose that $\phi, \psi \in \mathscr{C}(X_{x_1 \cdots x_{n-1}})$ and the set X has the form

$$\begin{aligned} X = \{&(x_1, \ldots, x_{n-1}, x_n) \in \mathbb{R}^n : (x_1, \ldots, x_{n-1}, 0) \in X_{x_1 \cdots x_{n-1}}, \\ &\phi(x_1, \ldots, x_{n-1}) \le x_n \le \psi(x_1, \ldots, x_{n-1})\}. \end{aligned} \tag{2.7}$$

In addition, suppose that $X_{x_1 \cdots x_{n-1}}$ is Jordan measurable. Then

$$\int_X f(x_1, \ldots, x_n) dx_1 \cdots dx_n = \int_{X_{x_1 \cdots x_{n-1}}} \left(\int_{\phi(x_1, \ldots, x_{n-1})}^{\psi(x_1, \ldots, x_{n-1})} f(x_1, \ldots, x_{n-1}, x_n) dx_n \right) dx_{n-1} \cdots dx_1. \tag{2.8}$$

If $X_{x_1 \cdots x_{n-1}}$ can be represented in a similar way in the form (2.7) of X, then the $(n-1)$-dimensional integral in the right-hand side of (2.8) can be reduced to an $(n-2)$-dimensional integral. Continuing this process, we get

$$\begin{aligned} &\int_X f(x_1, \ldots, x_n) dx_n \cdots dx_1 \\ &= \int_a^b \left(\int_{\phi_1(x_1)}^{\psi_1(x_1)} \left(\int_{\phi_2(x_1, x_2)}^{\psi_2(x_1, x_2)} \cdots \left(\int_{\phi_{n-1}(x_1, \ldots, x_{n-1})}^{\psi_{n-1}(x_1, \ldots, x_{n-1})} f(x_1, \ldots, x_{n-1}, x_n) dx_n \right) \cdots dx_3 \right) dx_2 \right) dx_1. \end{aligned}$$

Thus the integration of a function of n independent variables is reduced to n times integrations of functions of a single variable. Let $X_{x_1 \cdots x_m}$ be the projection of X onto $\mathbb{R}^m_{x_1 \cdots x_m}$, and let $X(x_1, \ldots, x_m)$ be the intersection of X with the $(n-m)$-dimensional hyperplane through the point $(x_1, \ldots, x_m, 0, \ldots, 0)$ and orthogonal to $\mathbb{R}^m_{x_1 \cdots x_m}$. Then

$$\int_X f(x_1, \ldots, x_n) dx_1 \cdots dx_n = \int_{X_{x_1 \cdots x_m}} \left(\int_{X(x_1, \ldots, x_m)} f(x_1, \ldots, x_n) dx_{m+1} \cdots dx_n \right) dx_1 \cdots dx_m.$$

If $f \equiv 1$ on X, then

$$\mu X = \int_{X_{x_1 \cdots x_m}} \mu X(x_1, \dots, x_m) dx_1 \cdots dx_m.$$

Example 2.8. We will find the volume of the n-dimensional ball

$$V_r^n = \{(x_1, \dots, x_n) \in \mathbb{R}^n : x_1^2 + \cdots + x_n^2 \le r^2\}.$$

For $n = 2$, we get

$$\begin{aligned}
\mu_2 V_t r^2 &= \int_{x_1^2 + x_2^2 \le r^2} dx_1 dx_2 \\
&= 2 \int_{-r}^{r} \sqrt{r^2 - x_1^2} dx_1 \\
&= 2 \int_{\pi}^{0} \sqrt{r^2 - r^2(\cos\theta)^2}(-r \sin\theta d\theta) \\
&= 2r^2 \int_0^{\pi} (\sin\theta)^2 d\theta \\
&= 2r^2 \int_0^{\pi} \frac{1 - \cos(2\theta)}{2} d\theta \\
&= r^2 \int_0^{\pi} d\theta - r^2 \int_0^{\pi} \cos(2\theta) d\theta \\
&= \pi r^2 - \frac{r^2}{2} \sin(2\theta) \Big|_{\theta=0}^{\theta=\pi} \\
&= \pi r^2.
\end{aligned}$$

For $n = 3$, we find

$$\begin{aligned}
\mu_3 V_r^3 &= 2 \int_0^r \mu_2 V^2_{\sqrt{r^2 - x_2^2}} dx_3 \\
&= 2\pi \int_0^r (r^2 - x_3^2) dx_3 \\
&= 2\pi r^2 \int_0^r dx_3 - 2\pi \int_0^r x_3^2 dx_3
\end{aligned}$$

$$= 2\pi r^3 - \frac{2}{3}\pi x_3^3 \Big|_{x_3=0}^{x_3=r}$$
$$= 2\pi r^3 - \frac{2}{3}\pi r^3$$
$$= \frac{4}{3}\pi r^3.$$

Assume that

$$\mu_{n-1} V_r^{n-1} = \kappa_{n-1} r^{n-1},$$

where κ_{n-1} is some constant. Then

$$\begin{aligned}
\mu V_r^n &= 2\int_0^r \mu_{n-1} V_{\sqrt{r^2-x_1^2}}^{n-1} dx_1 \\
&= 2\kappa_{n-1}\int_0^r \left(r^2 - x_1^2\right)^{\frac{n-1}{2}} dx_1 \\
&= 2\kappa_{n-1}\int_{\frac{\pi}{2}}^0 r^{n-1}\left((\sin\theta)^2\right)^{\frac{n-1}{2}} (-r\sin\theta d\theta) \\
&= 2\kappa_{n-1} r^n \int_0^{\frac{\pi}{2}} (\sin\theta)^n d\theta.
\end{aligned} \tag{2.9}$$

Let

$$I_n = \int_0^{\frac{\pi}{2}} (\sin t)^n dt, \quad n \in \mathbb{N}_0.$$

Then

$$\begin{aligned}
I_n &= \int_0^{\frac{\pi}{2}} (\sin t)^{n-1} d(-\cos t) \\
&= -(\sin t)^{n-1}\cos t\Big|_{t=0}^{t=\frac{\pi}{2}} + (n-1)\int_0^{\frac{\pi}{2}} (\sin t)^{n-2}(\cos t)^2 dt \\
&= (n-1)\int_0^{\frac{\pi}{2}} (\sin t)^{n-1}(1 - (\sin t)^2) dt
\end{aligned}$$

$$
\begin{aligned}
&= (n-1)\int_0^{\frac{\pi}{2}}(\sin t)^{n-2}dt - (n-1)\int_0^{\frac{\pi}{2}}(\sin t)^n dt \\
&= (n-1)I_{n-1} - (n-1)I_n,
\end{aligned}
$$

whereupon

$$nI_n = (n-1)I_{n-2},$$

or

$$I_n = \frac{n-1}{n}I_{n-2}.$$

Hence

$$
\begin{aligned}
I_{2k+1} &= \frac{2k}{2k+1}I_{2k-1} \\
&= \frac{2k}{2k+1}\frac{2k-2}{2k-3}I_{2k-3} \\
&\vdots \\
&= \frac{(2k)!!}{(2k+1)!!}I_1 \\
&= \frac{(2k)!!}{(2k+1)!!}\int_0^{\frac{\pi}{2}}\sin t dt \\
&= \frac{(2k)!!}{(2k+1)!!}\left(-\cos t\Big|_{t=0}^{t=\frac{\pi}{2}}\right) \\
&= \frac{(2k)!!}{(2k+1)!!}, \quad k \in \mathbb{N}_0,
\end{aligned}
$$

and

$$
\begin{aligned}
I_{2k} &= \frac{2k-1}{2k}I_{2k-2} \\
&= \frac{2k-1}{2k}\frac{2k-3}{2k-2}I_{2k-3} \\
&\vdots \\
&= \frac{(2k-1)!!}{(2k)!!}I_0 \\
&= \frac{(2k-1)!!}{(2k)!!}\int_0^{\frac{\pi}{2}}dt \\
&= \frac{(2k-1)!!}{(2k)!!}\frac{\pi}{2}, \quad k \in \mathbb{N}_0.
\end{aligned}
$$

Now applying (2.9), we find

$$\begin{aligned}\mu_n V_r^n &= \kappa_n r^n\\ &= 2\kappa_{n-1} r^n \begin{cases} \frac{\pi}{2}\frac{(n-1)!!}{n!!} & n \text{ is even},\\ \frac{(n-1)!!}{n!!} & n \text{ is odd},\end{cases}\end{aligned}$$

and

$$\kappa_n = 2\kappa_{n-1} \begin{cases} \frac{\pi}{2}\frac{(n-1)!!}{n!!} & n \text{ is even},\\ \frac{(n-1)!!}{n!!} & n \text{ is odd}.\end{cases}$$

By the last relation we get

$$\begin{aligned}\kappa_{2m} &= 2\kappa_{2m-1}\frac{\pi}{2}\frac{(2m-1)!!}{(2m)!!}\\ &= 2\frac{\pi}{2}\frac{(2m-1)!!}{(2m)!!}2\kappa_{2m-2}\frac{(2m-2)!!}{(2m-1)!!}\\ &= 2^2\frac{\pi}{2}\frac{(2m-2)!!}{(2m)!!}\kappa_{2m-2}\\ &= 2^2\frac{\pi}{2}\frac{(2m-2)!!}{(2m)!!}2\kappa_{2m-3}\frac{(2m-3)!!}{(2m-2)!!}\frac{\pi}{2}\\ &= 2^3\left(\frac{\pi}{2}\right)^2\frac{(2m-3)!!}{(2m)!!}\kappa_{2m-3}\\ &= 2^3\left(\frac{\pi}{2}\right)^2\frac{(2m-3)!!}{(2m)!!}2\frac{(2m-4)!!}{(2m-3)!!}\kappa_{2m-4}\\ &= 2^4\left(\frac{\pi}{2}\right)^2\frac{(2m-4)!!}{(2m)!!}\kappa_{2m-4}\\ &\;\vdots\\ &= 2^{2m-2}\left(\frac{\pi}{2}\right)^{m-1}\frac{2!!}{(2m)!!}\kappa_2\\ &= 2^{2m-2}\left(\frac{\pi}{2}\right)^{m-1}\frac{2}{2^m m!}\pi\\ &= \frac{\pi^m}{m!}, \quad m \in \mathbb{N}_0,\end{aligned}$$

and

$$\begin{aligned}\kappa_{2m+1} &= 2\kappa_{2m}\frac{(2m)!!}{(2m+1)!!}\\ &= 2\frac{(2m)!!}{(2m+1)!!}2\frac{\pi}{2}\kappa_{2m-1}\frac{(2m-1)!!}{(2m)!!}\end{aligned}$$

$$
\begin{aligned}
&= 2^2 \frac{\pi}{2} \frac{(2m-1)!!}{(2m+1)!!} \kappa_{2m-1} \\
&= 2^2 \frac{\pi}{2} \frac{(2m-1)!!}{(2m+1)!!} 2\kappa_{2m-2} \frac{(2m-2)!!}{(2m-3)!!} \\
&= 2^3 \frac{\pi}{2} \frac{(2m-2)!!}{(2m+1)!!} \kappa_{2m-2} \\
&= 2^3 \frac{\pi}{2} \frac{(2m-2)!!}{(2m+1)!!} 2 \frac{(2m-3)!!}{(2m-2)!!} \frac{\pi}{2} \kappa_{2m-3} \\
&= 2^4 \left(\frac{\pi}{2}\right)^2 \frac{(2m-3)!!}{(2m+1)!!} \kappa_{2m-3} \\
&\vdots \\
&= 2^{2m-2} \left(\frac{\pi}{2}\right)^{m-1} \frac{3!!}{(2m+1)!!} \kappa_3 \\
&= 2^{2m-2} \left(\frac{\pi}{2}\right)^{m-1} \frac{3}{(2m+1)!!!} \frac{4}{3}\pi \\
&= 4 \cdot 2^{m-1} \frac{\pi^m}{(2m+1)!!} \\
&= 2 \frac{(2\pi)^m}{(2m+1)!!}, \quad m \in \mathbb{N}_0.
\end{aligned}
$$

2.2 The Taylor formula

In this section, we deduce the Taylor formula for a function of a single variable with a remainder term represented with iterated integrals. We will start our investigations in this section with the following useful lemma.

Lemma 2.1. *Let $\phi \in \mathscr{C}([a,b])$. Then, for all $k \in \mathbb{N}$, we have*

$$
\int_a^t \int_a^{t_1} \cdots \int_a^{t_{k-1}} \phi(t_k) dt_k \cdots dt_2 dt_1 = \frac{1}{(k-1)!} \int_a^t \phi(s)(t-s)^{k-1} ds, \quad t \in [a,b]. \tag{2.10}
$$

Proof. We will prove our statement using the induction principle.

1. The case $k = 1$ is evident. Let $k = 2$. Set

$$
\begin{aligned}
g_2(t) &= \int_a^t \int_a^{t_1} \phi(t_2) dt_2 dt_1, \\
h_2(t) &= \int_a^t \phi(s)(t-s) ds, \\
f_2(t) &= g_2(t) - h_2(t), \quad t \in [a,b].
\end{aligned}
$$

Then

$$g_2'(t) = \int_a^t \phi(s)ds,$$

$$h_2'(t) = \int_a^t \phi(s)ds, \quad t \in [a,b],$$

and

$$\begin{aligned} f_2'(t) &= g_2'(t) - h_2'(t) \\ &= \int_a^t \phi(s)ds - \int_a^t \phi(s)ds \\ &= 0, \quad t \in [a,b]. \end{aligned}$$

Thus f_2 is a constant, and then

$$\begin{aligned} f_2(t) &= f_2(a) \\ &= g_2(a) - h_2(a) \\ &= 0, \quad t \in [a,b]. \end{aligned}$$

Thus

$$\int_a^t \int_a^{t_1} \phi(s)dsdt_1 = \int_a^t \phi(s)(t-s)ds, \quad t \in [a,b],$$

and the statement is valid for $k = 2$.

2. Assume that for some $k-1, k \in \mathbb{N}$, equality (2.10) holds.
3. We will prove that the statement is true for k, i. e., we will prove that

$$\int_a^t \int_a^{t_1} \cdots \int_a^{t_{k-1}} \int_a^{t_k} \phi(s)dsdt_k \cdots dt_2 dt_1 = \frac{1}{k!} \int_a^t \phi(s)(t-s)^k ds, \quad t \in [a,b]. \tag{2.11}$$

Set

$$g_{k+1}(t) = \int_a^t \int_a^{t_1} \cdots \int_a^{t_{k-1}} \int_a^{t_k} \phi(s)dsdt_k \cdots dt_2 dt_1,$$

$$h_{k+1}(t) = \frac{1}{k!} \int_a^t \phi(s)(t-s)^k ds,$$

$$f_{k+1}(t) = g_{k+1}(t) - h_{k+1}(t), \quad t \in [a,b].$$

Then

$$g'_{k+1}(t) = \int_a^t \int_a^{t_2} \cdots \int_a^{t_{n-1}} \int_a^{t_k} \phi(s)ds dt_k \cdots dt_2,$$

$$h_{k+1}(t) = \frac{1}{(k-1)!} \int_a^t \phi(s)(t-s)^{k-1} ds, \quad t \in [a,b],$$

and, applying (2.10), we arrive at the equalities

$$\begin{aligned} f'_{k+1}(t) &= g'_{k+1}(t) - h'_{k+1}(t) \\ &= 0, \quad t \in [a,b]. \end{aligned}$$

Thus f_{k+1} is a constant on $[a,b]$, and

$$\begin{aligned} f_{k+1}(t) &= f_{k+1}(a) \\ &= 0, \quad t \in [a,b], \end{aligned}$$

and (2.11) holds. Now applying the induction principle, we conclude that (2.10) holds for all $k \in \mathbb{N}$. This completes the proof. □

The Taylor formula reads as follows.

Theorem 2.4. *Let $f \in \mathscr{C}^n([a,b])$ and $x_0 \in [a,b]$. Then, for all $x \in [a,b]$, we have*

$$\begin{aligned} f(x) &= \sum_{k=0}^{n-1} \frac{f^{(k)}(x_0)}{k!}(x-x_0)^k + \int_{x_0}^{x} \int_{x_0}^{t_1} \cdots \int_{x_0}^{t_{k-1}} f^{(n)}(t_n) dt_n \cdots dt_2 dt_1 \\ &= \frac{1}{(n-1)!} \int_{x_0}^{x} f^{(n)}(t)(x-t)^{n-1} dt. \end{aligned}$$

Proof. We will prove the statment using the induction principle.

1. Let $n = 1$. Then by the Newton–Leibnitz formula we have

$$f(x) - f(x_0) = \int_{x_0}^{x} f'(s) ds, \quad x \in [a,b].$$

2. Assume that the statement is true for some $m \in \{1, \ldots, n-1\}$, i. e., assume that

$$f(x) = \sum_{k=0}^{m-1} \frac{f^{(k)}(x_0)}{k!}(x-x_0)^k + \int_{x_0}^{x} \int_{x_0}^{t_1} \cdots \int_{x_0}^{t_{m-1}} f^{(m)}(t_m) dt_m \cdots dt_2 dt_1. \tag{2.12}$$

3. We will prove the statement for $m+1$. By the Newton–Leibniz formula we have

$$f^{(m)}(x) - f^{(m)}(x_0) = \int_{x_0}^{x} f^{(m+1)}(t)dt.$$

Hence by (2.12) we get

$$\begin{aligned}
f(x) &= \sum_{k=0}^{m-1} \frac{f^{(k)}(x_0)}{k!}(x-x_0)^k + \int_{x_0}^{x}\int_{x_0}^{t_1}\cdots\int_{x_0}^{t_{m-1}} f^{(m)}(t_m)dt_m\cdots dt_2dt_1 \\
&= \sum_{k=0}^{m-1} \frac{f^{(k)}(x_0)}{k!}(x-x_0)^k + \int_{x_0}^{x}\int_{x_0}^{t_1}\cdots\int_{x_0}^{t_{m-1}} f^{(m)}(x_0)dt_m\cdots dt_2dt_1 \\
&\quad + \int_{x_0}^{x}\int_{x_0}^{t_1}\cdots\int_{x_0}^{t_{m-1}}\int_{x_0}^{t_m} f^{(m+1)}(t_{m+1})dt_{m+1}\cdots dt_2dt_1 \\
&= \sum_{k=0}^{m-1} \frac{f^{(k)}(x_0)}{k!}(x-x_0)^k + f^{(m)}(x_0)\int_{x_0}^{x}\int_{x_0}^{t_1}\cdots\int_{x_0}^{t_{m-1}} dt_m\cdots dt_2dt_1 \\
&\quad + \int_{x_0}^{x}\int_{x_0}^{t_1}\cdots\int_{x_0}^{t_{m-1}}\int_{x_0}^{t_m} f^{(m+1)}(t_{m+1})dt_{m+1}\cdots dt_2dt_1 \\
&= \sum_{k=0}^{m-1} \frac{f^{(k)}(x_0)}{k!}(x-x_0)^k + f^{(m)}(x_0)\frac{(x-x_0)^m}{m!} \\
&\quad + \int_{x_0}^{x}\int_{x_0}^{t_1}\cdots\int_{x_0}^{t_{m-1}}\int_{x_0}^{t_m} f^{(m+1)}(t_{m+1})dt_{m+1}\cdots dt_2dt_1 \\
&= \sum_{k=0}^{m} \frac{f^{(k)}(x_0)}{k!}(x-x_0)^k + \int_{x_0}^{x}\int_{x_0}^{t_1}\cdots\int_{x_0}^{t_{m-1}}\int_{x_0}^{t_m} f^{(m+1)}(t_{m+1})dt_{m+1}\cdots dt_2dt_1.
\end{aligned}$$

From here, by the induction principle we conclude that (2.12) is valid for all $m \in \{1, \dots, n\}$.

Now applying Lemma 2.1 to (2.12), we get the desired result. This completes the proof. □

2.3 Linear maps on measurable sets

Linear maps play a fundamental role in the study of linear algebra, calculus, differential equations, differential geometry, and various other mathematical disciplines. Moreover, their significance extends to numerous practical applications across different fields. Linear maps can be used to generate complex regions from basic regions and can also be used to rotate and translate regions to get simpler representations of regions that are useful when we evaluate integrals.

Here we will investigate the general case of arbitrary nonhomogeneous linear maps. A linear map $A : \mathbb{R}^n_x \to \mathbb{R}^m_y$ has the form

$$A(x) = B(x) + a,$$

where $B : \mathbb{R}^n_x \to \mathbb{R}^m_y$ is a homogeneous linear map,

$$\begin{aligned} x &= (x_1, \dots, x_m) \in \mathbb{R}^n_x, \\ y &= (y_1, \dots, y_n) \in \mathbb{R}^m_y, \\ a &= (a_1, \dots, a_m) \in \mathbb{R}^m. \end{aligned}$$

By P we will denote the translation

$$P(y) = y + a, \quad y \in \mathbb{R}^m_y.$$

Then we can rewrite the map A as follows:

$$A(x) = PB(x), \quad x \in \mathbb{R}^n_x.$$

Let (b_{ij}), $i \in \{1, \dots, m\}$, $j \in \{1, \dots, n\}$, be the matrix of the homogeneous linear map B. Then the linear map A can be represented as follows:

$$y_j(x) = \sum_{l=1}^{n} b_{jl} x_l + a_j, \quad j \in \{1, \dots, m\}.$$

Below we will consider the case where $n = m$. We will give a connection of the distance between two points and the distance between their images under a linear map.

Theorem 2.5. *Let*

$$c_0 = \max_{i,j \in \{1, \dots, n\}} |b_{ij}|.$$

Then for all points $x^1, x^2 \in \mathbb{R}^n_x$, we have the inequality

$$|A(x^1) - A(x^2)| \le n c_0 |x^1 - x^2|.$$

Proof. In the previous section, we have proved that

$$|B(x)| \le \|B\| |x|$$

for all $x \in \mathbb{R}^n_x$. Moreover, it is proved that

$$\|B\| \le \sqrt{\sum_{i,j=1}^{n} b_{ij}^2},$$

where (b_{ij}), $i, j \in \{1, \ldots, n\}$, is the matrix of the linear map B. Then, for all $x^1, x^2 \in \mathbb{R}^n_x$, we have

$$\begin{aligned} |A(x^1) - A(x^2)| &= |B(x^1) + a - B(x^2) - a| \\ &= |B(x^1) - B(x^2)| \\ &= |B(x^1 - x^2)| \\ &\le \|B\| |x^1 - x^2| \\ &\le \sqrt{\sum_{i,j=1}^{n} b_{ij}^2} |x^2 - x^2| \\ &\le nc_0 |x^1 - x^2|. \end{aligned}$$

This completes the proof. □

The next result gives a connection between the measure of a set and the measure of its image under a linear map.

Theorem 2.6. *Let J be the Jacobian matrix of the linear homogeneous map B, and let $X \subset \mathbb{R}^n_x$ be a measurable set. Then*

$$\mu(A(X)) = |J| \mu(X).$$

Proof. Since B is a linear homogeneous map, it can be represented in the form

$$B = CD,$$

where C is a nonnegative self-adjoint linear map, and D is an orthogonal map. Then we have

$$\begin{aligned} A &= PB \\ &= PCD. \end{aligned}$$

Because C is a nonnegative self-adjoint linear map, it can be represented in the form

$$C = \begin{pmatrix} \lambda_1 & 0 & \cdots & 0 \\ 0 & \lambda_2 & \cdots & 0 \\ \vdots & & & \\ 0 & 0 & \cdots & \lambda_n \end{pmatrix},$$

where $\lambda_j \geq 0, j \in \{1, \dots, n\}$. Let Q be an n-dimensional cube whose edges are parallel to the coordinate axis and have a length h. Then

$$\begin{aligned}\mu(C(Q)) &= \lambda_1 \cdots \lambda_n \mu(Q)\\ &= \det C\mu(Q).\end{aligned}$$

Note that

$$\det B = \det C \det D$$

and $\det D = \pm 1$. Thus

$$\det C = |J|.$$

Hence

$$\mu(C(Q)) = |J|\mu(Q).$$

Further, we have the inclusions

$$s_k(X) \subset X \subset S_k(X).$$

Then

$$\begin{aligned}C(s_k(X)) \subset C(X) \subset C(S_k(X)),\\ C(S_k(X))\backslash C(s_k(X)) \subset C(S_k(X)\backslash s_k(X)),\end{aligned}$$

and

$$\begin{aligned}\mu(C(S_k(X))\backslash C(s_k(X))) &\leq \mu(C(S_k(X)\backslash s_k(X)))\\ &= |J|\mu(S_k(X)\backslash s_k(X)).\end{aligned}$$

Since

$$\lim_{k\to\infty} \mu(S_k(X)\backslash s_k(X)) = 0,$$

we obtain that

$$\lim_{k\to\infty} \mu(C(S_k(X))\backslash C(s_k(X))) = 0.$$

Therefore $C(X)$ is a measurable set, and since

$$\mu(C(s_k(X))) = |J|\mu(s_k(X)),$$

we arrive at

$$\begin{aligned}\mu(C(X)) &= \lim_{k\to\infty} \mu(C(s_k(X)))\\ &= |J| \lim_{k\to\infty} \mu(s_k(X))\\ &= |J|\mu(X).\end{aligned}$$

Consequently,

$$\begin{aligned}\mu(A(X)) &= \mu(PCD(X))\\ &= \mu(C(X))\\ &= |J|\mu(X).\end{aligned}$$

This completes the proof. □

2.4 Metric properties of differentiable maps

In this section, we deduce some important properties of differentiable maps. We need them when we make a change of variables for the evaluation of multiple integrals.

Let X be an open set in $\mathbb{R}^n_x$, and let $f : X \to \mathbb{R}^n_y$ be a map given by

$$y = f(x), \quad x \in X.$$

The map f can be written in the coordinate form

$$y_j = y_j(x), \quad x \in X,\ j \in \{1, \dots, n\}.$$

Take $x^0 = (x_1^0, \dots, x_n^0) \in X$ and

$$\begin{aligned}y^0 &= f(x^0),\\ \Delta x &= x - x^0\\ &= (\Delta x_1, \dots, \Delta x_n).\end{aligned}$$

Assume that f is a differentiable map at the point x^0. Then it can be represented in the form

$$y_j = y_j^0 + \sum_{l=1}^{n} a_{jl}(x_l - x_l^0) + \varepsilon_j(x^0, \Delta x)|\Delta x|, \quad j \in \{1, \dots, n\},$$

where

$$a_{jl} = \frac{\partial y_j}{\partial x_l}(x^0), \quad j, l \in \{1, \dots, n\},$$

and

$$\lim_{|\Delta x|\to 0} \varepsilon_j(x^0, \Delta x) = 0, \quad j \in \{1, \ldots, n\}.$$

Let L_{x^0} be the linear map given by

$$y_j = y_j^0 + \sum_{l=1}^{n} a_{jl}(x_l - x_l^0), \quad j \in \{1, \ldots, n\}.$$

Then the map f can be written as follows:

$$f(x) = L_{x^0}(x) + \varepsilon|\Delta x|, \quad x \in X,$$

where

$$\lim_{\Delta x\to 0} \varepsilon = 0$$

and

$$\begin{aligned}\varepsilon &= \varepsilon(x^0, \Delta x)\\ &= (\varepsilon_1, \ldots, \varepsilon_n),\\ \varepsilon_j &= \varepsilon_j(x^0, \Delta x), \quad j \in \{1, \ldots, n\}.\end{aligned}$$

Moreover, we can write that map f as follows:

$$f(x) = L_{x^0}(x) + o(\Delta x) \quad \text{as } \Delta x \to 0.$$

A continuous image of a compact set of measure zero is a compact set of measure zero.

Theorem 2.7. *If the map f is continuously differentiable on X and $K \subset X$ is a compact set of zero measure, then $f(K)$ is a compact set of zero measure.*

Proof. We have that $f(K)$ is a compact set. Since K is a compact set, it is a closed set. Because X is an open set in $\mathbb{R}^n_x$, we have that $\mathbb{R}^n_x \backslash X$ is a closed set. Therefore there is a real number $\delta > 0$ such that

$$d(K, \mathbb{R}^n_x \backslash X) = \delta > 0.$$

Take k_0 such that the diameter of any cube of rank k_0 is less than δ. Then any cube of rank k_0 that intersects the set K is contained in X. The set $S_{k_0}(K)$ that contains such cubes is a compact set. Because K is a bounded set, the set $S_{k_0}(K)$ is a union of a finite number of cubes of rank k_0, each of which is a compact set. Thus $S_{k_0}(K)$ is a compact set. Set

$$c = \max_{\substack{x \in S_{k_0}(K) \\ i,j \in \{1,\dots,n\}}} \left|\frac{\partial y_i}{\partial x_j}(x)\right|.$$

Let $x, x^1 \in S_{k_0}(K)$ be chosen so that the segment with endpoints x and x^1 is contained in $S_{k_0}(K)$. Now applying the mean value theorem and the Cauchy–Schwarz inequality, we get

$$\begin{aligned}
|f(x^1) - f(x)| &= \sqrt{\sum_{j=1}^{n} (y_j(x^1) - y_j(x))^2} \\
&= \sqrt{\sum_{j=1}^{n} \left(\sum_{l=1}^{n} \frac{\partial y_j}{\partial x_l}(x + \theta_l \Delta x) \Delta x_l \right)^2} \\
&= \sqrt{\sum_{j=1}^{n} \left(\sum_{l=1}^{n} \left(\frac{\partial y_j}{\partial x_l}(x + \theta_l \Delta x) \right)^2 \right) \left(\sum_{l=1}^{n} \Delta x_l^2 \right)} \\
&\le c \sqrt{\sum_{j=1}^{n} n \left(\sum_{l=1}^{n} \Delta x_l^2 \right)} \\
&= cn|\Delta x| \\
&= cn|x - x^1|.
\end{aligned}$$

Therefore, for any convex set A, we have

$$\begin{aligned}
\operatorname{diam} f(A) &= \sup_{x, x^1 \in A} |f(x^1) - f(x)| \\
&\le cn \sup_{x, x^1 \in A} |x^1 - x| \\
&= cn \operatorname{diam} A.
\end{aligned}$$

If A is an n-dimensional cube Q_h with edge length h, we have

$$Q_h \subset \sqrt{\sum_{j=1}^{n} \left(\sum_{l=1}^{n} \frac{\partial y_j}{\partial x_l}(x + \theta_l \Delta x) \Delta x_l \right)^2}.$$

Note that

$$\operatorname{diam} Q_h = h\sqrt{n}$$

and

$$\mu Q_h = h^n.$$

Then

$$\begin{aligned}\operatorname{diam} f(Q_h) &\le cn \operatorname{diam} Q_h \\ &= cn^{\frac{3}{2}} h.\end{aligned}$$

Consequently,

$$\begin{aligned}\mu^* f(Q_h) &\le \left(2 \operatorname{diam} f(Q_h)\right)^n \\ &\le 2^n c^n n^{\frac{3n}{2}} h^n \\ &= 2^n c^n n^{\frac{3n}{2}} \mu Q_h.\end{aligned}$$

Note that

$$S_k(K) \subset S_{k_0}(K) \quad \text{for all } k \ge k_0.$$

Let $k \ge k_0$, and let

$$S_k(K) = \bigcup_{j=1}^{j_k} Q_j^k, \quad Q_j^k \cap Q_l^k = \emptyset,\ j \ne l,\ j, l \in \{1, \ldots, j_k\},$$

where $Q_j^k, j \in \{1, \ldots, j_k\}$, are cubes of rank k. Then

$$f\left(\bigcup_{j=1}^{j_k} Q_j^k\right) = \bigcup_{j=1}^{j_k} f(Q_j^k).$$

Since

$$K \subset S_k(K),$$

we have the inclusion

$$f(K) \subset f(S_k(K)).$$

Therefore

$$\begin{aligned}\mu^* f(K) &\le \mu^* f(S_k(K)) \\ &= \mu^* f\left(\bigcup_{j=1}^{j_k} Q_j^k\right) \\ &= \bigcup_{j=1}^{j_k} \mu^* f(Q_j^k)\end{aligned}$$

$$\le 2^n c^n n^{\frac{3n}{2}} \sum_{j=1}^{j_k} \mu^* Q_j^k$$
$$= 2^n c^n n^{\frac{3n}{2}} \mu S_k(K).$$

Since $\mu K = 0$, we get

$$\lim_{k\to 0} \mu S_k(K) = 0$$

and

$$\mu^* f(K) \le 2^n c^n n^{\frac{3n}{2}} \lim_{k\to\infty} \mu S_k(K)$$
$$= 0,$$

and thus $\mu f(K) = 0$. This completes the proof. □

Denote

$$J_f = J_f(x)$$
$$= \frac{\partial(y_1,\ldots,y_n)}{\partial(x_1,\ldots,x_n)}.$$

Now we will give an estimate of the upper Jordan measure of the continuous image of a cube using the matrix J_f.

Theorem 2.8. *Let f be a continuously differentiable map on X such that $J_f(x) \neq 0$ for all $x \in X$. Let $K \subset X$ be a compact set, let $Q_h \subset X$ be an n-dimensional cube with edge of length h, and let $x^0 \in X \cap Q_h$. Then*

$$\mu^* f(Q_h) \le |J_f(x^0)|\mu Q_h + \alpha(h)h^n, \quad \lim_{h\to 0} \alpha(h) = 0. \tag{2.13}$$

Proof. By Theorem 2.6 we have that

$$\mu L_{x^0}(Q_h) = |J_f(x^0)|\mu Q_h. \tag{2.14}$$

Since K is a compact set and $\mathbb{R}^n \backslash X$ is a bounded set, we have that there is $\delta > 0$ such that

$$d(K, \mathbb{R}^n \backslash X) = \delta > 0.$$

Suppose that

$$h\sqrt{n} < \delta.$$

Then any cube Q_h that intersects the compact set K is contained in X. Next, since f is continuously differentiable at the point x^0, we have

$$|f(x) - L_{x^0}(x)| = |\varepsilon(x^0, \Delta x)||\Delta x|,$$

where $x^0 \in K$, and

$$\begin{aligned}\Delta x &= x - x^0,\\ \varepsilon(x^0, \Delta x) &\to 0 \quad \text{as } \Delta x \to 0.\end{aligned}$$

Then

$$\varepsilon(\Delta x) = \sup_{x^0 \in K} |\varepsilon(x^0, \Delta x)|,$$

and

$$\lim_{\Delta x \to 0} \varepsilon(\Delta x) = 0,$$

and we have the inequality

$$|f(x) - L_{x^0}(x)| \le \varepsilon(\Delta x)|\Delta x|, \quad x^0 \in K,\ x = x^0 + \Delta x \in X. \tag{2.15}$$

If $x^0, x \in Q_h$, then

$$\begin{aligned}|\Delta x| &= |x - x^0|\\ &\le h\sqrt{n}.\end{aligned}$$

Hence by (2.15), for $x^0, x \in Q_h$, we have

$$|f(x) - L_{x^0}(x)| \le \varepsilon(\Delta x) h\sqrt{n}.$$

Set

$$\begin{aligned}\alpha_0 &= \alpha_0(h)\\ &= \sqrt{n} \sup_{|\Delta x| \le h} \varepsilon(\Delta x).\end{aligned}$$

Therefore

$$|f(x) - L_{x^0}(x)| \le \alpha_0 h, \quad x^0 \in K \cap Q_h,\ x = x^0 + \Delta x \in Q_h,$$

where

$$\lim_{h \to 0} \alpha_0(h) = 0.$$

Let

$$P = L_{x^0}(Q_h),$$

and let

$$U_{a_0h} = U(P, a_0h)$$

be the a_0h-neighborhood of P. We have

$$f(Q_h) \subset U_{a_0h}.$$

Let P_{a_0h} be an n-dimensional parallelogram such that $n-1$ edges of P_{a_0h} are parallel to $n-1$ edges of P and $P \subset P_{a_0h}$. Then $U_{a_0h} \subset P_{a_0h}$, and then $f(Q_h) \subset P_{a_0h}$. By $a_1, \ldots, a_n$ we denote the lengths of the edges of P, and by H_j, $j \in \{1, \ldots, n\}$, we denote the distance between two parallel $(n-1)$-dimensional boundaries of P containing edges of lengths $a_1, \ldots, a_{j-1}$, $a_{j+1}, \ldots, a_n$. Then the distance between two $(n-1)$-dimensional parallel boundaries of P_{a_0h} containing edges of lengths $a_1 + \Delta a_1, \ldots, a_{j-1} + \Delta a_{j-1}$, $a_{j+1} + \Delta a_{j+1}, \ldots, a_n + \Delta a_n$, is $H_j + 2a_0h$, $j \in \{1, \ldots, n\}$. In other words, H_j and $H_j + 2a_0h$, $j \in \{1, \ldots, n\}$, are the altitudes of P and P_{a_0h}. By $\phi_j \in [0, \frac{\pi}{2}]$, $j \in \{1, \ldots, n\}$, we denote the angle between the edge of length a_j, $j \in \{1, \ldots, n\}$, and the $(n-1)$-dimensional hyperplane containing the edges of length $a_1, \ldots, a_{j-1}$, $a_{j+1}, \ldots, a_n$. We have that $\phi_j \in [0, \frac{\pi}{2}]$, $j \in \{1, \ldots, n\}$, is the angle between the edge of length $a_j + \Delta a_j$, $j \in \{1, \ldots, n\}$, and the $(n-1)$-dimensional hyperplane containing the edges of length $a_1 + \Delta a_1, \ldots, a_{j-1} + \Delta a_{j-1}$, $a_{j+1} + \Delta a_{j+1}, \ldots, a_n + \Delta a_n$. Then we get

$$\begin{aligned} a_j &= \frac{H_j}{\sin \phi_j}, \\ a_j + \Delta a_j &= \frac{H_j + 2a_0h}{\sin \phi_j}, \quad j \in \{1, \ldots, n\}. \end{aligned}$$

Hence

$$\begin{aligned} \Delta a_j &= \frac{H_j + 2a_0h}{\sin \phi_j} - a_j \\ &= \frac{H_j + 2a_0h}{\sin \phi_j} - \frac{H_j}{\sin \phi_j} \\ &= \frac{2a_0h}{\sin \phi_j}, \quad j \in \{1, \ldots, n\}. \end{aligned}$$

Also, we have

$$\phi_j(x^0) > 0, \quad j \in \{1, \ldots, n\}.$$

Therefore, for any $j \in \{1, \ldots, n\}$, there is $c_j \in (0, \frac{\pi}{2}]$ such that

$$\min_{x \in K} \phi_j(x) = c_j, \quad j \in \{1, \ldots, n\}.$$

Denote

$$c_0 = \max\left\{\frac{1}{\sin c_1}, \dots, \frac{1}{\sin c_n}\right\}.$$

Then

$$\begin{aligned}\Delta a_j &= \frac{2a_0 h}{\sin \phi_j} \\ &\le \frac{2a_0 h}{\sin c_j} \\ &\le 2a_0 c_0 h, \quad j \in \{1, \dots, n\}.\end{aligned}$$

Note that $P_{a_0 h} \backslash P_{\text{int}}$ is the union of the parallelograms $P_{j,1}$ and $P_{j,2}$ with edges $a_1 + \Delta a_1, \dots, a_{j-1} + \Delta a_{j-1}, a_{j+1} + \Delta a_{j+1}, \dots, a_n + \Delta a_n$, and the length of its altitude is $a_0 h$. Therefore

$$P_{a_0 h} \subset P \cup \left(\bigcup_{j=1}^{n} P_{j,1}\right) \cup \left(\bigcup_{j=1}^{n} P_{j,2}\right).$$

The edges of lengths $a_j, j \in \{1, \dots, n\}$, of the parallelogram P are mapped by the map L_{x_0} to the edges of Q_h of length h. Now applying Theorem 2.5, we get

$$\begin{aligned} a_j &\le cnh, \\ c &= \max_{\substack{i,j \in \{1,\dots,n\} \\ x \in X}} \left| \frac{\partial y_i}{\partial x_j}(x) \right|. \end{aligned}$$

Next,

$$\begin{aligned}\mu P_{j,l} &\le (a_1 + \Delta a_1) \cdots (a_{j-1} + \Delta a_{j-1})(a_{j+1} + \Delta a_{j+1}), \dots, (a_n + \Delta a_n) a_0 h \\ &\le (cnh + 2c_0 a_0 h)^{n-1} a_0 h \\ &= (cn + 2c_0 a_0)^{n-1} a_0 h^n \\ &= (cn + 2c_0 a_0)^{n-1} a_0 \mu Q_h, \quad j \in \{1, \dots, n\},\ l \in \{1, 2\}.\end{aligned}$$

By (2.14) we obtain

$$\begin{aligned}\mu P &= \mu L_{x^0}(Q_h) \\ &= |J_f(x^0)| \mu Q_h.\end{aligned}$$

We have the following estimates:

$$\begin{aligned}\mu^* f(Q_h) &\le \mu P_{\alpha_0 h}\\ &\le \mu P_{\alpha_0 h} + \sum_{j=1}^{n} \mu P_{j,1} + \sum_{j=1}^{n} \mu P_{j,2}\\ &\le |J_f(x^0)|\mu Q_h + 2n(cn + 2c_0\alpha_0)^{n-1}\alpha_0 \mu Q_h.\end{aligned}$$

Set

$$\alpha(h) = 2n(cn + 2c_0\alpha_0)^{n-1}\alpha_0.$$

Then

$$\mu^* f(Q_h) \le |J_f(x^0)|\mu Q_h + \alpha(h)\mu Q_h, \quad \lim_{h\to 0} \alpha(h) = 0,$$

i. e., we get (2.13). This completes the proof. □

Now we will give a condition when a continuous image of a measurable set is a measurable set.

Theorem 2.9. *Let $f : X \to \mathbb{R}^n_y$ be a continuously differentiable one-to-one map with Jacobian $J_f(x) \neq 0$, $x \in X$, and let $Y \subset X$ such that $\overline{Y} \subset X$ be a measurable set. Then $f(Y)$ is a measurable set in $\mathbb{R}^n_y$.*

Proof. We have that $\overline{Y}$ is a compact set. Hence $f(\overline{Y})$ is a compact set, and

$$f(Y) \subset f(\overline{Y}).$$

Because Y is a measurable set, its boundary ∂Y is a compact measurable set of zero measure. By Theorem 2.7 we have that $f(\partial Y)$ is a compact set and

$$\mu f(\partial Y) = 0.$$

Next,

$$f(\partial Y) = \partial f(Y).$$

Then

$$\begin{aligned}\mu \partial f(Y) &= \mu f(\partial Y)\\ &= 0.\end{aligned}$$

Thus $f(Y)$ is a bounded set with boundary $\partial f(Y)$ of zero measure. Therefore $f(Y)$ is a measurable set. This completes the proof. □

Remark 2.1. Let $f : X \to \mathbb{R}_y^n$ be a continuously differentiable one-to-one map with Jacobian $J_f(x) \neq 0, x \in X$, and let $Y \subset X$ such that $\overline{Y} \subset X$ be a measurable set. Then (2.13) can be written as follows:

$$\mu f(Q_h) \leq |J_f(x^0)| \mu Q_h + \alpha(h)h^n, \quad \lim_{h\to 0} \alpha(h) = 0. \tag{2.16}$$

2.5 Change of variables

In calculus, we have the substitution rule

$$\int_a^b f(g(x))g'(x)dx = \int_c^d f(u)du,$$

where $u = g(x), g(a) = c, g(b) = d$. In essence, this is taking an integral in terms of x and changing it into that in terms of u. We want to do something similar for multiple integrals. Although often the reason for changing variables is to get an integral that we can do with the new variables, another reason for changing variables is to convert the region into a nicer region to work with. When we convert to polar, cylindrical, or spherical coordinates, we do not worry about this change since it was easy enough to determine the new limits based on the given region. That is not always the case, however. In this section, we consider the change of variables for multiple integrals in the general case.

Let $X \subset \mathbb{R}_x^n$ be an open set, and let $F : X \to \mathbb{R}_y^n$ be a continuously differentiable one-to-one map such that $J_F(x) \neq 0, x \in X$. The next result is the first that gives a formula for the change of variables for multiple integrals.

Theorem 2.10. *Let Y be a measurable set such that $\overline{Y} \subset X$. Let also, $f \in \mathscr{C}(\overline{F(X)})$. Then*

$$\int_{\overline{F(Y)}} f(y)dy = \int_{\overline{Y}} f(F(x))|J_F(x)|dx. \tag{2.17}$$

Remark 2.2. Note that if a bounded function on a set $Z \subset \mathbb{R}_x^n$ is integrable or non-integrable on Z, then it is integrable or non-integrable on $\overline{Z}$, and conversely. In the case where f is integrable on Z and $\overline{Z}$, we have that

$$\int_Z f(x)dx = \int_{\overline{Z}} f(x)dx.$$

In our case, the functions $f(\cdot)$ and $f(F(\cdot))|J_F(\cdot)|$ are continuous on $\overline{F(X)}$ and $\overline{X}$, respectively. Then they are bounded and integrable on $\overline{F(X)}$ and $\overline{X}$, respectively. Thus

$$\int_{\overline{F(Y)}} f(y)dy = \int_{F(Y)} f(y)dy,$$

and

$$\int_{\overline{Y}} f(F(x))|J_F(x)|dx = \int_{Y} f(F(x))|J_F(x)|dx.$$

Therefore equation (2.17) is equivalent to the equation

$$\int_{F(Y)} f(y)dy = \int_{Y} f(F(x))|J_F(x)|dx. \tag{2.18}$$

Proof of Theorem 2.10. Suppose that

$$f(y) \geq 0, \quad y \in \overline{f(X)}. \tag{2.19}$$

If f changes its sign in $\overline{f(X)}$, then since $f \in \mathscr{C}(\overline{F(X)})$ and $\overline{F(X)}$ is a compact set in $\mathbb{R}^n$, f is a bounded function on $\overline{F(X)}$. Then there is a constant $c > 0$ such that

$$f(y) > -c, \quad y \in \overline{F(X)}.$$

Set

$$f_1(y) = f(y) + c,$$
$$f_2(y) = c, \quad y \in \overline{F(X)}.$$

If Theorem 2.10 is true for nonnegative functions on $\overline{F(X)}$, then it will be true for the functions f_1 and f_2. Hence, using the linear and homogeneous properties of the integral, we conclude that Theorem 2.10 will be true for the function

$$f(y) = f_1(y) - f_2(y), \quad y \in \overline{F(X)}.$$

Therefore, without loss of generality, assume that (2.19) holds. Let $Q_j^k, j \in \{1, \ldots, j_k\}$, be cubes of rank k, and suppose

$$Y_j^k = \overline{Y} \cap Q_j^k \neq \emptyset, \quad j \in \{1, \ldots, j_k\}.$$

Then the system of sets

$$\tau_k = \{Y_j^k\}_{j=1}^{j_k}$$

is a partition of the compact set $\overline{Y}$ such that

$$\lim_{k \to \infty} |\tau_k| = 0.$$

Consider the system of sets

$$F(\tau_k) = \{F(Y_j^k)\}_{j=1}^{j_k}. \tag{2.20}$$

Note that $Y_j^k, j \in \{1, \dots, j_k\}$, are measurable compact sets that are obtained by the intersection of the compact sets $\overline{Y}$ and $Q_j^k, j \in \{1, \dots, j_k\}$, lying in the open set X. Therefore their images $F(Y_j^k), j \in \{1, \dots, j_k\}$, are measurable sets. Since

$$\overline{Y} = \bigcup_{j=1}^{j_k} Y_j^k,$$

we have that

$$\begin{aligned} F(\overline{Y}) &= F\left(\bigcup_{j=1}^{j_k} Y_j^k\right) \\ &= \bigcup_{j=1}^{j_k} F(Y_j^k) \end{aligned}$$

and

$$F(Y_j^k) \cap F(Y_l^k) = F(Y_j^k \cap Y_l^k), \quad j, l \in \{1, \dots, j_k\},$$

because F is a one-to-one map. Note that

$$\mu(Y_j^k \cap Y_l^k) = \emptyset, \quad j, l \in \{1, \dots, j_k\}, \ j \neq l.$$

Hence

$$\begin{aligned} \mu(F(Y_j^k) \cap F(Y_l^k)) &= \mu(F(Y_j^k \cap Y_l^k)) \\ &= 0, \quad j, l \in \{1, \dots, j_k\}, \ j \neq l. \end{aligned}$$

Since F is a continuous map on $\overline{Y}$, it is uniformly continuous on $\overline{Y}$. Take arbitrary $\varepsilon > 0$. Then there is $\delta > 0$ such that for any set $Z \subset \overline{Y}$ with $\operatorname{diam} Z < \delta$, we have the inequality

$$\operatorname{diam} F(Z) < \varepsilon.$$

Now we choose a rank k_0 such that

$$10^{-k_0 n} < \delta.$$

Then, for all $k \geq k_0$, we have

$$10^{-kn} < \delta.$$

Hence, for all $k > k_0$, we have

$$\begin{aligned}|F(\tau_k)| &= \max_{j\in\{1,\dots,j_k\}} \operatorname{diam} F(Y_j^k)\\ &< \varepsilon,\end{aligned}$$

whereupon

$$\lim_{k\to\infty} |F(\tau_k)| = 0.$$

Because $\overline{Y} \subset G$, its boundary $\partial\overline{Y}$ is a compact set of measure 0, and

$$F(\partial\overline{Y}) = \partial F(\overline{Y}).$$

Therefore

$$\mu\partial\overline{F(Y)} = 0.$$

Consider the integral sums $\sigma_{F(\tau_k)(\partial\overline{F(Y)})}$ and $\sigma_{\tau_k(\partial\overline{Y})}$ for the integrals

$$\int_{\overline{F(Y)}} f(y)dy \quad \text{and} \quad \int_{\overline{Y}} f(F(x))|J_F(x)|dx,$$

respectively. Choose $F(Y_j^k), j \in \{1,\dots,j_k\}$, that do not intersect the boundary $\partial\overline{F(Y)}$ of the set $\overline{F(Y)}$. This is equivalent to choose $Y_j^k, j \in \{1,\dots,j_k\}$, that do not intersect the boundary $\partial\overline{Y}$ of the set $\overline{Y}$. By the definition of τ_k it follows that the cubes $Q_j^k, j \in \{1,\dots,j_k\}$, contain a point x of $\overline{Y}$. If $Q_j^k \not\subset \overline{Y}$, then $x \in \partial\overline{Y}$, and

$$x \in \overline{Y} \cap Q_j^k = Y_j^k.$$

This is a contradiction with the choice of the sets Y_j^k. Therefore

$$\begin{aligned}Y_j^k &= \overline{Y} \cap Q_j^k\\ &= Q_j^k\\ &\subset \overline{Y}, \quad j \in \{1,\dots,j_k\},\end{aligned}$$

i. e., $Y_j^k, j \in \{1,\dots,j_k\}$, are cubes of rank k. Let now $\eta^{j,k} \in F(Y_j^k), j \in \{1,\dots,j_k\}$, be arbitrarily chosen. Let also, $\xi^{j,k}, j \in \{1,\dots,j_k\}$, be such that

$$\eta^{j,k} = F(\xi^{j,k}), \quad j \in \{1,\dots,j_k\}.$$

Then

$$\begin{aligned}\sigma_{F(\tau_k)(\partial\overline{F(Y)})} &= \sum_{F(Y_j^k)\cap\partial\overline{F(Y)}=\emptyset} f(\eta^{j,k})\mu F(Y_j^k)\\ &= \sum_{F(Y_j^k)\cap\partial\overline{F(Y)}=\emptyset} f(\eta^{j,k})\mu F(Q_j^k),\end{aligned}$$

and

$$\sigma_{\tau_k(\partial\overline{Y})} = \sum_{Q_j^k\subset\overline{Y}} f(F(\xi^{j,k}))|J_F(\xi^{j,k})|\mu Q_j^k.$$

Because $f \in \mathscr{C}(\overline{F(Y)})$ and $\overline{F(Y)}$ is a compact set, there is a constant c_1 such that

$$|f(y)| \le c_1, \quad y \in \overline{F(Y)}.$$

Next, applying (2.16), we get

$$\begin{aligned}\sigma_{F(\tau_k)(\partial\overline{F(Y)})} &= \sum_{F(Y_j^k)\cap\partial\overline{F(Y)}=\emptyset} f(\eta^{j,k})\mu F(Q_j^k)\\ &\le \sum_{Q_j^k\subset\overline{Y}} f(F(\xi^{j,k}))|J_F(\xi^{j,k})|\mu Q_j^k\\ &\quad + ca\left(\frac{1}{10^k}\right)\sum_{Q_j^k\subset\overline{Y}}\mu Q_j^k\\ &\le \sigma_{\tau_k(\partial\overline{Y})} + ca\left(\frac{1}{10^k}\right)\mu\overline{Y}\end{aligned}$$

and

$$\lim_{k\to\infty} ca\left(\frac{1}{10^k}\right)\mu\overline{Y} = 0.$$

Hence

$$\begin{aligned}\int\limits_{\overline{F(Y)}} f(y)dy &= \lim_{k\to\infty}\sigma_{F(\tau_k)(\partial\overline{F(Y)})}\\ &\le \lim_{k\to\infty}\sigma_{\tau_k(\partial\overline{Y})} + \lim_{k\to\infty} ca\left(\frac{1}{10^k}\right)\mu\overline{Y}\\ &= \int\limits_{\overline{Y}} f(F(x))|J_F(x)|dx,\end{aligned}$$

i. e.,

$$\int\limits_{\overline{F(Y)}} f(y)dy \le \int\limits_{\overline{Y}} f(F(x))|J_F(x)|dx. \tag{2.21}$$

Note that F^{-1} is a continuously differentiable one-to-one map with Jacobian $J_{F^{-1}}(y) \neq 0$, $y \in F(Y)$, and

$$J_{F^{-1}}(y) = \frac{1}{J_F(x)},$$
$$y = f(x),$$
$$F(F^{-1}(y)) = y, \quad y \in F(Y).$$

Now applying inequality (2.21) to the right-hand side of (2.21) and the map F^{-1}, we get

$$\begin{aligned}\int_{\overline{Y}} f(F(x))|J_F(x)|dx &= \int_{\overline{F^{-1}(F(Y))}} f(F(x))|J_F(x)|dx \\ &\le \int_{\overline{F(Y)}} f(F^{-1}(F(y)))|J_F(F^{-1}(Y))||J_{F^{-1}}(y)|dy \\ &= \int_{\overline{F(Y)}} f(y)dy.\end{aligned}$$

By the last inequality and (2.21) we obtain equation (2.17). This completes the proof. □

Now we will give a generalization of the last result.

Theorem 2.11. *Let X be an open measurable set, and let $F : X \to \mathbb{R}^n_y$ be a continuously differentiable one-to-one map. Suppose that F and its Jacobian J_F can be continuously extended to $\overline{X}$ and $f \in \mathscr{C}(\widetilde{X})$ can be continuously extended to $\overline{\widetilde{X}}$, where $\widetilde{X} = F(X)$. Then*

$$\int_{\widetilde{X}} f(y)dy = \int_X f(F(x))|J_F(x)|dx. \tag{2.22}$$

Proof. Since X is a measurable set, we have that $\overline{\widetilde{X}}$ is a measurable compact set. Therefore the continuous extension of f to $\overline{\widetilde{X}}$ is integrable on $\overline{\widetilde{X}}$, and f is integrable on $\widetilde{X}$. As above, the function $f(F(\cdot))|J_F(\cdot)|$ is integrable on X. Thus both integrals in (2.22) exist. Because X is an open measurable set, there is a system of open measurable sets X_k, $k \in \mathbb{N}$, such that

$$X_k \subset X_{k+1}, \quad X_k \subset \overline{X_k} \subset X, \quad k \in \mathbb{N},$$

and

$$X = \bigcup_{k=1}^{\infty} X_k.$$

By Theorem 2.10 it follows that

$$\int_{F(X_k)} f(y)dy = \int_{\overline{X_k}} f(F(x))|J_F(x)|dx, \quad k \in \mathbb{N}. \tag{2.23}$$

Note that

$$F(X_k) \subset F(X_{k+1}), \quad k \in \mathbb{N},$$
$$\bigcup_{k=1}^{\infty} F(X_k) = F(X)$$
$$= \widetilde{X}.$$

Therefore

$$\lim_{k\to\infty} \int_{F(X_k)} f(y)dy = \int_{F(X)} f(y)dy,$$

and

$$\lim_{k\to\infty} \int_{\overline{X_k}} f(F(x))|J_F(x)|dx = \int_{\overline{X}} f(F(x))|J_F(x)|dx.$$

By the last two equalities and (2.23) we obtain (2.22). This completes the proof. □

Double integrals are sometimes much easier to evaluate if we change rectangular coordinates to polar coordinates. A point $p \in \mathbb{R}^2$ with Cartesian coordinates (x, y) is also specified by its polar coordinates (r, θ), where r is the distance from the origin to p, and θ is the angle from the positive x-axis to p (see Figure 2.15). The angle θ is not defined for $p = (0, 0)$. The Cartesian coordinates (x, y) can be expressed in terms of (r, θ) as follows:

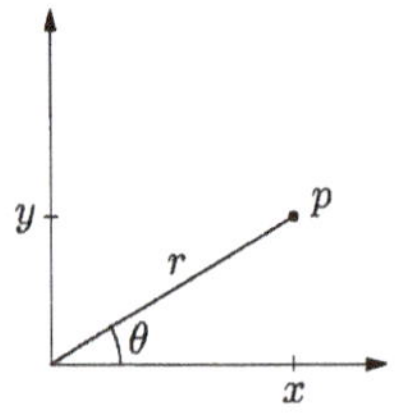

Figure 2.15: Polar coordinates.

$$x = r\cos\theta,$$
$$y = r\sin\theta.$$

The polar coordinates (r, θ) can be expressed by the Cartesian coordinates as follows:

$$r = \sqrt{x^2 + y^2},$$
$$\tan\theta = \frac{y}{x},$$

provided that $x \neq 0$. Note that the function arctan is not even a well-defined function until its rank is specified, i. e., as $(-\frac{\pi}{2}, \frac{\pi}{2})$. With this restriction, we have the following representation of θ:

$$\theta = \begin{cases} \arctan\frac{y}{x} & \text{if } x > 0,\ y > 0, \\ \frac{\pi}{2} & \text{if } x = 0,\ y > 0, \\ \arctan\frac{y}{x} + \pi & \text{if } x < 0, \\ \frac{3\pi}{2} & \text{if } x = 0,\ y < 0, \\ \arctan\frac{y}{x} + 2\pi & \text{if } x > 0,\ y < 0. \end{cases}$$

The change of variables mapping from polar to Cartesian coordinates is

$$\Phi : [0,\infty) \times [0, 2\pi] \to \mathbb{R}^2, \quad \Phi(r,\theta) = (r\cos\theta, r\sin\theta). \tag{2.24}$$

The mapping is injective except that the half-lines $[0,\infty) \times \{0\}$ and $[0,\infty) \times \{2\pi\}$ both map to the nonnegative x-axis, and the vertical segment $\{0\} \times [0, 2\pi]$ is squashed to the point $(0,0)$. Each horizontal half-line $[0,\infty) \times \{0\}$ maps to the ray of angle θ with the positive x-axis, and each vertical segment $\{r\} \times [0, 2\pi]$ maps to the circle of radius r (see Figure 2.16). We have that regions in the (x,y)-plane defined by radial or angular constraints are images under Φ of (r,θ)-regions defined by rectangular constraints. For example, the Cartesian disk

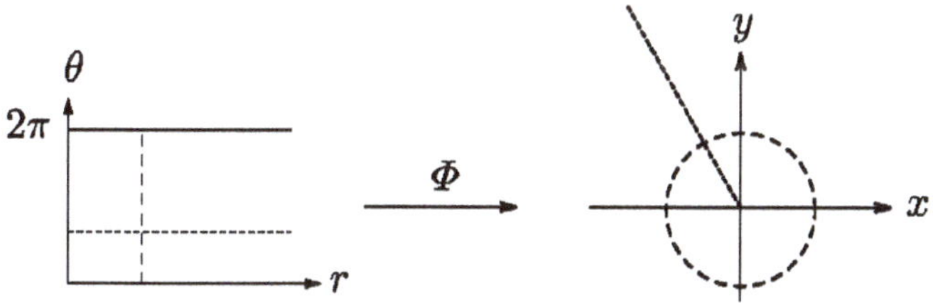

Figure 2.16: The polar coordinate mapping.

$$\{(x,y) \in \mathbb{R}^2 : x^2 + y^2 \le b^2\}$$

is the Φ-image of the polar rectangle

$$\{(r,\theta) : 0 \le r \le b,\ 0 \le \theta \le 2\pi\}$$

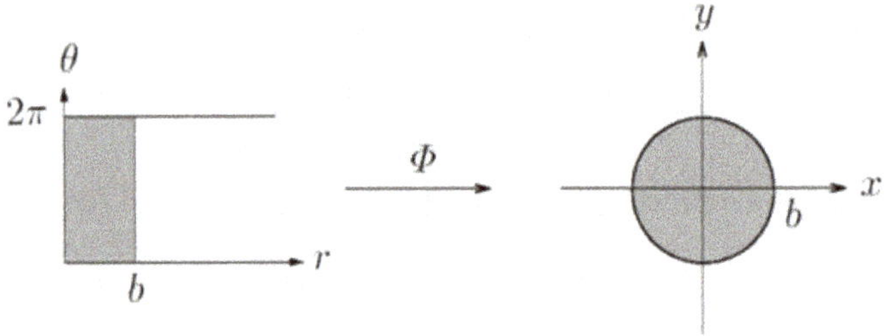

Figure 2.17: Rectangle to disk under polar coordinate mapping.

(see Figure 2.17). Similarly, the Cartesian annulus and quarter disk

$$\{(x,y) \in \mathbb{R}^2 : a^2 \leq x^2 + y^2 \leq b^2\},$$
$$\{(x,y) \in \mathbb{R}^2 : x \geq 0,\ y \geq 0,\ x^2 + y^2 \leq b^2\}$$

are the images of rectangles (see Figures 2.18 and 2.19). Notice that for the map (2.24), we have

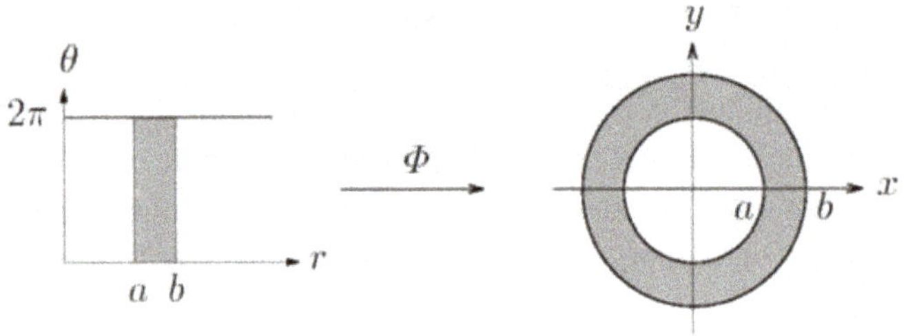

Figure 2.18: Rectangle to annulus under polar coordinate mapping.

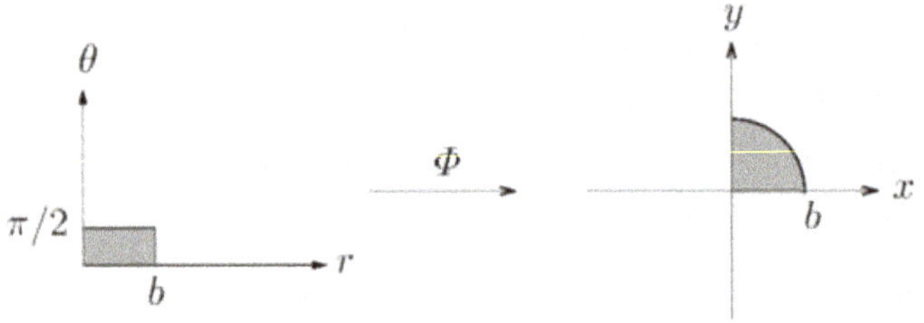

Figure 2.19: Rectangle to quarter disk under polar coordinate mapping.

$$\begin{aligned}\frac{\partial(x_1,x_2)}{\partial(r,\theta)} &= \det\begin{pmatrix}\cos\theta & \sin\theta\\ -r\sin\theta & r\cos\theta\end{pmatrix}\\ &= r(\cos\theta)^2 + r(\sin\theta)^2\\ &= r.\end{aligned}$$

If E is a region in $\mathbb{R}^2$, then applying Theorem 2.10, we get

$$\iint_E f(x,y)dxy = \int\int_{\Phi^{-1}(E)} f(r\cos\theta, r\sin\theta)rdrd\theta.$$

Example 2.9. We will compute the integral

$$I = \int_X \cos(\pi\sqrt{x^2+y^2})dxdy,$$

where

$$X = \{(x,y) \in \mathbb{R}^2 : x^2 + y^2 \le 1\}.$$

Consider the map Φ given by (2.24). We have

$$\begin{aligned}x^2 + y^2 &= (r\cos\theta)^2 + (r\sin\theta)^2\\ &= r^2((\cos\theta)^2 + (\sin\theta)^2)\\ &= r^2, \quad r \ge 0,\ \phi \in [0, 2\pi].\end{aligned}$$

Next, the inequality

$$x^2 + y^2 \le 1$$

holds if and only if $r \le 1$. Thus

$$\Phi(X) = \{(r,\theta) : r \in [0,1],\ \theta \in [0,2\pi]\}.$$

Applying Theorem 2.10, we get

$$\begin{aligned}I &= \int_0^1\int_0^{2\pi} \cos(\pi r) r d\theta dr\\ &= 2\pi\int_0^1 r\cos(\pi r)dr\\ &= 2\int_0^1 rd(\sin(\pi r))\end{aligned}$$

$$= 2r\sin(\pi r)\Big|_{r=0}^{r=1} - 2\int_0^1 \sin(\pi r)dr$$
$$= \frac{2}{\pi}\cos(\pi r)\Big|_{r=0}^{r=1}$$
$$= \frac{2}{\pi}(\cos\pi - \cos 0)$$
$$= -\frac{4}{\pi}.$$

Just as polar coordinates are convenient for radial symmetry in $\mathbb{R}^2$, cylindrical coordinates in $\mathbb{R}^3$ conveniently describe regions with symmetry about the z-axis. A point $p \in \mathbb{R}^3$ with Cartesian coordinates (r, θ, z), where (r, θ) are the polar coordinates for the point (x, y) (see Figure 2.20). The cylindrical change of variable mapping

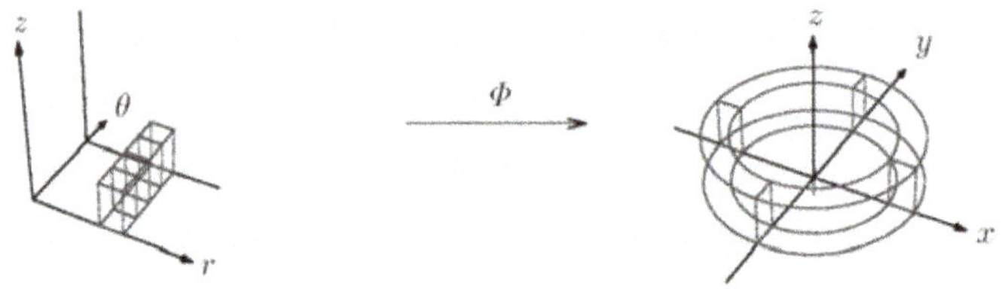

Figure 2.20: Cylindrical coordinates.

$$\Phi : [0, \infty) \times [0, 2\pi] \times \mathbb{R} \to \mathbb{R}^3$$

is given by

$$\Phi(r, \theta, z) = (r\cos\theta, r\sin\theta, z).$$

In this case, we have

$$\frac{\partial(x,y,z)}{\partial(r,\theta,z)} = \det\begin{pmatrix} \cos\theta & \sin\theta & 0 \\ -r\sin\theta & r\cos\theta & 0 \\ 0 & 0 & 1 \end{pmatrix}$$
$$= r(\cos\theta)^2 + r(\sin\theta)^2$$
$$= r.$$

Now applying Theorem 2.10, we get

$$\iiint_E f(x,y,z)dxdydz = \iiint_{\Phi^{-1}(E)} rf(r\cos\theta, r\sin\theta, z)drd\theta dz,$$

where E is a region in $\mathbb{R}^3$.

Example 2.10. We will evaluate the integral

$$I = \int\int_E x^2 z dxdydz,$$

where E is the cylinder

$$E = \{(x, y, z) \in \mathbb{R}^3 : x^2 + y^2 \leq 1,\ 0 \leq z \leq 2\}.$$

Using cylindrical coordinates, we find

$$\begin{aligned}
I &= \int_0^{2\pi}\int_0^1\int_0^2 r^3(\cos\theta)^2 z dz dr d\theta \\
&= \left(\int_0^{2\pi}(\cos\theta)^2 d\theta\right)\left(\int_0^1 r^3 dr\right)\left(\int_0^2 z dz\right) \\
&= \left(\int_0^{2\pi}\frac{1+\cos(2\theta)}{2}d\theta\right)\left(\frac{1}{4}r^4\Big|_{r=0}^{r=1}\right)\left(\frac{1}{2}z^2\Big|_{z=0}^{z=2}\right) \\
&= \left(\pi + \frac{1}{4}\sin(2\theta)\Big|_{\theta=0}^{\theta=2\pi}\right)\left(\frac{1}{4}\right)2 \\
&= \frac{\pi}{2}.
\end{aligned}$$

Spherical coordinates in $\mathbb{R}^3$ are designed to exploit symmetry about the origin. A point $p = (x, y, z) \in \mathbb{R}^3$ has spherical coordinates (ρ, θ, ϕ), where ρ is the distance from the origin to the point p, the longitude θ is the angle from the positive x-axis to the (x, y)-projection, and the colatitude ϕ is the angle from the positive z-axis to p. The spherical coordinate mapping

$$\Phi : [0, \infty) \times [0, 2\pi] \times [0, \pi] \to \mathbb{R}^3$$

is given by

$$\Phi(\rho, \theta, \phi) = (\rho\cos\theta\sin\phi, \rho\sin\theta\sin\phi, \rho\cos\phi).$$

In this case, we have

$$\begin{aligned}
\frac{\partial(x,y,z)}{\partial(\rho,\theta,\phi)} &= \begin{vmatrix} \cos\theta\sin\phi & \sin\theta\sin\phi & \cos\phi \\ -\rho\sin\theta\sin\phi & \rho\cos\theta\sin\phi & 0 \\ \rho\cos\theta\cos\phi & \rho\sin\theta\cos\phi & -\rho\sin\phi \end{vmatrix} \\
&= -\rho^2(\cos\theta)^2(\sin\phi)^3 - \rho^2(\sin\theta)^2\sin\phi(\cos\phi)^2 \\
&\quad - \rho^2(\cos\theta)^2(\cos\phi)^2\sin\phi - \rho^2(\sin\theta)^2(\sin\phi)^3
\end{aligned}$$

$$= -\rho^2(\sin\phi)^3 - \rho^2(\cos\phi)^2\sin\phi$$
$$= -\rho^2\sin\phi,$$

and if E is a region in $\mathbb{R}^3$, then applying Theorem 2.10, we find

$$\iiint\limits_E f(x,y,z)dxdydz = \iiint\limits_{\Phi^{-1}(E)} f(\rho\cos\theta\sin\phi, \rho\sin\theta\sin\phi, \rho\cos\phi)\rho^2\sin\phi d\rho d\theta d\phi.$$

Figure 2.21 shows the image under the spherical coordinate mapping of some (θ,ϕ)-rectangles, each having a fixed value ρ, and similarly, Figure 2.22 for some fixed values of θ, and Figure 2.23 for some fixed values of ϕ.

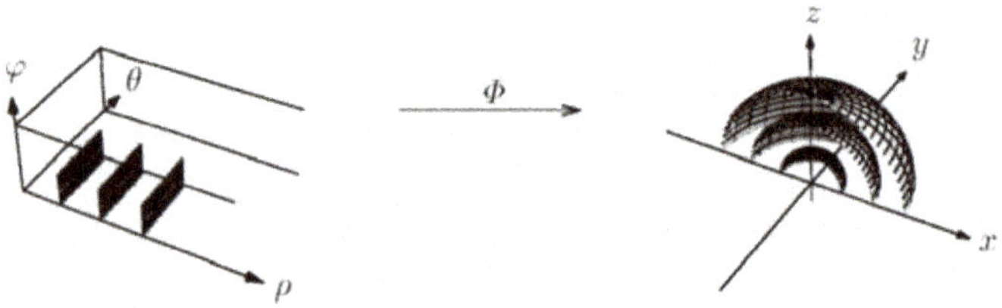

Figure 2.21: Spherical coordinates for some fixed spherical radii.

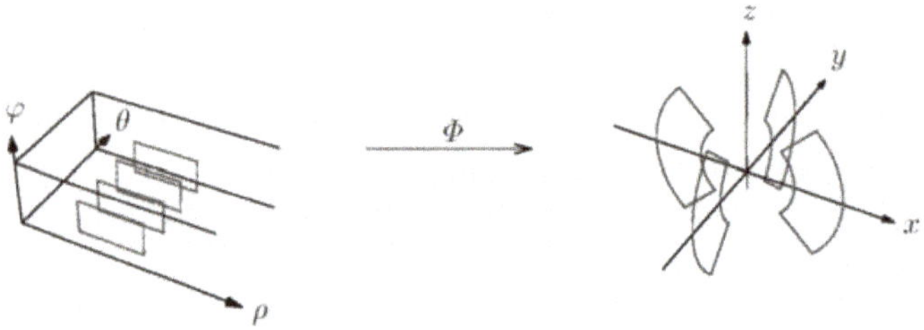

Figure 2.22: Spherical coordinates for some fixed longitudes.

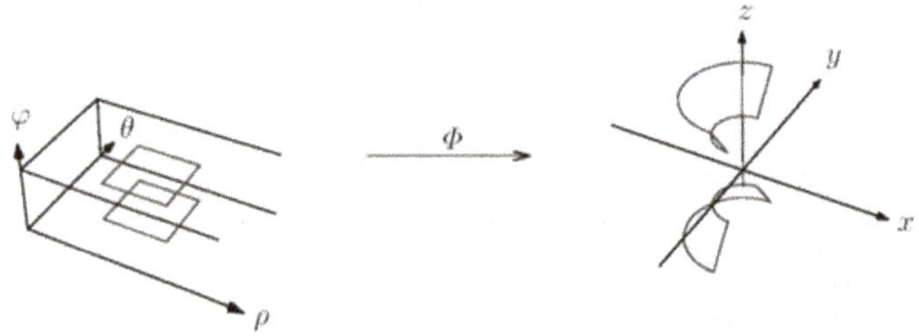

Figure 2.23: Spherical coordinates for some fixed colatitudes.

Example 2.11. We will compute the integral

$$I = \int\limits_X \sqrt{x^2 + y^2 + z^2}\,dxdydz,$$

where

$$X = \{(x, y, z) \in \mathbb{R}^3 : 1 \le x^2 + y^2 + z^2 \le 8\}.$$

We will use spherical coordinates. We have

$$\begin{aligned} x^2 + y^2 + z^2 &= (\rho\cos\theta\sin\phi)^2 + (\rho\sin\theta\sin\phi)^2 + (\rho\cos\phi)^2 \\ &= \rho^2(\cos\theta)^2(\sin\phi)^2 + \rho^2(\sin\theta)^2(\sin\phi)^2 + \rho^2(\cos\phi)^2 \\ &= \rho^2(\sin\phi)^2 + \rho^2(\cos\phi)^2 \\ &= \rho^2, \quad \rho \ge 0,\ \theta \in [0, 2\pi],\ \phi \in [0, \pi]. \end{aligned}$$

Note that the inequalities

$$1 \le x^2 + y^2 + z^2 \le 8$$

are equivalent to the inequalities

$$1 \le \rho^2 \le 8,$$

or $\rho \in [1, 2\sqrt{2}]$. Therefore

$$\Phi^{-1}(X) = \{(\rho, \theta, \phi) : \rho \in [1, 2\sqrt{2}],\ \theta \in [0, 2\pi],\ \phi \in [0, \pi]\}.$$

Now applying Theorem 2.10, we get

$$\begin{aligned} I &= \int\limits_1^{2\sqrt{2}} \int\limits_0^{2\pi} \int\limits_0^{\pi} \rho|-\rho^2 \sin\phi| d\phi d\theta d\rho \\ &= 2\pi\Bigg(\int\limits_1^{2\sqrt{2}} \rho^3 d\rho\Bigg)\Bigg(\int\limits_0^{\pi} \sin\phi d\phi\Bigg) \\ &= 2\pi\Bigg(\frac{\rho^4}{4}\Bigg|_{\rho=1}^{\rho=2\sqrt{2}}\Bigg)\Big(-\cos\phi\Big|_{\phi=0}^{\phi=\pi}\Big) \\ &= 2\pi\Bigg(\frac{16\cdot 4}{4} - \frac{1}{4}\Bigg)(1+1) \\ &= 4\pi\cdot\frac{63}{4} \\ &= 63\pi. \end{aligned}$$

Exercise 2.4. Compute the following integrals:

1.
$$\int_X \frac{dxdy}{x^2+y^2-1},$$

where

$$X = \{(x,y) \in \mathbb{R}^2 : 9 \le x^2+y^2 \le 25\}.$$

2.
$$\int_X |xy|dxdy,$$

where

$$X = \{(x,y) \in \mathbb{R}^2 : a^2 \le x^2+y^2 \le 4a^2\}, \quad a \in \mathbb{R},\ a \ne 0.$$

3.
$$\int_X xy^2 dxdy,$$

where

$$X = \{(x,y) \in \mathbb{R}^2 : x^2+y^2 \le a^2,\ x \ge 0\}, \quad a \in \mathbb{R},\ a \ne 0.$$

4.
$$\int_X y^2 e^{x^2+y^2} dxdy,$$

where

$$X = \{(x,y) \in \mathbb{R}^2 : x^2+y^2 \le 1,\ x \ge 0,\ y \ge 0\}.$$

5.
$$\int_X \frac{\log(x^2+y^2)}{x^2+y^2} dxdy,$$

where

$$X = \{(x,y) \in \mathbb{R}^2 : 1 \le x^2+y^2 \le a^2,\ y \ge 0\}, \quad a \in \mathbb{R},\ a \ge 1.$$

6.

$$\int_X (x^2 - y^2)dxdydz,$$

where

$$X = \{(x,y,z) : x^2 + y^2 + z^2 \leq a^2,\ y \geq 0,\ z \geq 0\}, \quad a \in \mathbb{R},\ a \neq 0.$$

7.

$$\int_X (x + z)dxdydz,$$

where

$$X = \{(x,y,z) : x^2 + y^2 + z^2 \leq R^2,\ z \leq \sqrt{x^2 + y^2}\}, \quad R \in \mathbb{R},\ R \neq 0.$$

8.

$$\int_X \frac{xyz}{(x^2 + y^2 + z^2)^{\frac{3}{2}}} dxdydz,$$

where

$$X = \{(x,y,z) \in \mathbb{R}^3 : (x^2 + y^2 + z^2)^{\frac{3}{2}} \leq 4xy,\ x \geq 0,\ y \geq 0,\ z \geq 0\}.$$

9.

$$\int_X z\sqrt{x^2 + y^2} dxdydz,$$

where

$$X = \{(x,y,z) \in \mathbb{R}^3 : x^2 + y^2 \leq 2x,\ 0 \leq z \leq y\}.$$

10.

$$\int_X xz^2 dxdydz,$$

where

$$X = \{(x,y,z) \in \mathbb{R}^3 : (3x - 4)^2 \leq y^2 + z^2 \leq x^2\}.$$

11.

$$\int_X z dx dy dz,$$

where

$$X = \{(x,y,z) \in \mathbb{R}^3 : 3(x^2 + y^2) \le z,\ 1 - x^2 - y^2 \ge z\}.$$

2.6 Some applications

We have already discussed applications of multiple integrals such as finding areas and volumes. In this section, we develop computational techniques for finding the center of mass and moments of inertia of several types physical objects using double integrals for lamina and triple integrals for a three-dimensional object with variable density. The density is supposed to be a constant number when the lamina or the object is homogeneous.

Consider the two-dimensional case. The center of mass is known as the center of gravity if the object is in a uniform gravitational field. If the object is homogeneous, then the center of mass is the geometric center of the object, called centroid. Figure 2.24 shows a point P as the center of mass of a lamina. To find the coordinates of the center of mass $P(\overline{x},\overline{y})$ of a lamina, we need to find the moment M_x of the lamina about the x-axis and the moment M_y about the y-axis. We also need to find the mass m of the lamina. Thus we have

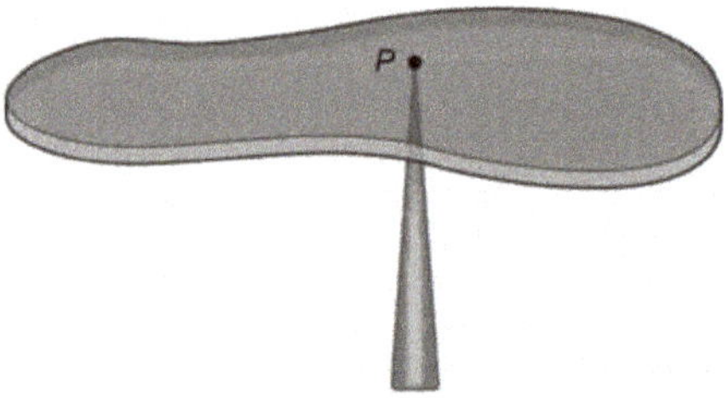

Figure 2.24: A lamina balanced about its center of mass.

$$\overline{x} = \frac{M_y}{m},$$
$$\overline{y} = \frac{M_x}{m}.$$

If the density function is constant, then $\overline{x}$ and $\overline{y}$ give the centroid of the lamina. Assume that the lamina occupies a region R in the (x,y)-plane, and let $\rho(x,y)$ be its density at any point (x,y). Then

$$\rho(x,y) = \lim_{\Delta A \to 0} \frac{\Delta m}{\Delta A},$$

where Δm and ΔA are the mass and area of a small rectangle containing the point (x,y), and the limit is taken as the dimensions of the rectangle go to 0 (see Figure 2.25). Now we divide the region R into tiny rectangles R_{ij} of area ΔA and choose (x_{ij}^*, y_{ij}^*) as sample points (see Figure 2.26). Then, the mass m_{ij} of each R_{ij} is equal to

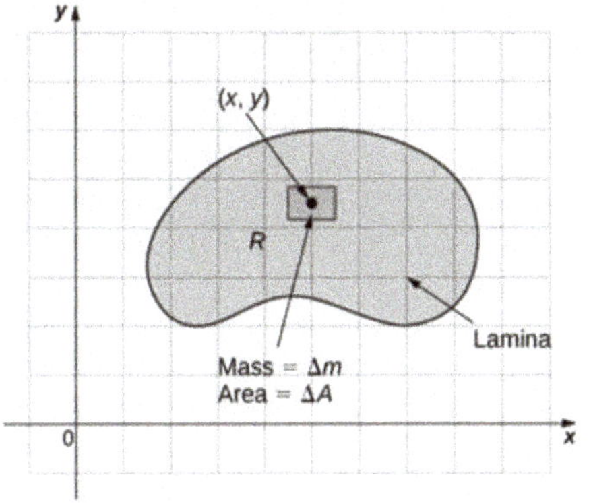

Figure 2.25: The density of a lamina at a point.

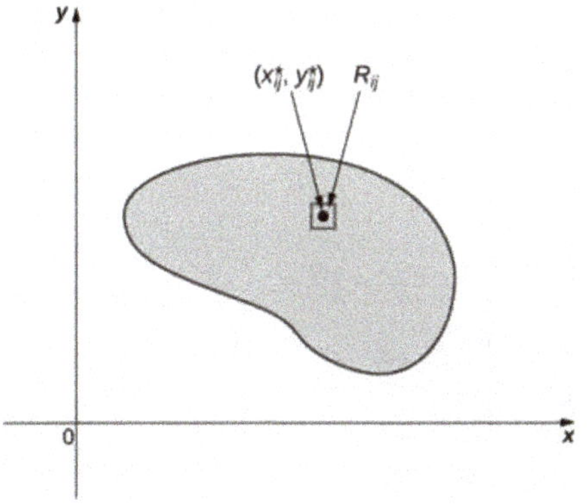

Figure 2.26: Subdividing the lamina into tiny rectangles.

$$\rho(x_{ij}^*, y_{ij}^*)\Delta A.$$

Let k and l be the number of subintervals in x and y, respectively. Then the mass of the lamina is

$$m = \lim_{k,l\to\infty} \sum_{i=1}^{k}\sum_{j=1}^{l} m_{ij}$$
$$= \lim_{k,l\to\infty} \sum_{i=1}^{k}\sum_{j=1}^{l} \rho(x_{ij}^*, y_{ij}^*)\Delta A$$
$$= \iint_R \rho(x,y)dxdy.$$

Example 2.12. Consider the triangular lamina R with vertices $(0,0)$, $(0,3)$, $(3,0)$ and density $\rho(x,y) = xy\,\mathrm{kg/m^2}$ (see Figure 2.27). Then R can be represented in the form

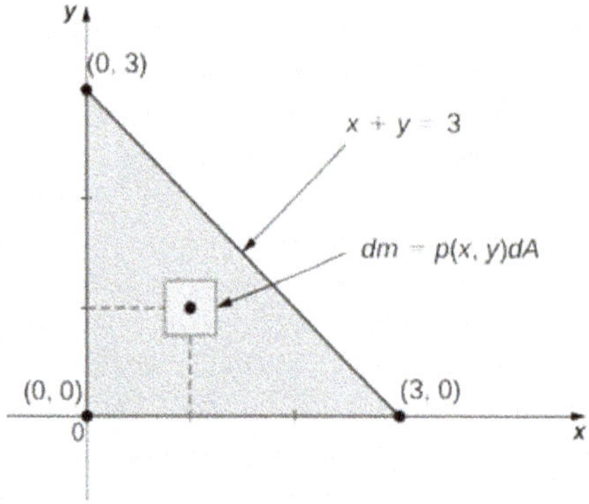

Figure 2.27: A lamina in the (x,y)-plane with density $\rho(x,y) = xy$.

$$R = \{(x,y) \in \mathbb{R}^2 : 0 \le x \le 3,\ 0 \le y \le 3 - x\},$$

and

$$m = \int_0^3 \int_0^{3-x} xydydx$$
$$= \int_0^3 x\left(\frac{1}{2}y^2\Big|_{y=0}^{y=3-x}\right)dx$$
$$= \frac{1}{2}\int_0^3 x(3-x)^2 dx$$
$$= \frac{1}{2}\int_0^3 x(x^2 - 6x + 9)dx$$
$$= \frac{1}{2}\int_0^3 (x^3 - 6x^2 + 9x)dx$$

$$= \frac{1}{2}\left(\frac{x^4}{4}\bigg|_{x=0}^{x=3} - 2x^3\bigg|_{x=0}^{x=3} + \frac{9}{2}x^2\bigg|_{x=0}^{x=3}\right)$$
$$= \frac{1}{2}\left(\frac{81}{4} - 54 + \frac{81}{2}\right)$$
$$= \frac{1}{2}\left(\frac{81 - 216 + 162}{4}\right)$$
$$= \frac{27}{8}.$$

Now we will find expressions for the moments. The moment M_x about the x-axis for R is the limit of the sums of moments of the regions R_{ij} about the x-axis. Then

$$M_x = \lim_{k,l\to\infty} \sum_{i=1}^{k}\sum_{j=1}^{l} y_{ij}^* m_{ij}$$
$$= \lim_{k,l\to\infty} \sum_{i=1}^{k}\sum_{j=1}^{l} y_{ij}^* \rho(x_{ij}^*, y_{ij}^*)\Delta A$$
$$= \int\int_R y\rho(x,y)dxdy.$$

Similarly,

$$M_y = \lim_{k,l\to\infty} \sum_{i=1}^{k}\sum_{j=1}^{l} x_{ij}^* m_{ij}$$
$$= \lim_{k,l\to\infty} \sum_{i=1}^{k}\sum_{j=1}^{l} x_{ij}^* \rho(x_{ij}^*, y_{ij}^*)\Delta A$$
$$= \int\int_R x\rho(x,y)dxdy.$$

Finally, for the center of the mass, we get

$$\overline{x} = \frac{M_y}{m}$$
$$= \frac{\int\int_R x\rho(x,y)dxdy}{\int\int_R \rho(x,y)dxdy}$$

and

$$\overline{y} = \frac{M_x}{m}$$
$$= \frac{\int\int_R y\rho(x,y)dxdy}{\int\int_R \rho(x,y)dxdy}.$$

Example 2.13. Consider the same triangular region R as in Example 2.12. Then

$$\begin{aligned}
M_x &= \int_0^3 \int_0^{3-x} xy^2 dydx \\
&= \frac{1}{3} \int_0^3 xy^3 \Big|_{y=0}^{y=3-x} dx \\
&= -\frac{1}{3} \int_0^3 x(x-3)^3 dx \\
&= -\frac{1}{3} \int_0^3 (x-3)^4 dx - \int_0^3 (x-3)^3 dx \\
&= -\frac{1}{15}(x-3)^5 \Big|_{x=0}^{x=3} - \frac{1}{4}(x-3)^4 \Big|_{x=0}^{x=3} \\
&= \frac{243}{15} - \frac{81}{4} \\
&= \frac{81}{5} - \frac{81}{4} \\
&= \frac{81}{20},
\end{aligned}$$

and

$$\begin{aligned}
\bar{y} &= \frac{M_x}{m} \\
&= \frac{\frac{81}{20}}{\frac{27}{8}} \\
&= \frac{\frac{3}{5}}{\frac{1}{2}} \\
&= \frac{6}{5}.
\end{aligned}$$

Exercise 2.5. Let R be the triangular region as in Example 2.12. Find M_y and $\bar{x}$.

The moment of inertia of a particle of mass m about an axis is mr^2, where r is the distance of the particle from the axis. From Figure 2.27 we can see that the moment of inertia of the subrectangle R_{ij} about the x-axis is

$$(y_{ij}^*)^2 \rho(x_{ij}^*, y_{ij}^*) \Delta A.$$

Similarly, the moment of inertia of the subrectangle R_{ij} about the y-axis is

$$(x_{ij}^*)^2 \rho(x_{ij}^*, y_{ij}^*) \Delta A.$$

The moment of inertia is related to the rotation of the mass. It measures the tendency of the mass to resist a change in rotational motion about an axis. The moment of inertia I_x about the x-axis for the region R is the limit of the sum of moments of inertia of the regions R_{ij} about the x-axis. Thus

$$\begin{aligned} I_x &= \lim_{k,l\to\infty} \sum_{i=1}^{k}\sum_{j=1}^{l} (y_{ij}^*)^2 m_{ij} \\ &= \lim_{k,l\to\infty} \sum_{i=1}^{k}\sum_{j=1}^{l} (y_{ij}^*)^2 \rho(x_{ij}^*, y_{ij}^*) \Delta A \\ &= \iint_R y^2 \rho(x,y)\,dxdy. \end{aligned}$$

Similarly, the moment of inertia I_y about the y-axis for R is

$$\begin{aligned} I_y &= \lim_{k,l\to\infty} \sum_{i=1}^{k}\sum_{j=1}^{l} (x_{ij}^*)^2 m_{ij} \\ &= \lim_{k,l\to\infty} \sum_{i=1}^{k}\sum_{j=1}^{l} (x_{ij}^*)^2 \rho(x_{ij}^*, y_{ij}^*) \Delta A \\ &= \iint_R x^2 \rho(x,y)\,dxdy. \end{aligned}$$

By I_0 we will denote the polar moment, which is obtained by adding the moments of inertia I_x and I_y. Then

$$I_0 = I_x + I_y.$$

Example 2.14. Consider the triangular region R given in Example 2.12 with the same density. Then

$$\begin{aligned} I_x &= \int_0^3 \int_0^{3-x} xy^3\,dydx \\ &= \frac{1}{4}\int_0^3 xy^4 \Big|_{y=0}^{y=3-x} dx \\ &= \frac{1}{4}\int_0^3 x(x-3)^4\,dx \end{aligned}$$

$$= \frac{1}{4}\int_0^1 (x-3)^5 dx + \frac{3}{4}\int_0^3 (x-3)^4 dx$$
$$= \frac{(x-3)^6}{24}\Bigg|_{x=0}^{x=3} + \frac{3}{20}(x-3)^5\Bigg|_{x=0}^{x=3}$$
$$= -\frac{729}{24} + \frac{729}{20}$$
$$= \frac{243}{40}.$$

All the expressions of double integrals discussed above can be formulated for triple integrals. Suppose that Q is a solid object with a density function $\rho(x,y,z)$ at any point (x,y,z). Then its mass is

$$m = \iiint_Q \rho(x,y,z)dxdydz.$$

Its moments about the (x,y)-, (x,z)-, and (y,z)-planes are

$$M_{xy} = \iiint_Q z\rho(x,y,z)dxdydz,$$
$$M_{xz} = \iiint_Q y\rho(x,y,z)dxdydz,$$
$$M_{yz} = \iiint_Q x\rho(x,y,z)dxdydz,$$

respectively. If the center of mass of the object is the point $(\overline{x},\overline{y},\overline{z})$, then

$$\overline{x} = \frac{M_{yz}}{m},$$
$$\overline{y} = \frac{M_{xz}}{m},$$
$$\overline{z} = \frac{M_{xy}}{m}.$$

Finally, the moments of inertia about the (y,z)-, (x,z)-, and (x,y)-planes are

$$I_x = \iiint_Q (y^2+z^2)\rho(x,y,z)dxdydz,$$
$$I_y = \iiint_Q (x^2+z^2)\rho(x,y,z)dxdydz,$$
$$I_z = \iiint_Q (x^2+y^2)\rho(x,y,z)dxdydz.$$

Example 2.15. Let Q be the solid region bounded by

$$x + 2y + 3z = 6 \tag{2.25}$$

and the coordinate planes, and let it have the density

$$\rho(x, y, z) = z.$$

The region Q is a tetrahedron meeting the axes at the points $(6, 0, 0)$, $(0, 3, 0)$, and $(0, 0, 2)$ (see Figure 2.28). By (2.25) we find

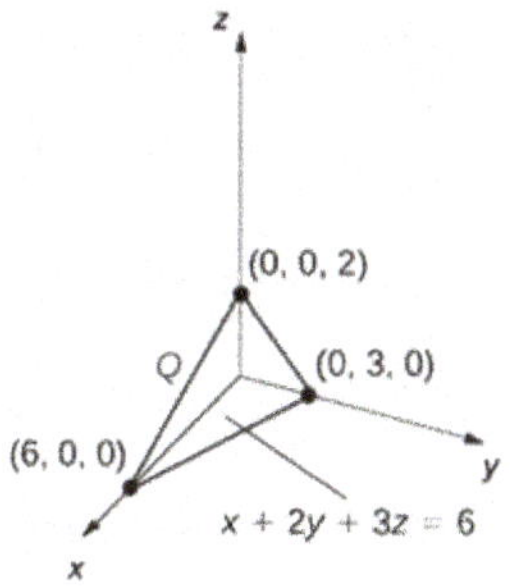

Figure 2.28: A tetrahedron.

$$z = \frac{1}{3}(6 - x - 2y),$$

and if $z = 0$, then the projection of Q onto the (x, y)-plane is bounded by the axes and the line

$$x + 2y = 6.$$

The region Q can be represented as follows:

$$Q = \Big\{(x, hy, z) \in \mathbb{R}^3 : 0 \le x \le 6,\ 0 \le y \le \frac{1}{2}(6 - x),$$
$$0 \le z \le \frac{1}{3}(6 - x - 2y)\Big\}.$$

Hence

$$m = \iiint\limits_Q \rho(x, y, z)\,dxdydz$$
$$= \int_0^6 \int_0^{\frac{1}{2}(6-x)} \int_0^{\frac{1}{3}(6-x-2y)} z\,dzdydx$$

$$
\begin{aligned}
&= \frac{1}{2}\int_0^6 \int_0^{\frac{1}{2}(6-x)} z^2\Big|_{z=0}^{z=\frac{1}{3}(6-x-2y)} dydx \\
&= \frac{1}{18}\int_0^6 \int_0^{\frac{1}{2}(6-x)} (6-x-2y)^2 dydx \\
&= -\frac{1}{108}\int_0^6 (6-x-2y)^3\Big|_{y=0}^{y=\frac{1}{2}(6-x)} dx \\
&= \frac{1}{864}\int_0^6 (6-x)^3 dx \\
&= -\frac{1}{3456}(6-x)^4\Big|_{x=0}^{x=6} \\
&= \frac{3}{8}.
\end{aligned}
$$

Exercise 2.6. Consider the region Q as in Example 2.15 with the same density function. Find M_{xy}, M_{xz}, M_{yz}, I_x, I_y, I_z, and the coordinates of the center of mass.

2.7 Advanced practical problems

Problem 2.1. Evaluate

1.
$$
\int_0^{\frac{\pi}{2}}\left(\int_1^{\cos\phi} r\sin\phi \log r dr\right)d\phi.
$$

2.
$$
\int_0^{\sin x}\left(\int_0^{y} \frac{uv}{\sqrt{u^2-v^2}}dv\right)du, \quad x \in \mathbb{R}.
$$

3.
$$
\int_0^{1}\left(\int_0^{\sqrt{1-x^2}} (1-y^2)^{\frac{3}{2}} dy\right)dx.
$$

4.
$$
\int_0^{a}\left(\int_x^{a} (a^2-y^2)^{\alpha} dy\right)dx, \quad a > 0, \quad \alpha > 0.
$$

5.

$$\int_{-1}^{1}\left(\int_{\sqrt[3]{|x|}}^{1}(1-y^2)^{\alpha}dy\right)dx, \quad \alpha > 0.$$

6.

$$\int_{0}^{1}\left(\int_{\sqrt{y}}^{\sqrt[5]{y}}\sqrt{1-x^3}dx\right)dy.$$

7.

$$\int_{0}^{1}\left(\int_{\sqrt[3]{x}}^{1}y^2\sqrt{y^4-x^2}dy\right)dx.$$

Problem 2.2. Compute

1.

$$\int\int_X xdxdy,$$

where

$$X = \{(x,y) \in \mathbb{R}^2 : x^2+y^2 \le 2,\ x^2-y^2 \le 1,\ x \ge 0,\ y \ge 0\}.$$

2.

$$\int\int_X ydxdy,$$

where

$$X = \{(x,y) \in \mathbb{R}^2 : 0 \le y \le 6,\ x < 6,\ xy > 3,\ y-x-2 < 0\}.$$

3.

$$\int\int_X (2y-x)dxdy,$$

where

$$X = \{(x,y) \in \mathbb{R}^2 : y(y-x) \le 2,\ x(x+y) \le 3\}.$$

4.

$$\iint_X x^2y^2\,dxdy,$$

where

$$X = \{(x,y) \in \mathbb{R}^2 : y > 0,\ xy < 1,\ x^2 - 3xy + 2y^2 < 0\}.$$

5.

$$\iint_X \frac{x}{x^2+y^2}\,dxdy,$$

where X is bounded by the lines

$$y = x\tan x, \quad y = x, \quad 0 \le x < \frac{\pi}{2}.$$

6.

$$\int_0^1\int_0^1 e^{x^2}\,dxdy.$$

7.

$$\int_0^1\int_{\sqrt{y}}^1 \sin(x^3-1)\,dxdy.$$

8.

$$\int_0^1\int_{\frac{x-1}{2}}^0 \tan(y^2+y)\,dydx.$$

9.

$$\int_0^1\int_0^{1-y} e^{2x-x^2}\,dxdy.$$

10.

$$\iint_X \sqrt{y^2-x^2}\,dxdy,$$

where X is bounded by the lines

$$y = 1, \quad y = x, \quad y = -x.$$

11.

$$\int\int_X (x^2 + y^2)dxdy,$$

where

$$X = \{(x,y) \in \mathbb{R}^2 : a \le |x| \le b,\ a \le |y| \le b\}, \quad 0 \le a \le b.$$

12.

$$\int\int_X |y| dxdy,$$

where

$$X = \left\{(x,y) \in \mathbb{R}^2 : \frac{x^2}{16} + \frac{y^2}{4} \le 1,\ x^2 + y^2 \ge 1\right\}.$$

13.

$$\int\int_X (2 - x - y)dxdy,$$

where

$$X = \{(x,y) \in \mathbb{R}^2 : 2y \le x^2 + y^2 \le 4\}.$$

14.

$$\int\int_X |xy| dxdy,$$

where

$$X = \{(x,y) \in \mathbb{R}^2 : a^2 \le x^2 + y^2 \le b^2\}, \quad 0 < a < b.$$

15.

$$\int\int_X \sqrt{|y - x^2|}dxdy,$$

where

$$X = [-1, 1] \times [0, 2].$$

16.
$$\iint_X \min\{x,y\}dxdy,$$
where
$$X = [0,a] \times [0,a], \quad a > 0.$$

17.
$$\iint_X \max\{\sin x, \sin y\}dxdy,$$
where
$$X = [0,\pi] \times [0,\pi].$$

18.
$$\iint_X \operatorname{sign}(2a - 2x - y)dxdy,$$
where
$$X = [0,a] \times [0,a], \quad a > 0.$$

19.
$$\iint_X \operatorname{sign}(x^2 - y^2 + 2)dxdy,$$
where
$$X = \{(x,y) \in \mathbb{R}^2 : x^2 + y^2 \le 4\}.$$

20.
$$\iint_X \left(\frac{x^2}{a^2} + \frac{y^2}{b^2}\right)dxdy,$$
where X is bounded by the lines
$$x = 0, \quad y = 0, \quad x = a\sin t, \quad y = b\cos t, \quad t \in \left[0, \frac{\pi}{2}\right].$$

21.

$$\iint_X (x-y)dxdy,$$

where X is bounded by the coordinate axes and the line

$$x = a(\cos t)^3, \quad y = a(\sin t)^3, \quad t \in \left[0, \frac{\pi}{2}\right].$$

22.

$$\iint_X ydxdy,$$

where X is bounded by the line

$$x = a(t - \sin t), \quad y = a(1 - \cos t), \quad t \in [0, 2\pi], \quad y = 0.$$

23.

$$\iint_X xdxdy,$$

where X is bounded by the line

$$x = a \sin t, \quad y = b \sin(2t), \quad t \in [0, \pi].$$

24.

$$\iint_X xydxdy,$$

where X is bounded by $x = 0$ and the line

$$x = 3t - \frac{t^3}{3}, \quad y = 2t - t^2, \quad t \in [0, 3].$$

25.

$$\iint_X (x+y)dxdy,$$

where X is bounded by $y = 0$ and the line

$$x = 2t(1-t), \quad y = 4t - t^3, \quad t \in [0, 2].$$

26.

$$\int\int_X y^2 dxdy,$$

where X is bounded by $y = 0$ and the line

$$x = \sin\left(\frac{3t}{2}\right), \quad y = \sin t, \quad t \in [0, \pi].$$

Problem 2.3. Let the function $y = \phi(x)$, $x \in [a, b]$, be defined parametrically

$$x = \alpha(t), \quad y = \beta(t), \quad t \in [t_1, t_2],$$

where $\alpha \in \mathscr{C}^1([t_1, t_2)$ is increasing, and

$$\alpha(t_1) = a, \quad \alpha(t_2) = b, \quad \beta(t) \geq 0, \quad t \in [t_1, t_2].$$

Let also,

$$X = \{(x,y) \in \mathbb{R}^2 : x \in [a, b],\ 0 \leq y \leq \phi(x)\}$$

and $f \in \mathscr{C}(X)$. Prove that

$$\int\int_X f(x,y)dxdy = \int_{t_1}^{t_2} \alpha'(t)\left(\int_0^{\beta(t)} f(\alpha(t),y)dy\right)dt.$$

If in addition, $\beta \in \mathscr{C}^1([t_1, t_2])$ and $\beta(t_1) = 0$, then prove that

$$\int\int_X f(x,y)dxdy = \int_{t_1}^{t_2} \alpha'(t)\left(\int_0^{\beta(t)} f(\alpha(t),\beta(s))\beta'(s)ds\right)dt.$$

Problem 2.4. Let X be bounded by $y = 0$ and tje line

$$x = a(t - \sin t), \quad y = a(1 - \cos t), \quad t \in [0, 2\pi].$$

Prove that

$$\int\int_X xy^n dxdy = \pi a \int\int_X y^n dxdy, \quad n \geq 0.$$

Problem 2.5. Compute

1.

$$\iiint_X y\,dxdydz,$$

where X is bounded by

$$x = 0, \quad y = 0, \quad z = 0, \quad 2x + y + z = 4.$$

2.

$$\iiint_X (1 + x + y + z)^{-3} dxdydz,$$

where X is bounded by

$$x + y + z = 1, \quad x = 0, \quad y = 0, \quad z = 0.$$

3.

$$\iiint_X (x + z) dxdydz,$$

where X is bounded by

$$x + y = 1, \quad x - y = 1, \quad x + z = 1, \quad z = 0, \quad x = 0.$$

4.

$$\iiint_X (x^2 - z^2) dxdydz,$$

where X is bounded by

$$y = -x, \quad z = x, \quad z = y, \quad z = 1.$$

5.

$$\iiint_X xy\,dxdydz,$$

where X is bounded by

$$x^2 + y^2 = 1, \quad z = 0, \quad z = 1, \quad x \geq 0, \quad y \geq 0.$$

6.

$$\iiint\limits_X xyz\,dxdydz,$$

where X is bounded by

$$x^2+y^2+z^2\le 1,\quad x,y,z\in\left[0,\frac{\pi}{2}\right].$$

7.

$$\iiint\limits_X \sqrt{x^2+y^2}\,dxdydz,$$

where X is bounded by

$$x^2+y^2+z^2=1,\quad z=1.$$

8.

$$\iiint\limits_X xy^2z^3\,dxdydz,$$

where X is bounded by

$$z=xy,\quad y=x,\quad x=1,\quad z=0.$$

9.

$$\iiint\limits_X xyz\,dxdydz,$$

where X is bounded by

$$y=x^2,\quad x=y^2,\quad z=xy,\quad z=0.$$

Problem 2.6. Let X be a Jordan-measurable domain that is symmetric with respect to Ox, and let $f:X\to\mathbb{R}$ be a Riemann-integrable function over X and odd with respect to the second variable. Prove that

$$\iint\limits_X f(x,y)dxdy=0.$$

Problem 2.7. Let X be a Jordan-measurable domain in $\mathbb{R}^3$ that is symmetric with respect to Ox, and let $f:X\to\mathbb{R}$ be a Riemann-integrable function over X such that

$$f(x, -y, -z) = -f(x, y, z), \quad (x, y, z) \in X.$$

Prove that

$$\iiint\limits_X f(x, y, z)dxdydz = 0.$$

Problem 2.8. Let X be a Jordan-measurable domain that is symmetric with respect to the origin, and let $f : X \to \mathbb{R}$ be a Riemann-integrable function over X such that

$$f(-x, -y, -z) = -f(x, y, z), \quad (x, y, z) \in X.$$

Prove that

$$\iiint\limits_X f(x, y, z)dxdydz = 0.$$

Problem 2.9. Compute the following integrals:

1.

$$\iint\limits_X (ax + by)dxdy,$$

where

$$X = \{(x, y) \in \mathbb{R}^2 : x^2 + y^2 \le R^2,\ x - y \le 0\}, \quad R \in \mathbb{R},\ R \ne 0.$$

2.

$$\iint\limits_X (x + y)dxdy,$$

where

$$X = \{(x, y) \in \mathbb{R}^2 : x^2 + y^2 \le R^2,\ y - kx > 0\}, \quad R, k \in \mathbb{R},\ R, k \ne 0.$$

3.

$$\iint\limits_X \operatorname{sign} y dxdy,$$

where

$$X = \{(x, y) \in \mathbb{R}^2 : x^2 + y^2 \le 1,\ y - kx > 0\}.$$

4.

$$\iint\limits_X \frac{y^2}{x^2+y^2}dxdy,$$

where

$$X = \{(x,y) \in \mathbb{R}^2 : x^2+y^2 \le ax\}, \quad a \in \mathbb{R},\ a > 0.$$

5.

$$\iint\limits_X ydxdy,$$

where

$$X = \{(x,y) \in \mathbb{R}^2 : x^2+y^2 \le 2x,\ x > y\}.$$

6.

$$\iint\limits_X \frac{y}{\sqrt{x^2+y^2}}dxdy,$$

where

$$X = \{(x,y) \in \mathbb{R}^2 : x^2+y^2 \le 1,\ x^2+y^2 \le 2y\}.$$

7.

$$\iint\limits_X \left(\frac{y}{x}\right)^2 dxdy,$$

where

$$X = \{(x,y) \in \mathbb{R}^2 : 1 \le x^2+y^2 \le 2x\}.$$

8.

$$\iint\limits_X xdxdy,$$

where

$$X = \{(x,y) \in \mathbb{R}^2 : ax \le x^2+y^2 \le 2ax,\ y \ge 0\}, \quad a \in \mathbb{R},\ a > 0.$$

9.

$$\iint_X \sqrt{a^2 - x^2 - y^2}dxdy,$$

where

$$X = \{(x,y) \in \mathbb{R}^2 : ay \le x^2 + y^2 \le a^2,\ y \ge 0,\ x \ge 0\}.$$

10.

$$\iint_X y^2 dxdy,$$

where

$$X = \{(x,y) \in \mathbb{R}^2 : 2x \le x^2 + y^2 \le 6x,\ y \le x\}.$$

Problem 2.10. Compute the following integrals:

1.

$$\iint_X \log(1 + x^2 + y^2)dxdy,$$

where

$$X = \{(x,y) \in \mathbb{R}^2 : 0 \le x \le 1,\ -\sqrt{1-x^2} \le y \le \sqrt{1-x^2}\}.$$

2.

$$\iint_X \sqrt{1 - x^2 - y^2}dxdy,$$

where

$$X = \{(x,y) \in \mathbb{R}^2 : -1 \le x \le 1,\ -\sqrt{1-x^2} \le y \le 0\}.$$

3.

$$\iint_X \frac{x^2}{x^2+y^2}dxdy,$$

where

$$X = \{(x,y) \in \mathbb{R}^2 : 0 \le x \le 2,\ 0 \le y \le \sqrt{2x-x^2}\}.$$

4.

$$\iint_X \sqrt{y}\,dxdy,$$

where

$$X = \{(x,y) \in \mathbb{R}^2 : 0 \le x \le 1,\ 1 - \sqrt{1-x^2} \le y \le 1 + \sqrt{1-x^2}\}.$$

5.

$$\iint_X \frac{1}{(x^2+y^2)^2}dxdy,$$

where X is bounded by the curves $x^2 - y^2 = 6$ and $x = 3$.

6.

$$\iint_X |xy|dxdy,$$

where

$$X = \{(x,y) \in \mathbb{R}^2 : (x^2+y^2)^2 \le x^2 - y^2,\ x \ge 0\}.$$

7.

$$\iint_X x^2 dxdy,$$

where

$$X = \{(x,y) \in \mathbb{R}^2 : (x^2+y^2)^2 \le 2xy,\ x \ge 0\}.$$

8.

$$\iint_X ydxdy,$$

where

$$X = \{(x,y) \in \mathbb{R}^2 : 0 \le x \le (x^2+y^2)^{\frac{3}{2}} \le 1,\ y \ge 0\}.$$

9.

$$\iint_X \frac{y}{\sqrt{x^2+y^2}}dxdy,$$

where

$$X = \left\{(x,y) \in \mathbb{R}^2 : \frac{3}{2}ay \le x^2 \le a^2 - y^2\right\}, \quad a \in \mathbb{R},\ a > 0.$$

10.

$$\iint_X \sqrt{x^2 + y^2}dxdy,$$

where

$$X = \{(x,y) \in \mathbb{R}^2 : ax \le x^2 + y^2 \le a(x + \sqrt{x^2 + y^2})\}, \quad a \in \mathbb{R},\ a > 0.$$

Problem 2.11. Compute the integrals

1.

$$\iiint_X (\sqrt{x^2 - y^2} + z)dxdydz,$$

where

$$X = \{(x,y,z) \in \mathbb{R}^3 : -1 \le x \le 3,\ 0 \le y \le x,\ 0 \le z \le \sqrt{x^2 - y^2}\}.$$

2.

$$\iiint_X x^3y^2zdxdydz,$$

where

$$X = \{(x,y,z) \in \mathbb{R}^3 : 0 \le x \le 1,\ 0 \le y \le 2x,\ 0 \le z \le \sqrt{xy}\}.$$

3.

$$\iiint_X (x + y + z)dxdydz,$$

where

$$X = \left\{(x,y,z) \in \mathbb{R}^3 : -1 \le x \le 1,\ \frac{x}{2} \le y \le 2x,\ x - y \le z \le x + y\right\}.$$

4.

$$\iiint_X ze^x \sin ydxdydz,$$

where

$$X = \{(x,y,z) \in \mathbb{R}^3 : 0 \leq x \leq a,\ 0 \leq y \leq b,\ 0 \leq z \leq c\}, \quad a,b,c \in \mathbb{R},\ a,b,c \geq 0.$$

5.

$$\iiint_X (x+y)e^{x-y}dxdydz,$$

where

$$X = \{(x,y,z) \in \mathbb{R}^3 : 0 \leq x \leq a,\ 0 \leq y \leq b,\ 0 \leq z \leq c\}, \quad a,b,c \in \mathbb{R},\ a,b,c \geq 0.$$

6.

$$\iiint_X xyzdxdydz,$$

where

$$X = \{(x,y,z) \in \mathbb{R}^3 : 0 \leq x \leq 1,\ -1 \leq y \leq 1,\ 0 \leq z \leq 1\}.$$

7.

$$\iiint_X (xy + yz + xz)dxdydz,$$

where

$$X = \{(x,y,z) \in \mathbb{R}^3 : 0 \leq x \leq 1,\ 0 \leq y \leq 1,\ 0 \leq z \leq 1\}.$$

8.

$$\iiint_X \sin(x+y+z)dxdydz,$$

where

$$X = \{(x,y,z) \in \mathbb{R}^3 : 0 \leq x \leq \pi,\ 0 \leq y \leq \pi,\ 0 \leq z \leq \pi\}.$$

9.

$$\iiint_X (x + 2y + 3z)dxdydz,$$

where X is bounded by the planes

$$y = 0, \quad z = 0, \quad z = 2, \quad x + y = 2, \quad 2x - y + 2 = 0.$$

10.

$$\iiint_X (xy)^2 dxdydz,$$

where

$$X = \{(x, y, z) \in \mathbb{R}^3 : 0 \le x \le y \le z \le 1\}.$$

11.

$$\iiint_X \frac{x}{R^4 + (x^2 + y^2 + z^2)^2} dxdydz,$$

where

$$X = \{(x, y, z) \in \mathbb{R}^3 : x^2 + y^2 + z^2 \le R^2,\ x \ge 0\}, \quad R \in \mathbb{R},\ R > 0.$$

12.

$$\iiint_X (x^2 + y^2 - z^2) dxdydz,$$

where

$$X = \{(x, y, z) \in \mathbb{R}^3 : 1 \le x^2 + y^2 + z^2 \le 4,\ x \ge 0,\ y \ge 0\}.$$

13.

$$\iiint_X (yz + xz) dxdydz,$$

where X is the domain in the first quadrant bounded by the surfaces

$$y = x, \quad x = 0, \quad z = 0, \quad x^2 + y^2 + z^2 = R^2.$$

14.

$$\iiint_X \frac{z}{\sqrt{x^2 + y^2 + z^2}} dxdydz,$$

where

$$X = \left\{(x, y, z) \in \mathbb{R}^3 : x^2 + y^2 + z^2 \le R^2,\ z \ge \frac{h}{a}\sqrt{x^2 + y^2}\right\}, \quad R, h, a \in \mathbb{R},\ R, h, a > 0.$$

15.

$$\iiint_X \sqrt{x^2+y^2+z^2}\,dxdydz,$$

where

$$X = \{(x,y,z) \in \mathbb{R}^3 : x^2+y^2+z^2 \le z\}.$$

16.

$$\iiint_X (x^2+y^2+z^2)\,dxdydz,$$

where

$$X = \{(x,y,z) \in \mathbb{R}^3 : x^2+y^2 \le R^2,\ 0 \le x,\ 0 \le z \le H\}, \quad H,R \in \mathbb{R},\ H,R > 0.$$

17.

$$\iiint_X (x+y+z)\,dxdydz,$$

where X is bounded by the surfaces

$$x^2+y^2 = 1, \quad z = 0, \quad x+y+z = 2.$$

18.

$$\iiint_X (x^2+y^2)\,dxdydz,$$

where

$$X = \left\{(x,y,z) \in \mathbb{R}^3 : \frac{x^2+y^2}{2} \le z \le 2\right\}.$$

19.

$$\iiint_X (z-x+y)\,dxdydz,$$

where X is bounded by the surfaces

$$ay = z^2+x^2, \quad y^2 = z^2+x^2, \quad a \in \mathbb{R},\ a > 0.$$

20.

$$\iiint_X \left(\frac{x^2}{a^2} + \frac{y^2}{b^2} + \frac{z^2}{c^2}\right) dxdydz,$$

where

$$X = \left\{(x, y, z) \in \mathbb{R}^3 : \frac{x^2}{a^2} + \frac{y^2}{b^2} + \frac{z^2}{c^2} \leq 1\right\}, \quad a, b, c \in \mathbb{R},\ a, b, c > 0.$$

21.

$$\iiint_X (x^2 + y^2) dxdydz,$$

where

$$X = \left\{(x, y, z) \in \mathbb{R}^3 : \frac{x^2}{a^2} + \frac{y^2}{b^2} + \frac{z^2}{c^2} \leq 1\right\}, \quad a, b, c \in \mathbb{R},\ a, b, c > 0.$$

22.

$$\iiint_X \sqrt{1 - \frac{x^2}{a^2} - \frac{y^2}{b^2} - \frac{z^2}{c^2}} dxdydz,$$

where

$$X = \left\{(x, y, z) \in \mathbb{R}^3 : \frac{x^2}{a^2} + \frac{y^2}{b^2} + \frac{z^2}{c^2} \leq 1\right\}, \quad a, b, c \in \mathbb{R},\ a, b, c > 0.$$

23.

$$\iiint_X z dxdydz,$$

where X is bounded by the surfaces

$$R^2 z^2 = h^2(x^2 + y^2), \quad z = h,\ R, h \in \mathbb{R},\ R, h > 0.$$

24.

$$\iiint_X \sqrt{y^2 + z^2} dxdydz,$$

where X is bounded by the surfaces

$$y^2 + z^2 = R^2, \quad x + y = R, \quad y - x = R, \quad R \in \mathbb{R},\ R > 0.$$

25.

$$\iiint\limits_X \frac{xyz}{(a^2+x^2+y^2+z^2)^3}dxdydz,$$

where

$$X = \{(x,y,z) \in \mathbb{R}^3 : x^2+y^2 \le a^2,\ y^2+z^2 \le a^2,\ x \ge 0,\ y \ge 0,\ z \ge 0\}, \quad a \in \mathbb{R},\ a > 0.$$

26.

$$\iiint\limits_X \sqrt{x^2+y^2}dxdydz,$$

where X is the domain in the first quadrant bounded by the surfaces

$$z = 0, \quad y = 0, \quad x^2+y^2 = a^2, \quad az = x^2 - y^2, \quad a \in \mathbb{R},\ a > 0.$$

27.

$$\iiint\limits_X z^2 dxdydz,$$

where

$$X = \{(x,y,z) \in \mathbb{R}^3 : x^2+y^2+z^2 \le R^2,\ x^2+y^2+z^2 \le 2Rz\}, \quad R \in \mathbb{R},\ R > 0.$$

28.

$$\iiint\limits_X \frac{|xy|}{x^2}dxdydz,$$

where

$$X = \{(xyz) \in \mathbb{R}^3 : \sqrt{x^2+y^2} \le z \le \sqrt{1-x^2-y^2}\}.$$

29.

$$\iiint\limits_X \frac{1}{(x+y)(x+y+z)}dxdydz,$$

where

$$X = \{(x,y,z) \in \mathbb{R}^3 : 1 \le x \le 2,\ 1 \le x+y \le 3,\ 1 \le x+y+z \le 5\}.$$

30.

$$\iiint_X (x^2 - y^2)(z + x^2 - y^2)dxdydz,$$

where

$$X = \{(x, y, z) \in \mathbb{R}^3 : x - 1 \le y \le x,\ 1 - x \le y \le 2 - x,\ 1 - x^2 + y^2 \le z \le y^2 - x^2 + 2x\}.$$

31.

$$\iiint_X xyzdxdydz,$$

where

$$X = \{(x, y, z) \in \mathbb{R}^3 : x \le yz \le 2x,\ y \le xz \le 2y,\ z \le xy \le 2z\}.$$

32.

$$\iiint_X x^2 dxdydz,$$

where X is bounded by the surfaces

$$z = ay^2, \quad z = by^2, \quad z = \alpha x, \quad z = \beta x, \quad z = h, \quad 0 < a < b,\ 0 < \alpha < \beta,\ h > 0.$$

33.

$$\iiint_X xyzdxdydz,$$

where X is the domain in the first quadrant bounded by the surfaces

$$mz = x^2 + y^2, \quad nz = x^2 + y^2, \quad xy = a^2, \quad xy = b^2, \quad y = \alpha x, \quad y = \beta x,$$

and

$$0 < a < b, \quad \alpha < \beta, \quad 0 < m < n.$$

Problem 2.12. Compute

$$\iiint_X \frac{\partial^3 f}{\partial x \partial y \partial z}(x, y, z)dxdydz,$$

where

$$X = \{(x,y,z) \in \mathbb{R}^3 : a_1 \le x \le a_2,\ b_1 \le y \le b_2,\ c_1 \le z \le c_2\},$$

$a_1, a_2, b_1, b_2, c_1, c_2 \in \mathbb{R}$, $a_1 \le a_2$, $b_1 \le b_2$, $c_1 \le c_2$, and $f \in \mathscr{C}^3(X)$.

Problem 2.13. Let

$$X = \{(x,y,z) \in \mathbb{R}^3 : x^2 + y^2 \le R^2,\ 0 \le z \le H\}, \quad R, H \in \mathbb{R},\ R > 0,\ H > 0,$$

and $f \in \mathscr{C}(X)$. Prove that the function

$$\iiint\limits_{x^2+y^2 \le z^2} f(x,y,z)dxdydz$$

is a continuous function on $[0, H]$.

Problem 2.14. Let $f \in \mathscr{C}([0,\infty))$. Find $\frac{dF}{dt}(t)$, $t \ge 0$, where

$$F(t) = \iiint\limits_{x^2+y^2+z^2 \le t^2} f(\sqrt{x^2+y^2+z^2})dxdydz, \quad t \ge 0.$$

Problem 2.15. Let

$$X = \{(x,y,z) \in \mathbb{R}^3 : z \ge \sqrt{x^2+y^2}\}$$

and $f \in \mathscr{C}(X)$. Find $\frac{dF}{dt}(t)$, $t \ge 0$, where

$$F(t) = \iiint\limits_{\sqrt{x^2+y^2} \le z \le t} f(x,y,z)dxdydz, \quad t \ge 0.$$

Problem 2.16. Let

$$X = \{(x,y,z) \in \mathbb{R}^3 : x \ge 0,\ y \ge 0,\ z \ge 0\},$$
$$Y(t) = \{(x,y,z) \in \mathbb{R}^3 : y \ge 0,\ z \ge 0,\ x + y + z \le t\}, \quad t \ge 0,$$

and $f \in \mathscr{C}(X)$. Find $\frac{dF}{dt}(t)$, $t \ge 0$, where

$$F(t) = \iiint\limits_{Y(t)} f(x,y,z)dxdydz, \quad t \ge 0.$$

Problem 2.17. Let

$$X = \{(x,y,z) : x \ge 0,\ y \ge 0,\ z \ge 0\},$$
$$Y(x,y,z) = \{(x_1,y_1,z_1) \in \mathbb{R}^3 : 0 \le x_1 \le x,\ 0 \le y_1 \le y,\ 0 \le z_1 \le z\}, \quad (x,y,z) \in X,$$

and $f \in \mathscr{C}(X)$. Find $\frac{\partial^3 F}{\partial x \partial y \partial z}(x,y,z)$, $(x,y,z) \in X$, where

$$F(x,y,z) = \int\limits_{Y(x,y,z)} f(\xi,\eta,\zeta)d\xi d\zeta d\eta, \quad (x,y,z) \in X.$$

Problem 2.18. Let

$$X = \{(x_1,x_2,x_3,x_4) \in \mathbb{R}^4 : x_1 + x_2 + x_3 + x_4 \le a,\ x_j \ge 0,\ j \in \{1,\dots,4\}\}.$$

Compute

1.
$$\int\limits_X dx_1 dx_2 dx_3 dx_4.$$

2.
$$\int\limits_X (x_1 + x_2 + x_3 + x_4)^\alpha dx_1 dx_2 dx_3 dx_4, \quad \alpha \ge 0.$$

Problem 2.19. Let

$$X = \{(x_1,x_2,x_3,x_4) \in \mathbb{R}^4 : 0 \le x_1 \le x_2 \le x_3 \le x_4 \le a\}, \quad a \in \mathbb{R},\ a \ge 0.$$

Compute

1.
$$\int\limits_X x_1x_2x_3x_4 dx_1 dx_2 dx_3 dx_4.$$

2.
$$\int\limits_X (x_1x_2 + x_3x_4) dx_1 dx_2 dx_3 dx_4.$$

Problem 2.20. Let

$$X = \{(x_1,x_2,x_3,x_4) \in \mathbb{R}^4 : x_1^2 + x_2^2 + x_3^2 \le R^2,\ 0 \le x_4 \le H\}, \quad R,H \in \mathbb{R},\ R \ge 0,\ H \ge 0.$$

Compute

1.
$$\int\limits_X dx_1 dx_2 dx_3 dx_4.$$

2.

$$\int_X (x_1^2 + x_2^2 + x_3^2 + x_4^2)dx_1dx_2dx_3dx_4.$$

Problem 2.21. Let

$$X = \{(x_1, x_2, x_3, x_4) \in \mathbb{R}^4 : x_1^2 + x_2^2 + x_3^2 \le a^2x_4^2,\ 0 \le x_4 \le H\}, \quad a, H \in \mathbb{R},\ a \ge 0,\ H \ge 0.$$

Compute

1.

$$\int_X dx_1dx_2dx_3dx_4.$$

2.

$$\int_X (x_1^2 + x_2^2 + x_3^2)dx_1dx_2dx_3dx_4.$$

Problem 2.22. Let

$$X = \{(x_1, x_2, x_3, x_4) \in \mathbb{R}^4 : x_1^2 + x_2^2 \le a^2,\ x_3^2 + x_4^2 \le b^2\}, \quad a, b \in \mathbb{R},\ a, b \ge 0.$$

Compute

1.

$$\int_X dx_1dx_2dx_3dx_4.$$

2.

$$\int_X (x_1 + x_2 + x_3 + x_4)^2 dx_1dx_2dx_3dx_4.$$

Problem 2.23. Let

$$X = [a, b],$$
$$X^2 = X \times X,$$
$$X^n = X^{n-1} \times X,$$

$K \in \mathscr{C}(X^2)$, and

$$K_2(x, y) = \int_X K(x, \xi)K(\xi, y)d\xi,$$

$$\vdots$$

$$K_{n+1}(x,y) = \int_X K_n(x,\xi)K(\xi,y)d\xi, \quad (x,y) \in X^2.$$

Prove that

$$K_{n+1}(x_1,x_2) = \int_{X^n} K(x_1,\xi_1)K(\xi_1,\xi_2)\cdots K(\xi_n,x_2)d\xi_1\cdots d\xi_n.$$

Problem 2.24. Let $X = [0,a]^n$, $n \geq 2$, $a \in \mathbb{R}$, $a > 0$. Compute the following integrals:

1.

$$\int_X x_k^p dx_1\cdots dx_n, \quad 1 \leq k \leq n,\ p \geq 0.$$

2.

$$\int_X \sum_{k=1}^n x_k dx_1\cdots dx_n.$$

3.

$$\int_X \sum_{k=1}^n x_k^p dx_1\cdots dx_n, \quad p \geq 0.$$

4.

$$\int_X \left(\sum_{k=1}^n x_k\right)^p dx_1\cdots dx_n, \quad p \geq 0.$$

5.

$$\int_X e^{c_1x_1+\cdots+c_nx_n} dx_1\cdots dx_n, \quad c_k \in \mathbb{R},\ c_k \neq 0,\ k \in \{1,\ldots,n\}.$$

6.

$$\int_X \left(\cos\left(\frac{\pi}{2an}(x_1+\cdots+x_n)\right)\right)^2 dx_1\cdots dx_n.$$

Problem 2.25. Let

$$X = \{(x_1,\ldots,x_n) : 0 \leq x_n \leq x_{n-1} \leq \cdots \leq x_2 \leq x_1 \leq a\}.$$

Compute the following integrals:

1.
$$\int_X dx_1 \cdots dx_n.$$

2.
$$\int_X x_1 \cdots x_n dx_1 \cdots dx_n.$$

3.
$$\int_X \sum_{k=1}^n x_k dx_1 \cdots dx_n.$$

Problem 2.26. Let

$$X = \left\{(x_1, \ldots, x_n) \in \mathbb{R}^n : \sum_{k=1}^n x_k \le a,\ x_k \ge 0,\ k \in \{1, \ldots, n\}\right\}.$$

Compute the following integrals:

1.
$$\int_X dx_1 \cdots dx_n.$$

2.
$$\int_X \sum_{k=1}^n x_k dx_1 \cdots dx_n.$$

3.
$$\int_X \sum_{k=1}^n x_k^2 dx_1 \cdots dx_n.$$

4.
$$\int_X \left(\sum_{k=1}^n x_k\right)^{\frac{1}{2}} dx_1 \cdots dx_n.$$

5.
$$\int_X \left(\sum_{k=1}^n x_k\right)^{p} dx_1 \cdots dx_n.$$

Problem 2.27. Find the volume of the n-dimensional parallelogram bounded by the planes

$$\sum_{j=1}^{n} a_{ij}x_j = \pm h_i, \quad h_i > 0,\ i \in \{1,\ldots,n\},$$

provided that $\det(a_{ij}) \neq 0$.

Problem 2.28. Find the volume of the n-dimensional pyramid

$$\sum_{j=1}^{n} \frac{x_j}{a_j} \leq 1, \quad x_j \geq 0,\ a_j \geq 0,\ j \in \{1,\ldots,n\}.$$

Problem 2.29. Consider the map F given by

$$\begin{aligned}
x_n &= r\sin\psi_{n-1},\\
x_{n-1} &= r\cos\psi_{n-1}\sin\psi_{n-2},\\
x_{n-2} &= r\cos\psi_{n-1}\cos\psi_{n-2}\sin\psi_{n-3},\\
&\vdots\\
x_2 &= r\cos\psi_{n-1}\cdots\cos\psi_2\sin\psi_1,\\
x_1 &= r\cos\psi_{n-1}\cdots\cos\psi_2\cos\psi_1,
\end{aligned}$$

where $\psi_1 \in [0,2\pi)$, $\psi_j[-\frac{\pi}{2},\frac{\pi}{2}]$, $j \in \{2,\ldots,n-1\}$. Find the Jacobian of this map.

Problem 2.30. Find the volume of the n-dimensional cylinder

$$\sum_{j=1}^{n-1} x_j^2 \leq R^2, \quad 0 \leq x_n \leq H,$$

where $H > 0$.

Problem 2.31. Find the volume of the n-dimensional cone

$$\sum_{j=1}^{n-1} x_j^2 \leq x^2 x_n^2, \quad 0 \leq x_n \leq H,$$

where $H > 0$.

Problem 2.32. Let

$$X = \left\{(x_1,\ldots,x_n) \in \mathbb{R}^n : \sum_{j=1}^{n} x_j^2 \leq R^2,\ 0 \leq x_n \leq H\right\},$$

where $R, H \in \mathbb{R}$, $R > 0$, $H > 0$. Compute the integral

$$\int\limits_X x_n^2 dx_1 \cdots dx_n.$$

Problem 2.33. Find the volume of

$$\sum_{j=1}^{n} \frac{x_j^2}{a_j^2} \le 1,$$

where $a_j \in \mathbb{R}$, $a_j \neq 0, j \in \{1, \dots, n\}$.

Problem 2.34. Let

$$X = \left\{ (x_1, \dots, x_n) \in \mathbb{R}^n : \sum_{j=1}^{n} x_j^2 \le R^2 \right\},$$

where $R > 0$. Compute the integral

$$\int\limits_X (R^2 - x_1^2 - \cdots - x_n^2)^{\frac{1}{2}} dx_1 \cdots dx_n.$$

Problem 2.35. Let $f \in \mathscr{C}([0, \infty))$, and let

$$X = \left\{ (x_1, \dots, x_n) \in \mathbb{R}^n : \sum_{j=1}^{n} x_j^2 \le R^2 \right\},$$

where $R > 0$. Reduce the integral

$$\int\limits_X f(\sqrt{x_1^2 + \cdots + x_n^2}) dx_1 \cdots dx_n$$

to a one-dimensional integral.

Problem 2.36. Let

$$X = [0, a]^n,$$
$$Y(x_1, \dots, x_n) = [0, x_1] \times \cdots \times [0, x_n], \quad x_k > 0, \ k \in \{1, \dots, n\}.$$

Let also, $f \in \mathscr{C}(X)$, and let

$$F(x_1, \dots, x_n) = \int\limits_{Y(x_1, \dots, x_n)} f(u_1, \dots, u_n) du_1 \cdots du_n.$$

Find

$$\frac{\partial^n F}{\partial x_1 \cdots \partial x_n}.$$

Problem 2.37. Let $a > 0, X = [0,a]^n, f \in \mathscr{C}([0,a])$ and f is positive on $[0,a]$. Compute the integral

$$\int_X \frac{\sum_{k=1}^m f(x_k)}{\sum_{k=1}^n f(x_k)} dx_1 \cdots dx_n, \quad 1 \le m \le n.$$

Problem 2.38. Let $f \in \mathscr{C}([0,1]), X_n = [0,1]^n$. Prove that

$$\lim_{n\to\infty} \int_{X_n} f(x_1,\ldots,x_n)dx_1\cdots dx_n = f(0).$$

Problem 2.39. Let

$$X = \left\{(x_1,\ldots,x_n) : \sum_{k=1}^n x_k \le 1,\ x_k \ge 0,\ k \in \{1,\ldots,n\}\right\},$$

and $p_k > 0$, $k \in \{1,\ldots,n\}$. Prove that

$$\int_X x_1^{p_1-1}\cdots x_n^{p_n-1}dx_1\cdots dx_n = \frac{\Gamma(p_1)\cdots\Gamma(p_n)}{\Gamma(p_1+\cdots+p_n+1)}.$$

Here Γ is the gamma function.

Problem 2.40. Let

$$X = \left\{(x_1,\ldots,x_n) : \sum_{k=1}^n x_k \le 1,\ x_k \ge 0,\ k \in \{1,\ldots,n\}\right\},$$

$p_k > 0$, $k \in \{1,\ldots,n\}$, and $f \in \mathscr{C}([0,\infty))$. Prove that

$$\int_X f(x_1+\cdots+x_n)x_1^{p_1-1}\cdots x_n^{p_n-1}dx_1\cdots dx_n$$

$$= \frac{\Gamma(p_1)\cdots\Gamma(p_n)}{\Gamma(p_1+\cdots+p_n)}\int_0^1 f(t)t^{p_1+\cdots+p_n-1}dt.$$

Here Γ is the gamma function (see the appendix of this book).

3 Improper multiple integrals

In this chapter, we introduce improper multiple integrals. Some criteria for their convergence are deduced. Absolute convergence of improper multiple integrals is defined and investigated.

3.1 Basic definitions

In single-variable calculus, an improper integral arises when attempting to integrate a function on an unbounded region or a function on an interval where it is discontinuous. Improper integrals can be also defined for functions of several variables. The definition given below is different from that in the one-dimensional case and is valid only for the multidimensional case. Improper multiple integrals have various real-life applications such as in the Laplace and Fourier transforms, quantum mechanics, and probability theory.

Let $n \geq 2$, and let $X \subset \mathbb{R}^n$ be an open set.

Definition 3.1. We say that the sequence $\{X_k\}_{k=1}^{\infty}$ of open sets of $\mathbb{R}^n$ monotonically exhausts the open set X if

1. $\overline{X_k} \subset X_{k+1}$, $k \in \mathbb{N}$, and
2. $\bigcup_{k=1}^{\infty} X_k = X$.

We will show that for any open set X, there is a monotonically exhausting sequence of open sets. Let

$$Q_k = \{x \in \mathbb{R}^n : |x| < k\}, \quad k \in \mathbb{N},$$

and let

$$s_k = s_k(X \cap Q_k), \quad k \in \mathbb{N},$$

be the set of all cubes of rank k in $X \cap Q_k$. Since for all $m \in \mathbb{N}$,

$$\operatorname{dist}(s_{k_m}, \partial X) > 0$$

and the diameters of all cubes lying in s_k tend to 0 as $k \to \infty$, there is $m \in \mathbb{N}$ such that

$$s_{k_m} \subset (s_{k_{m+1}})_{\text{int}}.$$

Denote

$$X_m = (s_{k_m})_{\text{int}}, \quad m \in \mathbb{N}.$$

https://doi.org/10.1515/9783112219607-003

Then

$$\overline{X_m} \subset X_{m+1}, \quad m \in \mathbb{N}.$$

Now since

$$X = \bigcup_{k=1}^{\infty} s_k(X),$$

we obtain

$$X = \bigcup_{m=1}^{\infty} X_m.$$

Definition 3.2. Let f be a Riemann-integrable function on any Jordan-measurable set Y such that $\overline{Y} \subset X$. The function f is said to be an integrable function in improper sense on the open set X if for any sequence $\{X_k\}_{k=1}^{\infty}$ that monotonically exhausts the set X, the limit

$$\lim_{k\to\infty} \int_{X_k} f(x)dx$$

does not depend on the choice of the sequence $\{X_k\}_{k=1}^{\infty}$. This limit is called the improper integral of the function f on the set X and is denoted by

$$\int_X f(x)dx. \tag{3.1}$$

Thus

$$\int_X f(x)dx = \lim_{k\to\infty} \int_{X_k} f(x)dx.$$

If the integral (3.1) exists, then we say that it is convergent. Otherwise, we say that it is divergent.

Example 3.1. The integral

$$\int_{-1}^{2} \int_{-\infty}^{\infty} e^{-x^2-y^2} dydx$$

is an improper integral.

Remark 3.1. Definition 3.2 is not equivalent to the corresponding definition of the improper integral in the case $n = 1$. Thus the defined improper integral in Definition 3.2 is valid only in the case $n > 1$.

Remark 3.2. If the open set X is Jordan measurable and the function f is integrable on X, then the improper integral (3.1) coincides with the usual Riemann integral.

Remark 3.3. Definition 3.2 allows us to transfer some properties of proper integrals to improper integrals (3.1), such as additivity, linearity, integration of inequalities, reducing to iterated integrals, change of variables, and others.

Remark 3.4. By the additivity of the proper integrals it follows that if $X_0 \subset X$ is Jordan measurable, then

$$\int_X f(x)dx - \int_{X_0} f(x)dx = \int_{X \setminus X_0} f(x)dx.$$

Thus the integral (3.1) is convergent if and only if for any sequence $\{X_k\}_{k=1}^{\infty}$ that monotonically exhausts the set X, there exist the integrals

$$\int_{X \setminus X_k} f(x)dx,$$

and

$$\lim_{k\to\infty} \int_{X \setminus X_k} f(x)dx = 0.$$

3.2 Improper integrals of nonnegative functions

In this section, we give a criterion for convergence of improper multiple integrals of nonnegative functions. The main result reads as follows.

Theorem 3.1. *Let f be a nonnegative function on X. Then for any sequence $\{X_k\}_{k=1}^{\infty}$ of Jordan-measurable sets that monotonically exhausts the open set X, the limit*

$$\lim_{k\to\infty} \int_{X_k} f(x)dx$$

exists, does not depend on the choice of the sequence $\{X_k\}_{k=1}^{\infty}$, and is finite or equal to ∞. In the case where the integral

$$\int_X f(x)dx \tag{3.2}$$

exists, we have that

$$\lim_{k\to\infty} \int_{X_k} f(x)dx = \int_X f(x)dx.$$

If

$$\lim_{k\to\infty} \int_{X_k} f(x)dx = \infty,$$

then the integral (3.2) *does not exist, and we write*

$$\int_X f(x)dx = \infty.$$

Proof. Let $\{X_k\}_{k=1}^{\infty}$ be a sequence of Jordan-measurable sets that monotonically exhausts the open set X. Then, for any $k \in \mathbb{N}$, we have

$$\overline{X_k} \subset X_{k+1},$$

and hence

$$\int_{X_k} f(x)dx \le \int_{X_{k+1}} f(x)dx.$$

In the last inequality, we have used that the function f is nonnegative on X. Therefore the sequence

$$\left\{\int_{X_k} f(x)dx\right\}_{k=1}^{\infty}$$

has a finite or infinite limit. Denote

$$\lim_{k\to\infty} \int_{X_k} f(x)dx = I_1.$$

Suppose that $\{Y_k\}_{k=1}^{\infty}$ is another sequence of Jordan measurable sets that monotonically exhausts the set X. Denote

$$\lim_{k\to\infty} \int_{Y_k} f(x)dx = I_2.$$

We will prove that $I_1 = I_2$. Note that for any $k \in \mathbb{N}$, there is $k_0 = k_0(k) \in \mathbb{N}$ such that

$$\overline{X_k} \subset Y_{k_0}. \tag{3.3}$$

Since $\overline{X_k}$ is a Jordan-measurable set, it is bounded, and thus it is a compact set. By (3.3) it follows that the system $\{Y_l\}_{l=1}^{\infty}$ is an open cover of the compact set $\overline{X_k}$. By the Heine–Borel theorem it follows that there is a finite system

$$\{Y_1, Y_2, \dots, Y_{k_0}\} \subset \{Y_l\}_{l=1}^{\infty}$$

such that

$$Y_1 \subset Y_2 \subset \cdots \subset Y_{k_0}$$

and (3.3) holds. Applying (3.3) again and using that $f \geq 0$ on X, we arrive at the following chain of inequalities:

$$\begin{aligned}\int_{X_k} f(x)dx &\leq \int_{Y_{k_0}} f(x)dx\\ &\leq I_2.\end{aligned}$$

Hence

$$\begin{aligned}I_1 &= \lim_{k\to\infty} \int_{X_k} f(x)dx\\ &\leq I_2,\end{aligned}$$

i. e.,

$$I_1 \leq I_2. \tag{3.4}$$

As above, exchanging the roles of X_k and Y_k, we obtain

$$I_2 \leq I_1.$$

By the last inequality and inequality (3.4) we get $I_1 = I_2$. This completes the proof. □

Example 3.2. Consider the integral

$$I = \int_{R^2} e^{-x^2-y^2}dxdy.$$

Let

$$X_k = \{(x,y) \in \mathbb{R}^2 : x^2 + y^2 < k^2\}, \quad k \in \mathbb{N}.$$

We have that

$$\overline{X_k} \subset X_{k+1}, \quad k \in \mathbb{N},$$

and

$$\bigcup_{k=1}^{\infty} X_k = \mathbb{R}^2.$$

Thus the system $\{X_k\}_{k=1}^{\infty}$ is a sequence of Jordan-measurable sets that monotonically exhausts $\mathbb{R}^2$. Then applying Theorem 3.1, we find

$$\begin{aligned}
I &= \lim_{k\to\infty} \int_{X_k} e^{-x^2-y^2} dxdy \\
&= \lim_{k\to\infty} \int_0^{2\pi}\int_0^{k} e^{-r^2} rdrd\phi \\
&= \lim_{k\to\infty} \left(2\pi \cdot \frac{1}{2} \int_0^{r} e^{-r^2} dr^2 \right) \\
&= \pi \lim_{k\to\infty} \left(-e^{-r^2}\Big|_{r=0}^{r=k}\right) \\
&= \pi \lim_{k\to\infty} (1 - e^{-k^2}) \\
&= \pi,
\end{aligned}$$

where we have used the polar coordinates

$$\begin{aligned}
x &= r\cos\phi, \\
y &= r\sin\phi, \quad r > 0,\ \phi \in [0, 2\pi].
\end{aligned}$$

Exercise 3.1. Let I be the integral in Example 3.2, and let

$$X_k = \{(x,y) \in \mathbb{R}^2 : |x| \le k,\ |y| \le k\}, \quad k \in \mathbb{N}.$$

1. Prove that $\{X_k\}_{k=1}^{\infty}$ is a sequence of Jordan-measurable sets that monotonically exhausts $\mathbb{R}^2$.
2. Compute the integral I and compare the obtained result with the result in Example 3.2.
3. Compute the integral

$$\int_{-\infty}^{\infty} e^{-x^2} dx$$

using the computations in item 2.

Example 3.3. Consider the integral

$$I = \int_X \frac{1}{(x^2 + y^2 + z^2)^{\frac{\alpha}{2}}} dxdydz,$$

where $\alpha \in \mathbb{R}$, and

$$X = \{(x,y,z) \in \mathbb{R}^3 : x^2 + y^2 + z^2 < 1\}.$$

Let

$$X_k = \left\{(x,y,z) \in \mathbb{R}^3 : \frac{1}{k^2} < x^2 + y^2 + z^2 < \left(1 - \frac{1}{k}\right)^2\right\}, \quad k \in \mathbb{N},\ k \geq 2.$$

We have

$$\overline{X_k} \subset X_{k+1}, \quad k \in \mathbb{N},\ k \geq 2,$$

and

$$X = \bigcup_{k=2}^{\infty} X_k.$$

Thus $\{X_k\}_{k=2}^{\infty}$ is a sequence of Jordan-measurable sets that monotonically exhausts X. By Theorem 3.1 we get

$$\begin{aligned}
I &= \lim_{k\to\infty} \int_{X_k} \frac{1}{(x^2+y^2+z^2)^{\frac{\alpha}{2}}} dxdydz \\
&= \lim_{k\to\infty} \int_0^{2\pi}\int_0^{\pi}\int_{\frac{1}{k}}^{1-\frac{1}{k}} \frac{1}{r^\alpha}|-r^2 \sin\psi| dr d\psi d\phi \\
&= 2\pi \lim_{k\to\infty} \int_0^{\pi}\int_{\frac{1}{k}}^{1-\frac{1}{k}} \frac{1}{r^{\alpha-2}} \sin\psi dr d\psi \\
&= 2\pi \lim_{k\to\infty} \left(\int_0^{\pi} \sin\psi d\psi\right)\left(\int_{\frac{1}{k}}^{1-\frac{1}{k}} \frac{1}{r^{\alpha-2}} dr\right) \\
&= 2\pi(-\cos\psi|_{\psi=0}^{\psi=\pi}) \lim_{k\to\infty}\left(\int_{\frac{1}{k}}^{1-\frac{1}{k}} \frac{1}{r^{\alpha-2}} dr\right) \\
&= 4\pi \begin{cases} \lim_{k\to\infty}\left(\frac{r^{3-\alpha}}{3-\alpha}\Big|_{r=\frac{1}{k}}^{r=1-\frac{1}{k}}\right) & \text{if } \alpha \neq 3, \\ \lim_{k\to\infty}\left(\log r\Big|_{r=\frac{1}{k}}^{r=1-\frac{1}{k}}\right) & \text{if } \alpha = 3 \end{cases} \\
&= 4\pi \begin{cases} \frac{1}{3-\alpha}\lim_{k\to\infty}((1-\frac{1}{k})^{3-\alpha} - (\frac{1}{k})^{3-\alpha}) & \text{if } \alpha \neq 3, \\ \lim_{k\to\infty}(\log(1-\frac{1}{k}) - \log(\frac{1}{k})) & \text{if } \alpha = 3 \end{cases} \\
&= \begin{cases} \frac{4\pi}{3-\alpha} & \text{if } \alpha < 3, \\ \text{does not exist} & \text{if } \alpha \geq 3, \end{cases}
\end{aligned}$$

where we have used the spherical coordinates

$$x = r\cos\phi\sin\psi,$$
$$y = r\sin\phi\sin\psi,$$
$$z = r\cos\psi, \quad r > 0,\ \phi \in [0, 2\pi],\ \psi \in [0, \pi].$$

Example 3.4. Consider the integral

$$I = \int_{\mathbb{R}^2} |\sin((x^2+y^2)^2)| dxdy.$$

Let

$$X_k = \{(x,y) \in \mathbb{R}^2 : x^2 + y^2 < k^2\}, \quad k \in \mathbb{N}.$$

We have

$$\overline{X_k} \subset X_{k+1}, \quad k \in \mathbb{N},\ k \geq 2,$$

and

$$X = \bigcup_{k=2}^{\infty} X_k.$$

Thus the sequence $\{X_k\}_{k=2}^{\infty}$ is a sequence of Jordan-measurable sets that monotonically exhausts $\mathbb{R}^2$. Let now

$$I_k = \int_{X_k} \sin((x^2+y^2)^2) dxdy, \quad k \in \mathbb{N}.$$

Then

$$\begin{aligned} I_k &= \int_0^{2\pi}\int_0^k |\sin(r^4)| r dr d\phi \\ &= 2\pi \int_0^k |\sin(r^4)| r dr \\ &= 2\pi \int_0^{k^4} |\sin t| t^{\frac{1}{4}} \frac{dt}{4t^{\frac{3}{4}}} \\ &= \frac{\pi}{2} \int_0^{k^4} \frac{|\sin t|}{\sqrt{t}} dt, \quad k \in \mathbb{N}, \end{aligned}$$

where we have used the polar coordinates

$$x = r\cos\phi,$$
$$y = r\sin\phi, \quad r > 0,\ \phi \in [0, 2\pi].$$

Now applying Theorem 3.1, we get

$$\begin{aligned} I &= \lim_{k\to\infty} I_k \\ &= \lim_{k\to\infty} \int_0^{k^4} \frac{|\sin t|}{\sqrt{t}} dt \\ &= \int_0^{\infty} \frac{|\sin t|}{\sqrt{t}} dt. \end{aligned}$$

Since the integral

$$\int_0^{\infty} \frac{|\sin t|}{\sqrt{t}} dt$$

is divergent, we conclude that I is divergent.

Exercise 3.2. Investigate for convergence the following integrals:

1. $$\iint_X \frac{1}{(x^2+y^2)^{\frac{\alpha}{2}}} dxdy, \quad \alpha \in \mathbb{R},$$

   where

   $$X = \{(x,y) \in \mathbb{R}^2 : x^2 + y^2 > 1\}.$$

2. $$\iint_{\mathbb{R}^2} \frac{1}{(1+x^2+xy+y^2)^{\alpha}} dxdy, \quad \alpha \in \mathbb{R}.$$

3. $$\iint_{\mathbb{R}^2} \frac{1}{(1+x^4+y^4)^{\alpha}} dxdy, \quad \alpha \in \mathbb{R}.$$

3.3 A comparison criterion

Sometimes, we can determine if an improper multiple integral is convergent or not without its evaluation. This can be done by comparisons with other integrals for which we know that they are convergent or divergent. To do this, we have a test for convergence or divergence, which can help us answer the question of convergence for an improper multiple integral.

Let $n \geq 2$, and let $X \subset \mathbb{R}^n$ be an open set.

Theorem 3.2. *Let $f, g : X \to \mathbb{R}$ be nonnegative functions such that*

$$f(x) \leq g(x), \quad x \in X. \tag{3.5}$$

Then from the convergence of the integral

$$\int_X g(x)dx \tag{3.6}$$

it follows the convergence of the integral

$$\int_X f(x)dx, \tag{3.7}$$

and from the divergence of the integral (3.7) *it follows the divergence of the integral* (3.6).

Proof. Let $\{X_k\}_{k=1}^{\infty}$ be a sequence of Jordan-measurable sets that monotonically exhausts the open set X. Then by (3.5) we get

$$\int_{X_k} f(x)dx \leq \int_{X_k} g(x)dx. \tag{3.8}$$

Suppose that (3.6) is convergent. Then

$$\lim_{k\to\infty} \int_{X_k} g(x)dx < \infty.$$

Now applying (3.8), we get

$$\lim_{k\to\infty} \int_{X_k} f(x)dx \leq \lim_{k\to\infty} \int_{X_k} g(x)dx$$
$$< \infty.$$

So the integral (3.7) is convergent. Let now the integral (3.7) be divergent. Then

$$\lim_{k\to\infty} \int_{X_k} f(x)dx = \infty.$$

From this and from (3.8) we arrive at

$$\begin{aligned}\infty &= \lim_{k\to\infty} \int_{X_k} f(x)dx \\ &\leq \lim_{k\to\infty} \int_{X_k} g(x)dx.\end{aligned}$$

Therefore (3.6) is divergent. This completes the proof. □

Example 3.5. Consider the integral

$$\int_{\mathbb{R}^2} \frac{e^{-x^2-y^2}}{1+x^2+y^2} dxdy. \tag{3.9}$$

We have that

$$\frac{e^{-x^2-y^2}}{1+x^2+y^2} \leq e^{-x^2-y^2}, \quad (x,y) \in \mathbb{R}^2.$$

By Example 3.2 we have that the integral

$$\int_X e^{-x^2-y^2} dxdy$$

is convergent. Applying Theorem 3.2, we conclude that the integral (3.9) is convergent.

Example 3.6. Consider the integral

$$\int_X \frac{2+(\sin x)^2}{(x^2+y^2+z^2)^{\frac{\alpha}{2}}} dxdydz, \quad \alpha \in \mathbb{R}, \tag{3.10}$$

where X is the set in Example 3.3. We have that

$$\frac{2+(\sin x)^2}{(x^2+y^2+z^2)^{\frac{\alpha}{2}}} \geq \frac{1}{(x^2+y^2+z^2)^{\frac{\alpha}{2}}}, \quad (x,y,z) \in X.$$

By Example 3.3 it follows that the integral

$$\int_X \frac{1}{(x^2+y^2+z^2)^{\frac{\alpha}{2}}} dxdydz$$

is divergent for $\alpha \geq 3$. Now applying Theorem 3.2, we conclude that the integral (3.10) is divergent for $\alpha \geq 3$.

Example 3.7. Consider the integral

$$\int_{\mathbb{R}^2} (4 + |\sin((x^2 + y^2)^2)|)dxdy. \tag{3.11}$$

We have that

$$4 + |\sin((x^2 + y^2)^2)| \geq |\sin((x^2 + y^2)^2)|, \quad (x, y) \in \mathbb{R}^2.$$

By Example 3.4 we have that the integral

$$\int_{\mathbb{R}^2} |\sin((x^2 + y^2)^2)|dxdy$$

is divergent. Now applying Theorem 3.2, we conclude that the integral (3.11) is divergent.

Exercise 3.3. Let $f \in \mathscr{C}(\mathbb{R}^2)$ be such that

$$0 < m \leq f(x, y) \leq M, \quad (x, y) \in \mathbb{R}^2,$$

for some positive constants m and M such that $m \leq M$. Investigate for convergence the following integrals:

1.

$$\int_{\mathbb{R}^2} f(x, y)e^{-x^2-y^2}dxdy.$$

2.

$$\int_X \frac{f(x, y)}{(x^2 + y^2 + z^2)^{\frac{\alpha}{2}}}dxdydz, \quad \alpha \in \mathbb{R},$$

where

$$X = \{(x, y, z) \in \mathbb{R}^3 : x^2 + y^2 + z^2 < 1\}.$$

3.

$$\int_{\mathbb{R}^2} f(x, y)|\sin((x^2 + y^2)^2|dxdy.$$

3.4 Absolute convergence

Since most of the tests of convergence for improper multiple integrals are only valid for nonnegative functions, it is legitimate to wonder what happens to improper multiple integrals involving non-positive functions. First, notice that there is a very natural way of generating a positive number from a given number: just take the absolute value of the number. So consider a function $f(x)$ (not necessarily positive) defined on a set $X \subset \mathbb{R}^n$. Then let us consider the positive function $|f(x)|$ still defined on X. It is easy to see that both functions $f(x)$ and $|f(x)|$ exhibit the same kind of improper behavior. Therefore we may naturally ask what conclusion we can draw if we know something about the integral

$$\int_X f(x)dx.$$

To answer this question, in this section, we introduce the concept of absolute convergence for improper multiple integrals.

Suppose that $n \geq 2$ and $X \subset \mathbb{R}^n$ is an open set.

Definition 3.3. The improper integral

$$\int_X f(x)dx \tag{3.12}$$

is said to be absolutely convergent if the integral

$$\int_X |f(x)|dx \tag{3.13}$$

is convergent.

For a function $f : X \to \mathbb{R}$ and $x \in X$, define

$$f_+(x) = \begin{cases} f(x) & \text{if } f(x) \geq 0, \\ 0 & \text{if } f(x) < 0, \end{cases}$$

$$f_-(x) = \begin{cases} -f(x) & \text{if } f(x) \geq 0, \\ 0 & \text{if } f(x) > 0. \end{cases}$$

For any $x \in X$, we have

$$\begin{aligned} 0 &\leq f_+(x) \\ &\leq |f(x)|, \\ 0 &\leq f_-(x) \\ &\leq |f(x)|, \\ f(x) &= f_+(x) - f_-(x), \\ |f(x)| &= f_+(x) + f_-(x). \end{aligned}$$

Theorem 3.3. *If the integral* (3.12) *is convergent, then the integral* (3.13) *is convergent.*

Proof. Assume that the integral (3.12) is absolutely divergent. Then for any sequence $\{X_k\}_{k=1}^{\infty}$ of Jordan-measurable sets that monotonically exhausts the set X, we have

$$\lim_{k\to\infty} \int\limits_{X_k} |f(x)|dx = \infty.$$

Without loss of generality, suppose that

$$\int\limits_{X_{k+1}} |f(x)|dx > 3 \int\limits_{X_k} |f(x)|dx + 2k, \quad k \in \mathbb{N}. \tag{3.14}$$

Set

$$A_k = X_{k+1} \backslash \overline{X_k}, \quad k \in \mathbb{N}.$$

Then the A_k, $k \in \mathbb{N}$, are open sets, and

$$X_{k+1} = A_k \cup \overline{X_k}, \quad k \in \mathbb{N},$$

and

$$\int\limits_{X_{k+1}} |f(x)|dx = \int\limits_{A_k} |f(x)|dx + \int\limits_{X_k} |f(x)|dx.$$

Now applying (3.14), we get

$$\begin{aligned}\int\limits_{A_k} |f(x)|dx + \int\limits_{X_k} |f(x)|dx &= \int\limits_{X_{k+1}} |f(x)|dx \\ &> 3 \int\limits_{X_k} |f(x)|dx + 2k, \quad k \in \mathbb{N},\end{aligned}$$

whereupon

$$\int\limits_{A_k} |f(x)|dx > 2 \int\limits_{X_k} |f(x)|dx + 2k, \quad k \in \mathbb{N},$$

or

$$\int\limits_{A_k} f_+(x)dx + \int\limits_{A_k} f_-(x)dx > 2 \int\limits_{X_k} |f(x)|dx + 2k, \quad k \in \mathbb{N}.$$

Now since

$$\int_{A_k} f_+(x)dx \geq \int_{A_k} f_-(x)dx, \quad k \in \mathbb{N},$$

we obtain

$$2\int_{A_k} f_+(x)dx > 2\int_{X_k} |f(x)|dx + 2k, \quad k \in \mathbb{N},$$

or

$$\int_{A_k} f_+(x)dx > \int_{X_k} |f(x)|dx + k, \quad k \in \mathbb{N}.$$

Let now $\tau_k = \{X_{jk}\}_{k=1}^{j_{\tau_k}}$ be a partitions of the sets A_k, $k \in \mathbb{N}$, such that $X_{jk}, j \in \{1, \ldots, j_{\tau_k}\}$, $k \in \mathbb{N}$, are Jordan-measurable sets and $\mu X_{jk} > 0, j \in \{1, \ldots, j_{\tau_k}\}$, $k \in \mathbb{N}$. Then

$$\sum_{j=1}^{j_{\tau_k}} f_+(\xi_{jk})\mu X_{jk} > \int_{X_k} |f(x)|dx + k, \quad \xi_{jk} \in X_{jk}, \ j \in \{1, \ldots, j_{\tau_k}\}, \ k \in \mathbb{N}. \tag{3.15}$$

Let X_{jk}^*, $j \in \{1, \ldots, j_{\tau_k}\}$, $k \in \mathbb{N}$, be the sets X_{jk}, $j \in \{1, \ldots, j_{\tau_k}\}$, $k \in \mathbb{N}$, by the partition τ_k, $k \in \mathbb{N}$, for which $f_+(\xi) > 0$ for all $\xi \in X_{jk}$, $j \in \{1, \ldots, j_{\tau_k}\}$, $k \in \mathbb{N}$. Let also,

$$\tau_k^* = \{X_{jk}^*\}_{j=1}^{j_{\tau_k}}, \quad k \in \mathbb{N}.$$

If

$$Y_j \notin \tau_k^*, \quad \mu Y_j > 0, \quad \text{then take} \quad \xi_j \in Y_j \quad \text{such that} \quad f(\xi_j) = 0, \tag{3.16}$$

$j \in \{1, \ldots, j_{\tau_k}\}$, $k \in \mathbb{N}$. Let $B_k \subset \{1, \ldots, j_{\tau_k}\}$, $k \in \mathbb{N}$, be such that $X_{jk} = X_{jk}^*$ for all $j \in B_k$. Then applying (3.15), we get

$$\sum_{j \in B_k} f_+(\xi_{jk})\mu X_{jk} > \int_{X_k} |f(x)|dx + k, \quad \xi_{jk} \in X_{jk}, \ j \in \{1, \ldots, j_{\tau_k}\}, \ k \in \mathbb{N}.$$

Set

$$C_k = \bigcup_{j \in B_k} X_{jk}^*, \quad k \in \mathbb{N}.$$

We have that C_k, $k \in \mathbb{N}$, are Jordan-measurable sets, $C_k \subset A_k$, $k \in \mathbb{N}$, and τ_k^*, $k \in \mathbb{N}$, are their partitions. On C_k, $k \in \mathbb{N}$, we have that $f_+ > 0$ and $f_+ = f$. By $s_{\tau_k}^*$, $k \in \mathbb{N}$, we will denote the lower Darboux sums on C_k. Then

$$s^*_{\tau_k} \geq \int\limits_{X_k} f(x)dx + k,$$

whereupon, taking $|\tau_k| \to 0$, we obtain

$$\int\limits_{C_k} f(x)dx \geq \int\limits_{X_k} f(x)dx + k, \quad k \in \mathbb{N}. \tag{3.17}$$

Because $f \geq -|f|$, we have

$$\int\limits_{X_k} f(x)dx \geq -\int\limits_{X_k} |f(x)|dx, \quad k \in \mathbb{N}.$$

Now applying (3.17), we arrive at

$$\int\limits_{C_k} f(x)dx + \int\limits_{X_k} f(x)dx \geq k, \quad k \in \mathbb{N}. \tag{3.18}$$

Set

$$D_k = C_k \cup X_k, \quad k \in \mathbb{N}.$$

We have that D_k, $k \in \mathbb{N}$, are Jordan-measurable sets and

$$X_k \subset D_k \subset X_{k+1}, \quad k \in \mathbb{N}.$$

Note that

$$C_k \cap X_k = \emptyset, \quad k \in \mathbb{N}.$$

Hence, applying (3.18), we get

$$\lim_{k\to\infty} \int\limits_{D_k} f(x)dx = \infty.$$

From here we conclude that

$$\int\limits_{X} f(x)dx = \infty.$$

This completes the proof. □

Remark 3.5. By Theorem 3.3 it follows that for $n \geq 2$, the convergence and absolute convergence for improper integrals are equivalent.

Example 3.8. By Theorem 3.3 we obtain that the integral

$$\int_{\mathbb{R}^2} \sin((x^2+y^2)^2)\,dxdy$$

is divergent, because by Example 3.4 we have that the integral

$$\int_{\mathbb{R}^2} |\sin((x^2+y^2)^2)|\,dxdy$$

is divergent.

Exercise 3.4. Investigate for convergence the following integrals:

1.
$$\iint_X \frac{\sin x \cos y}{(x-y)^p}\,dxdy, \quad p \in \mathbb{R},$$

   where

$$X = \{(x,y) \in \mathbb{R}^2 : x - y > 1\}.$$

2.
$$\iint_{R^2} \sin(x^4+y^4)\,dxdy.$$

3.
$$\iint_X \sin((x^2+y^2)^\alpha)\,dxdy, \quad \alpha \in \mathbb{R},$$

   where

$$X = \{(x,y) \in \mathbb{R}^2 : x^2+y^2 \geq 1\}.$$

4.
$$\iint_X \frac{1}{(x^2+y^2)^{\frac{\alpha}{2}}}\,dxdy, \quad \alpha \in \mathbb{R},$$

   where

$$X = \{(x,y) \in \mathbb{R}^2 : x^2+y^2 < 1\}.$$

5.

$$\iint_X \frac{1}{(x^2 - xy + y^2)^\alpha} dxdy, \quad \alpha \in \mathbb{R},$$

where

$$X = \{(x,y) \in \mathbb{R}^2 : |x| + |y| < 1\}.$$

6.

$$\iint_X \frac{1}{(1 - x^2 - y^2)^\alpha} dxdy, \quad \alpha \in \mathbb{R},$$

where

$$X = \{(x,y) \in \mathbb{R}^2 : x^2 + y^2 < 1\}.$$

7.

$$\iint_X \frac{1}{(a - x)^\alpha (x - y)^\beta} dxdy, \quad a, \alpha, \beta \in \mathbb{R},$$

where

$$X = \{(x,y) \in \mathbb{R}^2 : 0 < y < x < a\}.$$

8.

$$\iint_X \frac{1}{x^\alpha + y^\beta} dxdy, \quad \alpha, \beta \in \mathbb{R},\ \alpha, \beta > 0,$$

where

$$X = \{(x,y) \in \mathbb{R}^2 : x > 0,\ y > 0,\ x^\alpha + y^\beta < 1\}.$$

9.

$$\iint_X \frac{1}{(x^\alpha + y^\beta)^p} dxdy, \quad \alpha, \beta, p \in \mathbb{R},\ \alpha > 0,\ \beta > 0,\ p > 0,$$

where

$$X = \{(x,y) \in \mathbb{R}^2 : x > 0,\ y > 0,\ x + y < 1\}.$$

10.

$$\iint_X \frac{1}{(1-x^\alpha-y^\beta)^p}dxdy, \quad \alpha,\beta,p \in \mathbb{R},\ \alpha,\beta,p > 0,$$

where

$$X = \{(x,y) \in \mathbb{R}^2 : x > 0,\ y > 0,\ x^\alpha + y^\beta < 1\}.$$

3.5 Advanced practical problems

Problem 3.1.

1. Investigate for convergence the integral

$$\iint_X e^{-xy} \sin x dxdy,$$

where $X = [0,\infty)^2$.

2. Prove that the integrals

$$\int_0^\infty \left(\int_0^\infty e^{-xy} \sin x dx\right)dy \quad \text{and} \quad \int_0^\infty \left(\int_0^\infty e^{-xy} \sin x dy\right)dx$$

are convergent.

3. Prove that the limit

$$\lim_{a,b\to\infty} \int_0^a \int_0^b e^{-xy} \sin x dxdy$$

exists, find this limit, and compare it with the results in item 2.

Problem 3.2.

1. Investigate for convergence the integral

$$\iint_X \sin(x^2+y^2)dxdy,$$

where $X = [0,\infty)^2$.

2. Prove that the integrals

$$\int_0^\infty \left(\int_0^\infty \sin(x^2+y^2)dx\right)dy \quad \text{and} \quad \int_0^\infty \left(\int_0^\infty \sin(x^2+y^2)dy\right)dx$$

are convergent.

3. Prove that the limit

$$\lim_{a,b\to\infty}\int_0^a\int_0^b \sin(x^2+y^2)\,dxdy$$

exists, find this limit, and compare it with the results in item 2.
4. Find the limit

$$\lim_{k\to\infty}\iint_{X_k}\sin(x^2+y^2)\,dxdy,$$

where

$$X_k = \{(x,y)\in\mathbb{R}^2 : x^2+y^2<2\pi k,\ x>0,\ y>0\},\quad k\in\mathbb{N}.$$

Compare the obtained result with the result in item 3.

Problem 3.3.
1. Investigate for convergence the integral

$$\iint_X \frac{x^2-y^2}{(x^2+y^2)^2}\,dxdy,$$

where $X=[1,\infty)^2$.
2. Prove that the integrals

$$\int_1^\infty\left(\int_1^\infty \frac{x^2-y^2}{(x^2+y^2)^2}\,dx\right)dy \quad\text{and}\quad \int_1^\infty\left(\int_1^\infty \frac{x^2-y^2}{(x^2+y^2)^2}\,dy\right)dx$$

are convergent.

Problem 3.4. Investigate for convergence the following integrals:
1.

$$\iint_{\mathbb{R}^2}\frac{1}{(1+|x|^\alpha)(1+|y|^\beta)}\,dxdy,\quad \alpha,\beta\in\mathbb{R}.$$

2.

$$\iint_X \frac{1}{x^\alpha+y^\beta}\,dxdy,$$

where

$$X = \{(x,y) \in \mathbb{R}^2 : x > 0,\ y > 0,\ x^\alpha + y^\beta > 1\}, \quad \alpha, \beta > 0.$$

3.

$$\iint_X \frac{1}{(|x|^\alpha + |y|^\beta)^p} dxdy,$$

where

$$X = \{(x,y) \in \mathbb{R}^2 : |x| + |y| > 1\}, \quad \alpha, \beta > 0.$$

4.

$$\iint_X \frac{x^2}{(1 + x^2 + y^2)^p} dxdy, \quad p \in \mathbb{R},$$

where

$$X = \{(x,y) \in \mathbb{R}^2 : |y| < 1\}.$$

5.

$$\iint_X \frac{1}{(x+y)^p} dxdy, \quad p \in \mathbb{R},$$

where

$$X = \{(x,y) \in \mathbb{R}^2 : y > 1 + x^2\}.$$

6.

$$\iint_X \frac{1}{(x+y)^p} dxdy, \quad p \in \mathbb{R},$$

where

$$X = \{(x,y) \in \mathbb{R}^2 : x > 0,\ y > 0,\ x - y > 1\}.$$

Problem 3.5. Let $f \in \mathscr{C}(\mathbb{R}^2)$ be such that

$$0 < m \le f(x,y) \le M, \quad (x,y) \in \mathbb{R}^2,$$

for some positive constants m and M. Investigate for convergence the following integrals:

1.
$$\iint_{\mathbb{R}^2} \frac{f(x,y)}{(x^2+y^2+1)^{\alpha}} dxdy, \quad \alpha \in \mathbb{R}.$$

2.
$$\iint_X \frac{f(x,y)}{(x^4+y^4)^{\alpha}} dxdy, \quad \alpha \in \mathbb{R},$$

where
$$X = \{(x,y) \in \mathbb{R}^2 : 1 < y < 2\}.$$

3.
$$\iint_X \frac{f(x,y)}{(|x|+y)^{\alpha}} dxdy, \quad \alpha \in \mathbb{R},$$

where
$$X = \{(x,y) \in \mathbb{R}^2 : 1 + x^2 < y < 2 + x^2\}.$$

Problem 3.6. Let $f \in \mathscr{C}(\mathbb{R}^2)$ be such that

$$0 < m \le |f(x,y)| \le M, \quad (x,y) \in \mathbb{R}^2,$$

for some positive constants m and M such that $m \le M$. Investigate for convergence the following integrals:

1.
$$\iint_X \frac{f(x,y)}{|x-y|^p} dxdy, \quad p \in \mathbb{R},$$

where
$$X = \{(x,y) \in \mathbb{R}^2 : 0 \le x \le 1,\ 0 \le y \le 1\}.$$

2.
$$\iint_X \frac{f(x,y)}{(1-x-y)^p} dxdy,$$

where
$$X = \{(x,y) \in \mathbb{R}^2 : |x| + |y| < 1\}.$$

3.

$$\iint_X \frac{f(x,y)}{(1-x^2-y^2)^p}dxdy,$$

where

$$X = \{(x,y) \in \mathbb{R}^2 : x^2+y^2 < 1\}.$$

Problem 3.7. Let $f \in \mathscr{C}([a,b]\times[c,d])$ and $g \in \mathscr{C}([a,b])$. Prove that the integral

$$\int_a^b \int_c^d \frac{f(x,y)}{|y-g(x)|^p}dxdy$$

is convergent for $p < 1$.

Problem 3.8. Investigate for convergence the following integrals:

1.

$$\iint_X \frac{1}{(x^2+y^2)^\alpha}dxdy, \quad \alpha \in \mathbb{R},$$

where

$$X = \{(x,y) \in \mathbb{R}^2 : x > 0,\ x^2+y^2 < 1,\ 0 < y < x^2\}.$$

2.

$$\iint_X \frac{1}{(1-x^2-y^2)^\alpha}dxdy, \quad \alpha \in \mathbb{R},$$

where

$$X = \{(x,y) \in \mathbb{R}^2 : x > 0,\ y > 0,\ x+y < 1\}.$$

3.

$$\iint_X \frac{1}{(1-x^2-y^2)^\alpha}dxdy, \quad \alpha \in \mathbb{R},$$

where

$$X = \{(x,y) \in \mathbb{R}^2 : \sqrt{x}+\sqrt{y} < 1\}.$$

4.

$$\iint\limits_X \frac{y^2 - x^2}{(x^2 + y^2)^2} dxdy,$$

where

$$X = \{(x,y) \in \mathbb{R}^2 : 0 < x < y < 1\}.$$

5.

$$\iint\limits_X \frac{y^2 - x^2}{(x^2 + y^2)^2} dxdy,$$

where

$$X = \{(x,y) \in \mathbb{R}^2 : 1 < x < y < \infty\}.$$

6.

$$\iint\limits_X \frac{y^2 - x^2}{(x^2 + y^2)^2} dxdy,$$

where

$$X = \{(x,y) \in \mathbb{R}^2 : 0 < x < 1,\ 1 < y < \infty\}.$$

7.

$$\iint\limits_X \sin((x^2 + y^2)^\alpha) dxdy, \quad \alpha \in \mathbb{R},$$

where

$$X = \{(x,y) \in \mathbb{R}^2 : x^2 + y^2 \le 1\}.$$

Problem 3.9. Prove that the integral

$$\int\limits_X \frac{1}{|x|^\alpha} dx_1 \dots dx_n, \quad x = (x_1, \dots, x_n),\ \alpha \in \mathbb{R},$$

where

$$X = \{x = (x_1, \dots, x_n) \in \mathbb{R}^n : |x| > 1\},$$

is convergent for $\alpha > n$ and divergent for $\alpha \le n$.

Problem 3.10. Prove that the integral

$$\int\limits_X \frac{1}{|x|^\alpha} dx_1 \dots dx_n, \quad x = (x_1, \dots, x_n),\ \alpha \in \mathbb{R},$$

where

$$X = \{x = (x_1, \dots, x_n) \in \mathbb{R}^n : |x| < 1\},$$

is convergent for $\alpha < n$ and divergent for $\alpha \geq n$.

Problem 3.11. Let $f \in \mathscr{C}(\mathbb{R}^3)$ and

$$0 < m \leq |f(x, y, z)| \leq M, \quad (x, y, z) \in \mathbb{R}^3,$$

for some positive constants m and M such that $m \leq M$. Investigate for convergence the following integrals:

1.

$$\iiint\limits_X \frac{f(x, y, z)}{(x^2 + y^2 + z^2)^p} dxdydz, \quad p \in \mathbb{R},$$

where

$$X = \{(x, y, z) \in \mathbb{R}^3 : x^2 + y^2 + z^2 > 1\}.$$

2.

$$\iiint\limits_X \frac{1}{|x|^p + |y|^q + |z|^r} dxdydz, \quad p, q, r \in \mathbb{R},$$

where

$$X = \{(x, y, z) \in \mathbb{R}^3 : |x| + |y| + |z| > 1\}.$$

3.

$$\iiint\limits_X \frac{f(x, y, z)}{(x^2 + y^2 + z^2)^p} dxdydz, \quad p \in \mathbb{R},$$

where

$$X = \{(x, y, z) \in \mathbb{R}^3 : x^2 + y^2 + z^2 < 1\}.$$

4.

$$\iiint\limits_X \frac{f(x,y,z)}{(1-x^2-y^2-z^2)^p} dxdydz, \quad p \in \mathbb{R},$$

where

$$X = \{(x,y,z) \in \mathbb{R}^3 : x^2 + y^2 + z^2 < 1\}.$$

5.

$$\iiint\limits_X \frac{1}{|x-y-z|^p} dxdydz, \quad p \in \mathbb{R},$$

where

$$X = \{(x,y,z) \in \mathbb{R}^3 : |x| < 1,\ |y| < 1,\ |z| < 1,\ x \neq y + z\}.$$

6.

$$\iiint\limits_X \frac{1}{|t^2-x^2-y^2-z^2|^p} dxdydz, \quad t,p \in \mathbb{R},\ t > 0,$$

where

$$X = \{(x,y,z) \in \mathbb{R}^3 : \sqrt{x^2+y^2+z^2} < t < T < \infty\}$$

for some nonnegative constant T.

Problem 3.12. Let $f \in \mathscr{C}(X)$, $X = [0,a]^3$, $\phi, \psi \in \mathscr{C}([0,a])$, and $p < 1$. Prove that the integral

$$\iiint\limits_X \frac{f(x,y,z)}{(|y-\phi(x)|^2 + |z-\psi(x)|^2)^p} dxdydz$$

is convergent.

4 Line integrals

Line integrals or curvilinear integrals have many applications to engineering and physics. They also allow us to make several useful generalizations of the fundamental theorem of calculus and are closely connected to the properties of vector fields. A line integral gives enables us to integrate multivariable functions and vector fields over arbitrary curves in a plane or space.

In this chapter, we introduce line integrals of the first and second kinds. Some of their basic properties are deduced. The Green theorem is formulated and proved.

4.1 Line integrals of the first kind

A line integral enables us to integrate multivariable functions over arbitrary curves in a plane or space. There are two types line integrals: line integrals of the first kind and line integrals of the second kind.

For a line integral of the first kind, let C be a smooth curve in a plane or space, and let f be a function with a domain including the curve C. We divide the curve C into small pieces. For each piece, we choose a point P in that piece and evaluate f at P. We multiply $f(P)$ by the arc length of the piece Δs, add the product $f(P)\Delta s$ over all the pieces, and then let the arc length of the pieces shrink to zero by taking a limit.

For a formal description of a line integral of the first kind, let C be a smooth curve in space given by the parameterization

$$C : r(t) = (x(t), y(t), z(t)), \quad t \in [a, b].$$

Let $f = f(x, y, z)$ be a function with a domain including the curve C and divide the parameter interval $[a, b]$ into n subintervals $[t_{j-1}, t_j], j \in \{1, \dots, n\}$, where $t_0 = a$ and $t_n = b$ (see Figure 4.1). Let $t_j^* \in [t_{j-1}, t_j], j \in \{1, \dots, n\}$. Denote the end points of $r(t_0), r(t_1), \dots, r(t_n)$ by

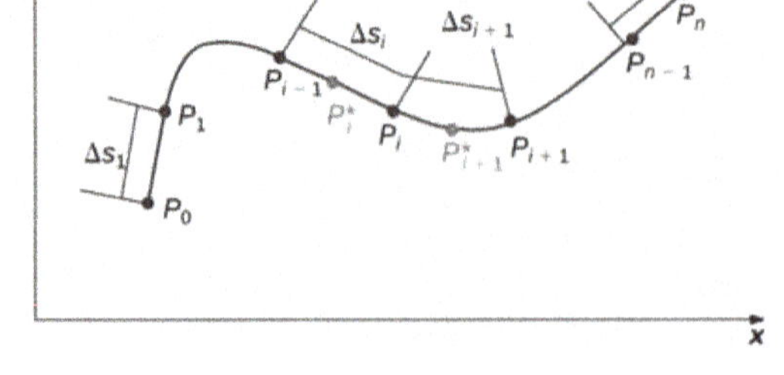

Figure 4.1: Curve C divided into n pieces.

https://doi.org/10.1515/9783112219607-004

$P_0, P_1, \ldots, P_n$, respectively. The points $P_j, j \in \{0, 1, \ldots, n\}$, divide the curve C into n pieces $C_1, C_2, \ldots, C_n$ with arc lengths $\Delta s_1, \Delta s_2, \ldots, \Delta s_n$, respectively. Let $P_j^* = r(t_j^*), j \in \{1, \ldots, n\}$. Now we form the sum

$$\sum_{j=1}^{n} f(P_j^*)\Delta s_j.$$

Definition 4.1. If the limit

$$\lim_{n\to\infty} \sum_{j=1}^{n} f(P_j^*)\Delta s_j$$

exists, it is called the line integral of the first kind and is denoted by

$$\int_C f(x,y,z)ds. \tag{4.1}$$

If C is a smooth plane curve represented by the parametric equations

$$x = x(t), \quad y = y(t), \quad t \in [a,b],$$

and $f(x,y)$ is a function of two variables, then the line integral of the first kind of f along C is defined similarly as

$$\int_C f(x,y)ds = \lim_{n\to\infty} \sum_{j=1}^{n} f(P_j^*)\Delta s_j, \tag{4.2}$$

if this limit exists.

Suppose that C is a smooth plane or space curve. If the integrals (4.1) and (4.2) exist, then they can be represented as follows:

$$\int_C f(x,y,z)ds = \int_a^b f(x(t),y(t),z(t))\sqrt{(x'(t))^2 + (y'(t))^2 + (z'(t))^2}dt$$

and

$$\int_C f(x,y)ds = \int_a^b f(x(t),y(t))\sqrt{(x'(t))^2 + (y'(t))^2}dt,$$

respectively.

Now we will examine the geometry captured by these integrals. Suppose that $f(x,y) \geq 0$ for all points (x,y) on a smooth plane curve. We project it “up” to the surface defined by $f(x,y)$. Thus we create a new curve C' that lies in the graph of $f(x,y)$ (see Figure 4.2).

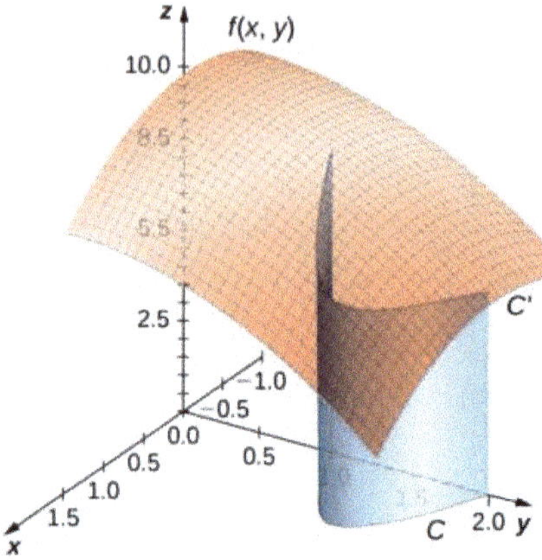

Figure 4.2: The area of the blue sheet is $\int_C f(x,y)ds$.

We drop a "sheet" from C' down to the (x,y)-plane. The area of this sheet is the integral (4.2). If $f(x,y) \leq 0$ for some points of C, then the value of (4.2) is the area above the (x,y)-plane minus the area below the (x,y)-plane. We see that the line integral (4.2) does not depend on the parameterization $r(t)$ of the curve C. Since the curve is traversed exactly once by the parameterization, the area of the sheet formed by the function and the curve is the same. This same kind of geometric argument can be extended to show that the line integral of a three-variable function over a curve in space does not depend on the parameterization of the curve.

Example 4.1. We will evaluate the integral

$$I = \int_C 4ds,$$

where C is the upper half of the unit circle (see Figure 4.3). Here

$$C : r(t) = (\cos t, \sin t), \quad t \in [0, \pi],$$

and

$$f(x,y) = 4.$$

Then

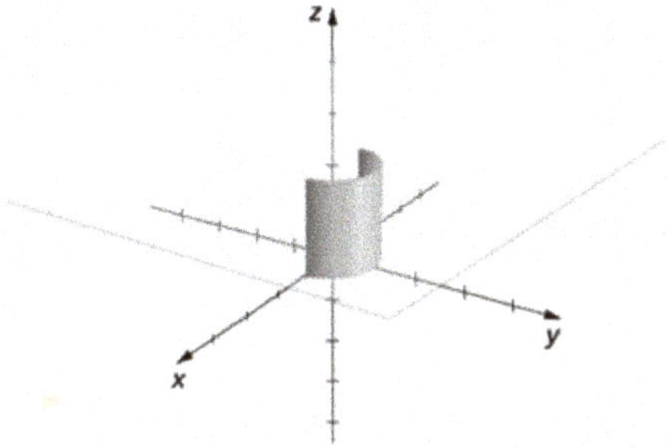

Figure 4.3: The upper half of the unit circle.

$$\begin{aligned} x(t) &= \cos t, \\ y(t) &= \sin t, \quad t \in [0, \pi], \\ x'(t) &= -\sin t, \\ y'(t) &= \cos t, \quad t \in [0, \pi], \end{aligned}$$

and

$$\begin{aligned} I &= 4 \int_0^{\pi} \sqrt{(-\sin t)^2 + (\cos t)^2} dt \\ &= 4 \int_0^{\pi} dt \\ &= 4\pi. \end{aligned}$$

Example 4.2. Consider the integral

$$I = \int_C \frac{1}{\sqrt{x^2 + y^2 + 4}} ds,$$

where C is the segment with end points $(0, 0)$ and $(1, 2)$. Note that the line through the points $(0, 0)$ and $(1, 2)$ has the equation

$$x = \frac{y}{2},$$

or

$$y = 2x.$$

Hence C is given by

$$y = 2x, \quad x \in [0, 1].$$

Therefore

$$\begin{aligned} I &= \int_0^1 \frac{1}{\sqrt{x^2+4x^2+4}} \sqrt{1+4}dx \\ &= \int_0^1 \frac{1}{\sqrt{4+(\sqrt{5}x)^2}} d(\sqrt{5}x) \\ &= \log(\sqrt{5}x + \sqrt{5x^2+4})\Big|_{x=0}^{x=1} \\ &= \log(\sqrt{5}+3) - \log 2 \\ &= \log\left(\frac{\sqrt{5}+3}{2}\right). \end{aligned}$$

Example 4.3. We will compute the integral

$$I = \int_C xyds,$$

where C is the boundary of the square with edges $A(1,0)$, $B(0,1)$, $C(-1,0)$, and $D(0,-1)$. The lines through the points A and B, B and C, C and D, and A and D have the following equations:

$$\begin{aligned} \frac{x-1}{-1} &= \frac{y-0}{1}, \\ \frac{x-0}{-1} &= \frac{y-1}{-1}, \\ \frac{x+1}{1} &= \frac{y-0}{-1}, \end{aligned}$$

and

$$\frac{x-1}{-1} = \frac{y-0}{-1},$$

respectively, or

$$\begin{aligned} y &= -x+1, \\ y &= x+1, \\ y &= -x-1, \end{aligned}$$

and

$$y = x-1,$$

respectively. Let

$$
\begin{aligned}
C_1 &= \{(x,y) \in \mathbb{R}^2 : y = -x+1,\ x \in [0,1]\},\\
C_2 &= \{(x,y) \in \mathbb{R}^2 : y = x+1,\ x \in [-1,0]\},\\
C_3 &= \{(x,y) \in \mathbb{R}^2 : y = -x-1,\ x \in [-1,0]\},\\
C_4 &= \{(x,y) \in \mathbb{R}^2 : y = x-1,\ x \in [0,1]\},
\end{aligned}
$$

and

$$C = C_1 \cup C_2 \cup C_3 \cup C_4.$$

Therefore

$$
\begin{aligned}
I &= \int_{C_1} xyds + \int_{C_2} xyds + \int_{C_3} xyds\\
&\quad + \int_{C_4} xyds\\
&= \sqrt{2}\int_0^1 x(-x+1)dx + \sqrt{2}\int_{-1}^0 x(x+1)dx + \sqrt{2}\int_{-1}^0 x(-x-1)dx\\
&\quad + \sqrt{2}\int_0^1 x(x-1)dx\\
&= \sqrt{2}\int_0^1 (-x^2+x+x^2-x)dx + \sqrt{2}\int_{-1}^0 (x^2+x-x^2-x)dx_1\\
&= 0.
\end{aligned}
$$

Example 4.4. We will compute the integral

$$I = \int_C \left(\sqrt{x^2+y^2} + z\right)ds,$$

where the curve C is given by the equations

$$
\begin{aligned}
x &= t\cos t,\\
y &= t\sin t,\\
z &= t, \quad t \in [0, 2\pi].
\end{aligned}
$$

We have

$$
\begin{aligned}
x'(t) &= \cos t - t\sin t,\\
y'(t) &= \sin t + t\cos t,
\end{aligned}
$$

$$z'(t) = 1, \quad t \in [0, 2\pi],$$

and

$$\begin{aligned}(x'(t))^2 + (y'(t))^2 + (z'(t))^2 &= (\cos t - t\sin t)^2 + (\sin t + t\cos t)^2 + 1\\ &= (\cos t)^2 - 2t\sin t\cos t + t^2(\sin t)^2\\ &\quad + (\sin t)^2 + 2t\sin t\cos t + t^2(\cos t)^2 + 1\\ &= 2 + t^2,\\ (x(t))^2 + (y(t))^2 &= (t\cos t)^2 + (t\sin t)^2\\ &= t^2(\cos t)^2 + t^2(\sin t)^2\\ &= t^2, \quad t \in [0, 2\pi].\end{aligned}$$

Therefore

$$\begin{aligned}I &= \int_0^{2\pi} \left(\sqrt{(x(t))^2 + (y(t))^2} + z(t)\right)\sqrt{(x'(t))^2 + (y'(t))^2 + (z'(t))^2}\,dt\\ &= \int_0^{2\pi} (\sqrt{t^2} + t)\sqrt{2 + t^2}\,dt\\ &= 2\int_0^{2\pi} t\sqrt{2 + t^2}\,dt\\ &= \int_0^{2\pi} \sqrt{2 + t^2}\,d(2 + t^2)\\ &= \frac{2}{3}(2 + t^2)^{\frac{3}{2}}\Big|_{t=0}^{t=2\pi}\\ &= \frac{2}{3}(2 + 4\pi^2)^{\frac{3}{2}} - \frac{2^{\frac{5}{2}}}{3}\\ &= \frac{2^{\frac{5}{2}}}{3}\left((1 + 2\pi^2)^{\frac{3}{2}} - 1\right).\end{aligned}$$

Exercise 4.1. Compute the following line integrals of the first kind:

1.

$$\int_C ds,$$

where C is the segment with end points $(0, 0)$ and $(1, 2)$.

2.

$$\int_C (2x + y)ds,$$

where C is the curve $ABOA$, and $A(1,0)$, $B(0,2)$, $O(0,0)$.

3.

$$\int_C \frac{z^2}{x^2+y^2}ds,$$

where C is the first loop of the helix

$$x = a\cos t,$$
$$y = a\sin t,$$
$$z = bt, \quad a, b > 0.$$

4.

$$\int_C zds,$$

where the curve C is given by

$$x = t\cos t,$$
$$y = t\sin t,$$
$$z = t, \quad t \in [0, 2\pi].$$

5.

$$\int_C zds,$$

where C is the arc of the curve

$$x^2 + y^2 = z^2,$$
$$y^2 = ax$$

from the point $(0,0,0)$ to the point $(a, a, a\sqrt{2})$, $a > 0$.

4.2 Line integrals of the second kind

To define the second type of line integrals, suppose that

$$F(x,y,z) = P(x,y,z)e_1 + Q(x,y,z)e_2 + R(x,y,z)e_3$$

is a continuous vector field in $\mathbb{R}^3$ that represents a force on a particle, and let C be a smooth curve in $\mathbb{R}^3$ contained in the domain of F. We will evaluate the work done by F in moving a particle along C. For this aim, assume that $r(t)$ is a parameterization of C for $t \in [a,b]$ such that the curve is traversed exactly once by the particle and the particle moves in the positive direction along C. Divide the parameter interval $[a,b]$ into n subintervals $[t_{j-1}, t_j]$, $j \in \{1,\dots,n\}$, of equal width. Denote the endpoints of $r(t_0)$, $r(t_1),\dots,r(t_n)$ by $P_0, P_1,\dots,P_n$, respectively. The points P_j, $j \in \{1,\dots,n\}$, divide the curve into n pieces. Denote the arc length of the piece from P_{j-1} to P_j by Δs_j, $j \in \{1,\dots,n\}$. For each $j \in \{1,\dots,n\}$, choose $t_j^* \in [t_{j-1}, t_j]$. The endpoint of $r(t_j^*)$ is a point in the piece of C between P_{j-1} and P_j (see Figure 4.4). If Δs_j, $j \in \{1,\dots,n\}$, are small, then as the particle moves from P_{j-1} to P_j along C, it moves approximately in the direction of $T(P_j)$, the unit tangent vector at the endpoint of $r(t_j^*)$. Let P_j^*, $j \in \{1,\dots,n\}$, denote the endpoint of $r(t_j^*)$. Then the work done by the force vector field in moving the particle from P_{j-1} to P_j is

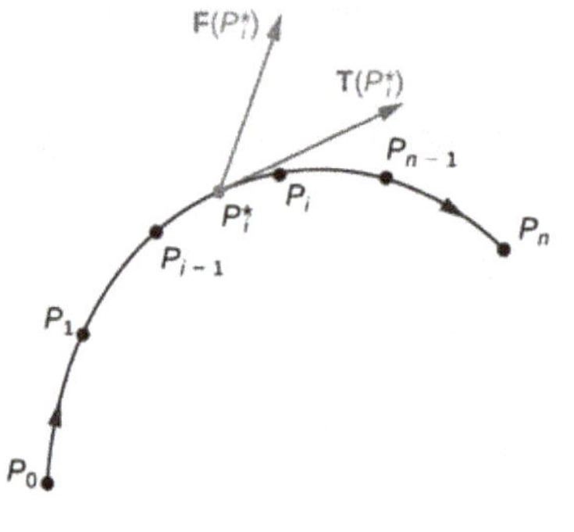

Figure 4.4: Curve C divided into n pieces.

$$\langle F(P_j^*), T(P_j^*)\Delta s_j\rangle, \quad j \in \{1,\dots,n\}.$$

So the total work is given by

$$\sum_{j=1}^{n}\langle F(P_j^*), T(P_j^*)\Delta s_j\rangle, \quad j \in \{1,\dots,n\}.$$

Letting the arc length of pieces of C get arbitrarily small by taking the limit as $n \to \infty$ gives the work done by the field in moving the particle along C. Therefore the work done by F in moving the particle in the positive direction along C is defined as follows:

$$W = \int_C \langle F, T \rangle ds,$$

which gives the concept of a line integral of the second type.

Definition 4.2. The line integral of the second type of the vector field F along an oriented smooth curve C is given by

$$\int_C \langle F, T \rangle ds = \lim_{n \to \infty} \sum_{j=1}^{n} \langle F(P_j^*), T(P_j^*) \Delta s_j \rangle, \quad j \in \{1, \ldots, n\},$$

if the limit exists. Sometimes, the line integrals of the second type are denoted by

$$\int_C P(x,y,z)dx + Q(x,y,z)dy + R(x,y,z)dz.$$

Let

$$\begin{aligned} r(t) &= (x(t), y(t), z(t)), \quad t \in [a, b], \\ \langle F, T \rangle ds &= \langle F(r(t)), T(t) \rangle |r'(t)| dt \\ &= \langle F(r(t)), r'(t) \rangle dt \\ &= (P(x(t), y(t), z(t))x'(t) + Q(x(t), y(t), z(t))y'(t) + R(x(t), y(t), z(t))z'(t))dt, \end{aligned}$$

and

$$\int_C \langle F, T \rangle ds = \int_a^b (P(x(t), y(t), z(t))x'(t) + Q(x(t), y(t), z(t))y'(t) + R(x(t), y(t), z(t))z'(t))dt.$$

Example 4.5. We will compute the following line integral of the second kind:

$$I = \int_C \frac{y}{x} dx + dy,$$

where Γ is the curve

$$y = \log x, \quad x \in [1, e].$$

Let

$$\begin{aligned} I_1 &= \int_C \frac{y}{x} dx, \\ I_2 &= \int_C dy. \end{aligned}$$

We find

$$\begin{aligned} I_1 &= \int_1^e \frac{\log x}{x} dx \\ &= \int_1^e \log x d \log x \\ &= \frac{(\log x)^2}{2} \Big|_{x=1}^{x=e} \\ &= \frac{1}{2} \end{aligned}$$

and

$$\begin{aligned} I_2 &= \int_1^e \frac{1}{x} dx \\ &= \log x \Big|_{x=1}^{x=e} \\ &= 1. \end{aligned}$$

Therefore

$$\begin{aligned} I &= I_1 + I_2 \\ &= \frac{1}{2} + 1 \\ &= \frac{3}{2}. \end{aligned}$$

Example 4.6. We will compute the integral

$$I = \int_C (x^2 + y^2) dx + (x^2 - y^2) dy,$$

where C is the curve

$$y = 1 - |x - 1|, \quad x \in [0, 2].$$

We have

$$\begin{aligned} y &= 1 - x + 1 \\ &= 2 - x, \quad x \in [1, 2], \end{aligned}$$

and

$$\begin{aligned} y &= 1 + x - 1 \\ &= x, \quad x \in [0, 1]. \end{aligned}$$

Let

$$I_1 = \int_C (x^2 + y^2)dx,$$
$$I_2 = \int_C (x^2 - y^2)dy.$$

Then

$$\begin{aligned} I_1 &= \int_0^1 (x^2 + x^2)dx + \int_1^2 (x^2 + (2 - x)^2)dx \\ &= 2\int_0^1 x^2dx + \int_1^2 (x^2 + x^2 - 4x + 4)dx \\ &= \frac{2}{3}x^3\Big|_{x=0}^{x=1} + \int_1^2 (2x^2 - 4x + 4)dx \\ &= \frac{2}{3} + \frac{2}{3}x^3\Big|_{x=1}^{x=2} - 2x^2\Big|_{x=1}^{x=2} + 4(2 - 1) \\ &= \frac{2}{3} + \frac{2}{3}(8 - 1) - 2(4 - 1) + 4 \\ &= \frac{2}{3} + \frac{14}{3} - 6 + 4 \\ &= \frac{16}{3} - 2 \\ &= \frac{10}{3}, \end{aligned}$$

and

$$\begin{aligned} I_2 &= \int_0^1 (x^2 - x^2)dx - \int_1^2 (x^2 - (2 - x)^2)dx \\ &= -\int_1^2 (x^2 - 4 + 4x - x^2)dx \\ &= -4\int_1^2 xdx + 4(2 - 1) \\ &= -2x^2\Big|_{x=1}^{x=2} + 4 \end{aligned}$$

$$
\begin{aligned}
&= -2(4-1) + 4 \\
&= -6 + 4 \\
&= -2.
\end{aligned}
$$

Therefore

$$
\begin{aligned}
I &= I_1 + I_2 \\
&= \frac{10}{3} - 2 \\
&= \frac{4}{3}.
\end{aligned}
$$

Example 4.7. Now we will compute the integral

$$
I = \int_C x dx + (x+y)dy + (x+y+z)dz,
$$

where C is the curve

$$
\begin{aligned}
x &= a \sin t, \\
y &= a \cos t, \\
z &= a(\sin t + \cos t), \quad t \in [0, 2\pi].
\end{aligned}
$$

Let

$$
\begin{aligned}
I_1 &= \int_C x dx, \\
I_2 &= \int_C (x+y)dy, \\
I_3 &= \int_C (x+y+z)dz.
\end{aligned}
$$

Then

$$
\begin{aligned}
I_1 &= a^2 \int_0^{2\pi} \sin t d \sin t \\
&= \frac{a^2}{2} (\sin t)^2 \Big|_{t=0}^{t=2\pi} \\
&= 0, \\
I_2 &= a^2 \int_0^{2\pi} (\sin t + \cos t) d \cos t
\end{aligned}
$$

$$\begin{aligned}
&= -a^2 \int_0^{2\pi} (\sin t)^2 dt + a^2 \int_0^{2\pi} \cos t d \cos t \\
&= -\frac{a^2}{2} \int_0^{2\pi} (1 - \cos(2t)) dt + a^2 \frac{(\cos t)^2}{2} \Big|_{t=0}^{t=2\pi} \\
&= -\frac{a^2}{2}(2\pi) + \frac{a^2}{4} \sin(2t) \Big|_{t=0}^{t=2\pi} \\
&= -a^2\pi,
\end{aligned}$$

and

$$\begin{aligned}
I_3 &= a^2 \int_0^{2\pi} (\sin t + \cos t + \sin t + \cos t) d(\sin t + \cos t) \\
&= 2a^2 \int_0^{2\pi} (\sin t + \cos t) d(\sin t + \cos t) \\
&= a^2 (\sin t + \cos t)^2 \Big|_{t=0}^{t=2\pi} \\
&= 0.
\end{aligned}$$

Therefore

$$\begin{aligned}
I &= I_1 + I_2 + I_3 \\
&= 0 - a^2\pi + 0 \\
&= -a^2\pi.
\end{aligned}$$

Exercise 4.2. Compute the following line integrals of the second kind:

1.

$$\int_C xy dx,$$

where Γ is the curve

$$y = \sin x, \quad x \in [0, \pi].$$

2.

$$\int_C \left(x - \frac{1}{y}\right) dy,$$

where C is the curve

$$y = x^2, \quad x \in [1, 2].$$

3.

$$\int_C x dy - y dx,$$

where C is the curve

$$y = x^3, \quad x \in [0, 2].$$

4.

$$\int_C 2xy dx + x^2 dy,$$

where C is the curve

$$y = \frac{x^2}{4}, \quad x \in [0, 2].$$

5.

$$\int_C 2xy dx - x^2 dy,$$

where C is the curve

$$y = \sqrt{\frac{x}{2}}, \quad x \in [0, 2].$$

6.

$$\int_C \cos y dx - \sin y dy,$$

where C is the curve

$$y = -x, \quad x \in [-2, 2].$$

7.

$$\int_C (xy - y^2) dx + x dy,$$

where C is the curve

$$y = 2\sqrt{x}, \quad x \in [0, 1].$$

8.

$$\int_C (x^2 - 2xy)dx + (y^2 - 2xy)dy,$$

where C is the curve

$$y = x^2, \quad x \in [-1, 1].$$

9.

$$\int_C ydx + zdy + xdz,$$

where C is the curve

$$\begin{aligned} x &= a\cos t, \\ y &= a\sin t, \\ z &= bt, \quad t \in [0, 2\pi], \; a, b > 0. \end{aligned}$$

10.

$$\int_C (y^2 - z^2)dx + 2yzdy - x^2dz,$$

where C is the curve

$$\begin{aligned} x &= t, \\ y &= t^2, \\ z &= t^3, \quad t \in [0, 1]. \end{aligned}$$

4.3 The Green formula

In calculus, the Green theorem relates a line integral around a simple closed curve C to a double integral over a plane region D bounded by C. The Green theorem extends the fundamental theorem of calculus to calculating double integrals. The Green theorem finds many applications in physics. One is solving two-dimensional flow integrals, stating that the sum of fluid outflowing from a volume is equal to the total outflow summed about an enclosing area. In plane geometry, and in particular, area surveying, the Green theorem can be used to determine the area and centroid of plane figures solely by integrating over the perimeter.

Let G be a bounded domain in $\mathbb{R}^2$ with boundary Γ that is a simple closed curve.

Definition 4.3. If the orientation of the curve Γ is chosen so that for each movement along the curve Γ corresponding to the chosen orientation, the domain G remains on the left, then we say that the orientation is positive. Otherwise, we say that the orientation is negative.

In Figure 4.5, a positive orientation is shown. In Figure 4.6, a negative orientation is shown. If Γ is positive oriented, then we will use the notation Γ^+. If Γ is negative oriented, then we will use the notation Γ^-.

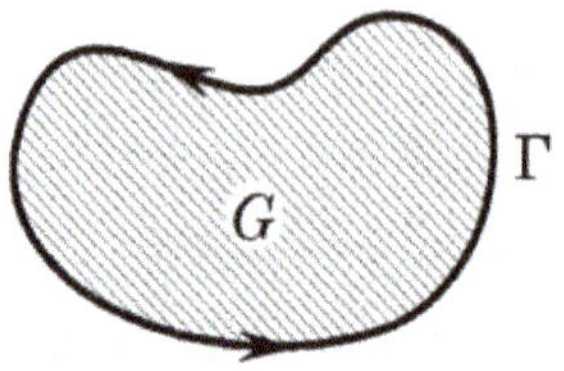

Figure 4.5: Positive orientation.

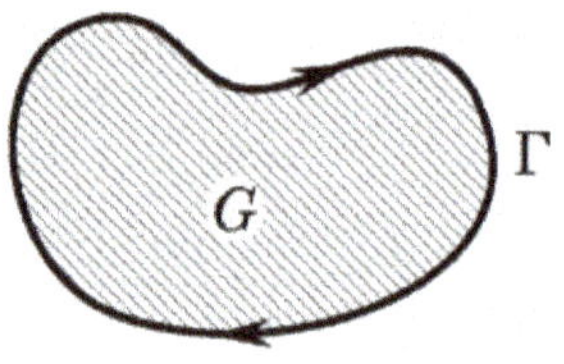

Figure 4.6: Negative orientation.

Definition 4.4. If the boundary Γ of the domain G can be represented as a union of continuous functions in x and segments parallel to Oy and continuous functions in y and segments parallel to Ox, then we say that G is an elementary domain.

Now we are ready to state and prove the Green theorem.

Theorem 4.1 (The Green theorem). *Suppose that the domain G is a union of finite number of elementary domains and its boundary Γ is a single closed curve. If $P, Q \in \mathscr{C}(\overline{G})$, $\frac{\partial P}{\partial y}$ and $\frac{\partial Q}{\partial x}$ exist on $\overline{G}$, and $\frac{\partial P}{\partial y}, \frac{\partial Q}{\partial x} \in \mathscr{C}(\overline{G})$, then we have the Green[1] formula*

1 George Green (14 July 1793–31 May 1841) was a British mathematical physicist who wrote An Essay on the Application of Mathematical Analysis to the Theories of Electricity and Magnetism in 1828. The essay introduced several important concepts, among them a theorem similar to the modern Green theorem, the idea of potential functions as currently used in physics, and the concept of what are now called the Green functions. Green was the first person to create a mathematical theory of electricity and magnetism,

$$\int_{\Gamma^+} Pdx + Qdy = \iint_G \left(\frac{\partial Q}{\partial x} - \frac{\partial P}{\partial y}\right)(x,y)dxdy. \tag{4.3}$$

Proof. Suppose that G is an elementary domain. Then its boundary Γ^+ is represented as a union of graphs of two continuous functions ϕ_1, ϕ_2 with respect to $x \in [a,b]$ and two segments $x = a, x = b$, and it can be represented as a union of two graphs of continuous functions ψ_1, ψ_2 with respect to $y \in [c,d]$ and two segments $y = c, y = d$ (see Figure 4.7). Then we have

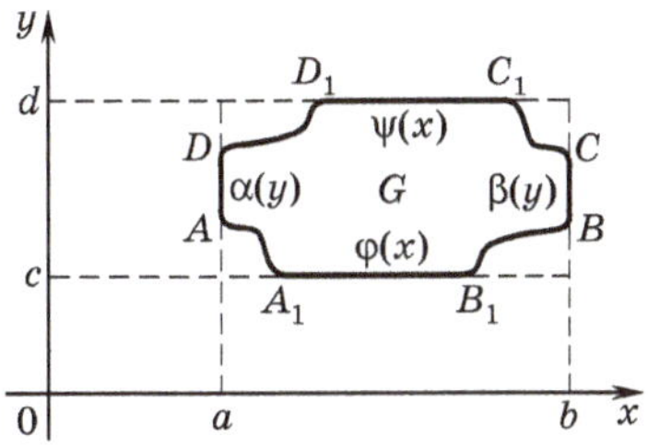

Figure 4.7: The domain G when it is an elementary domain.

$$\begin{aligned}\iint_G \frac{\partial P}{\partial y}(x,y)dxdy &= \int_a^b \left(\int_{\phi_1(x)}^{\phi_2(x)} \frac{\partial P}{\partial y}(x,y)dy\right)dx \\ &= \int_a^b (P(x,\phi_2(x)) - P(x,\phi_1(x)))dx \\ &= \int_{\widehat{DC}} P(x,y)dx - \int_{\widehat{AB}} P(x,y)dx \\ &= -\int_{\widehat{CD}} P(x,y)dx - \int_{\widehat{AB}} P(x,y)dx,\end{aligned}$$

i. e.,

$$\iint_G \frac{\partial P}{\partial y}(x,y)dxdy = -\int_{\widehat{CD}} P(x,y)dx - \int_{\widehat{AB}} P(x,y)dx. \tag{4.4}$$

Note that

---

and his theory formed the foundation for the work of other scientists such as James Clerk Maxwell, William Thomson, and others. His work on potential theory ran parallel to that of Carl Friedrich Gauss.

$$\int_{\widehat{BC}} P(x,y)dx = 0,$$

$$\int_{\widehat{DA}} P(x,y)dx = 0.$$

Now applying (4.4), we get

$$\begin{aligned}\iint_G \frac{\partial P}{\partial y}(x,y)dxdy &= -\int_{\widehat{CD}} P(x,y)dx - \int_{\widehat{AB}} P(x,y)dx \\ &\quad - \int_{\widehat{BC}} P(x,y)dx - \int_{\widehat{DA}} P(x,y)dx \\ &= -\int_{\Gamma^+} P(x,y)dx,\end{aligned}$$

whereupon

$$\int_{\Gamma^+} P(x,y)dx = -\iint_G \frac{\partial P}{\partial y}(x,y)dxdy. \tag{4.5}$$

Next,

$$\begin{aligned}\iint_G \frac{\partial Q}{\partial x}(x,y)dxdy &= \int_c^d \left(\int_{\psi_1(y)}^{\psi_2(y)} \frac{\partial Q}{\partial x}(x,y)dx \right) dy \\ &= \int_c^d (Q(\psi_2(y),y) - Q(\psi_1(y),y))dy \\ &= \int_c^d Q(\psi_2(y),y)dy - \int_c^d Q(\psi_1(y),y)dy \\ &= \int_{\widehat{B_1C_1}} Q(x,y)dy + \int_{\widehat{D_1A_1}} Q(x,y)dy \\ &= \int_{\widehat{B_1C_1}} Q(x,y)dy + \int_{\widehat{D_1A_1}} Q(x,y)dy \\ &\quad + \int_{\widehat{A_1B_1}} Q(x,y)dy + \int_{\widehat{C_1D_1}} Q(x,y)dy \\ &= \int_{\Gamma^+} Q(x,y)dy,\end{aligned}$$

where we have used that

$$\int_{\widehat{A_1B_1}} Q(x,y)dy = 0,$$

$$\int_{\widehat{C_1D_1}} Q(x,y)dy = 0.$$

Hence, applying (4.5), we get (4.3). Now suppose that

$$G = \bigcup_{j=1}^{k} G_j,$$

where $G_j, j \in \{1, \ldots, k\}$, are elementary domains with boundaries $\Gamma_j, j \in \{1, \ldots, k\}$. By the above computations we get

$$\int_{\Gamma_j^+} Pdx + Qdy = \int\int_{G_j} \left(\frac{\partial Q}{\partial x} - \frac{\partial P}{\partial y}\right)(x,y)dxdy, \quad j \in \{1, \ldots, k\}.$$

From here we get

$$\sum_{j=1}^{k} \int\int_{G_j} \left(\frac{\partial Q}{\partial x} - \frac{\partial P}{\partial y}\right)(x,y)dxdy = \sum_{j=1}^{k} \int_{\Gamma_j^+} Pdx + Qdy. \tag{4.6}$$

By the additivity property of the double integral we get

$$\sum_{j=1}^{k} \int\int_{G_j} \left(\frac{\partial Q}{\partial x} - \frac{\partial P}{\partial y}\right)(x,y)dxdy = \int_{G} \left(\frac{\partial Q}{\partial x} - \frac{\partial P}{\partial y}\right)(x,y)dxdy. \tag{4.7}$$

Now we consider the right-hand side of (4.6), where there are line integrals that appear twice on all interior parts of the boundaries Γ_j of the domains $G_j, j \in \{1, \ldots, k\}$, with opposite orientations, but they do not appear in the boundary of G. Then their sum equals 0 (see Figure 4.8). Hence, in the right-hand side of (4.7), there remain only positively oriented parts of the boundary Γ of the domain G, and their sum equals

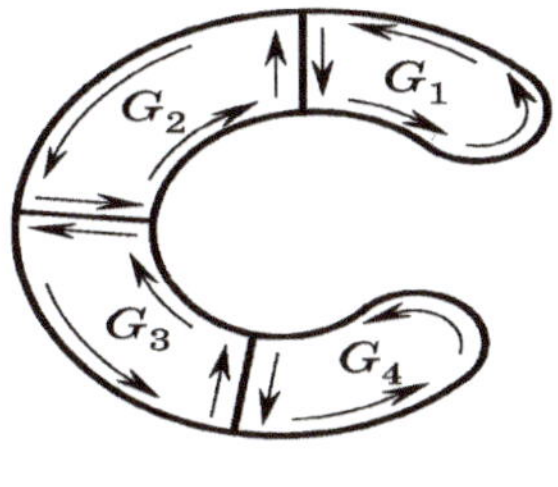

Figure 4.8: The domain G when it is an union of four elementary domains.

$$\int_{\Gamma^+} Pdx + Qdy.$$

Now applying (4.6) and (4.7), we obtain (4.3). This completes the proof. □

Corollary 4.1. *Suppose that all conditions of Theorem 4.1 hold. If in addition, Γ is partially smooth and $(\cos\alpha, \sin\alpha)$ is the unit tangent vector at each point of Γ, then we have*

$$\int_{\Gamma^+} (P\cos\alpha + Q\sin\alpha)ds = \int\int_G \left(\frac{\partial Q}{\partial x} - \frac{\partial P}{\partial y}\right)(x,y)dxdy,$$

where s is the arc length parameter of Γ.

Example 4.8. Using the Green formula, we will compute the integral

$$I = \int_\Gamma x^2ydx - xy^2dy,$$

where Γ is the circle

$$x^2 + y^2 = R^2, \quad R > 0.$$

Here

$$G = \{(x,y) \in \mathbb{R}^2 : x^2 + y^2 \le R^2\},$$

and

$$P(x,y) = x^2y,$$
$$Q(x,y) = -xy^2, \quad (x,y) \in G.$$

Then

$$\frac{\partial P}{\partial y}(x,y) = x^2,$$
$$\frac{\partial Q}{\partial x}(x,y) = -y^2, \quad (x,y) \in G.$$

Then applying the Green formula, we get

$$I = \int\int_G \left(\frac{\partial Q}{\partial x}(x,y) - \frac{\partial P}{\partial x}(x,y)\right)dxdy$$
$$= -\int\int_G (x^2 + y^2)dxdy$$

$$= -\int_0^{2\pi}\left(\int_0^R \rho^3 d\rho\right)d\phi$$

$$= -2\pi \frac{\rho^4}{4}\Big|_{\rho=0}^{\rho=R}$$

$$= -\frac{\pi R^4}{2},$$

where we have used the polar coordinates

$$x = \rho\cos\phi,$$
$$y = \rho\sin\phi, \quad \rho \in [0,R],\ \phi \in [0,2\pi].$$

Exercise 4.3. Using the Green formula, compute the following line integrals:

1.

$$\int_\Gamma (xy + x + y)dx + (xy + x - y)dy,$$

where Γ is the closed curve

$$\frac{x^2}{a^2} + \frac{y^2}{b^2} = 1, \quad a, b > 0,$$

oriented in positive direction.

2.

$$\int_\Gamma (2xy - y)dx + x^2 dy,$$

where Γ is the closed curve

$$\frac{x^2}{a^2} + \frac{y^2}{b^2} = 1, \quad a, b > 0,$$

oriented in positive direction.

3.

$$\int_\Gamma \frac{xdy + ydx}{x^2 + y^2},$$

where Γ is the closed curve

$$(x-1)^2 + (y-1)^2 = 1$$

oriented in positive direction.

4.

$$\int_{\Gamma} (x+y)^2 dx - (x^2+y^2) dy,$$

where Γ is the closed curve oriented in positive direction that is the boundary of the triangle with edges

$$(1,1), \quad (3,2), \quad (2,5).$$

5.

$$\int_{\Gamma} (y-x^2) dx + (x+y^2) dy,$$

where Γ is the closed curve oriented in positive direction that is the boundary of

$$\left\{(r,\phi) : 0 < r < R,\ 0 < \phi < \alpha \leq \frac{\pi}{2}\right\},$$

and (r,ϕ) are polar coordinates.

4.4 Applications of line integrals

Line integrals of the first kind have many applications. They can be used to compute the length or mass of a wire, the surface area of a sheet of a given height, or the electric potential of a charged wire given a linear charge density. Line integrals of the second kind are extremely useful in physics. They can be used to compute the work done on a particle as it moves through a force field, or the flow rate of a fluid across a curve.

Assume that a piece of wire is modeled by a curve C in space. The mass per unit length of the wire is a continuous function $\rho(x,y,z)$. The mass is density multiplied by length, and therefore the density of a small piece of the wire can be approximated by

$$\rho(x^*,y^*,z^*)\Delta s$$

for some point (x^*,y^*,z^*) in the piece. Letting the length of the pieces shrink to zero with a limit yields the line integral of the first kind

$$\int_C \rho(x,y,z) ds.$$

Example 4.9. We will compute the mass I of a spring in the shape of a curve parameterized by

$$(t, 2\cos t, 2\sin t), \quad t \in \left[0, \frac{\pi}{2}\right],$$

with density

$$\rho(x, y, z) = 2yz \text{ kg/m}$$

(see Figure 4.9). To do this, we must to compute the integral

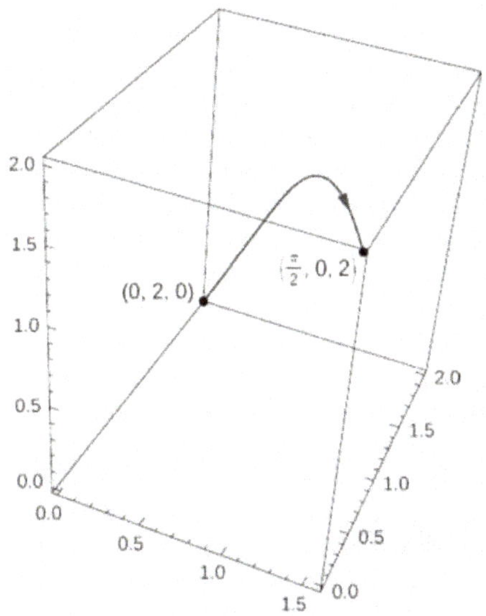

Figure 4.9: The wire from Example 4.9.

$$I = 2 \int_C yz ds.$$

Here

$$x(t) = t,$$
$$y(t) = 2\cos t,$$
$$z(t) = 2\sin t, \quad t \in \left[0, \frac{\pi}{2}\right].$$

Then

$$x'(t) = 1,$$
$$y'(t) = -2\sin t,$$
$$z'(t) = 2\cos t, \quad t \in \left[0, \frac{\pi}{2}\right],$$

and

$$\begin{aligned}\sqrt{(x'(t))^2 + (y'(t))^2 + (z'(t))^2} &= \sqrt{1 + (-2\sin t)^2 + (2\cos t)^2}\\ &= \sqrt{1 + 4((\sin t)^2 + (\cos t)^2)}\\ &= \sqrt{1+4}\\ &= \sqrt{5}.\end{aligned}$$

Thus

$$\begin{aligned}I &= 4\sqrt{5}\int_0^{\frac{\pi}{2}} \sin t \cos t dt\\ &= -4\sqrt{5}\int_0^{\frac{\pi}{2}} \cos t d(\cos t)\\ &= -2\sqrt{5}(\cos t)^2\Big|_{t=0}^{t=\frac{\pi}{2}}\\ &= 2\sqrt{5}.\end{aligned}$$

Exercise 4.4. Compute the mass of a spring in the shape of a helix parameterized by

$$(\cos t, \sin t, t), \quad t \in [0, 6\pi],$$

with a density function

$$\rho(x, y, z) = 3x + 2y + 4z.$$

Recall that if an object moves along a curve C parameterized by $r(t)$, $t \in [a, b]$, then the work required to move the object is given by

$$W = \int_a^b \langle F(r(t)), r'(t)\rangle dt.$$

Example 4.10. We will find how much work W is required to move an object in the vector force field

$$F = (yz, xy, xz)$$

along the path C parameterized by

$$r(t) = (t^2, t, t^4), \quad t \in [0, 1]$$

(see Figure 4.10). We have

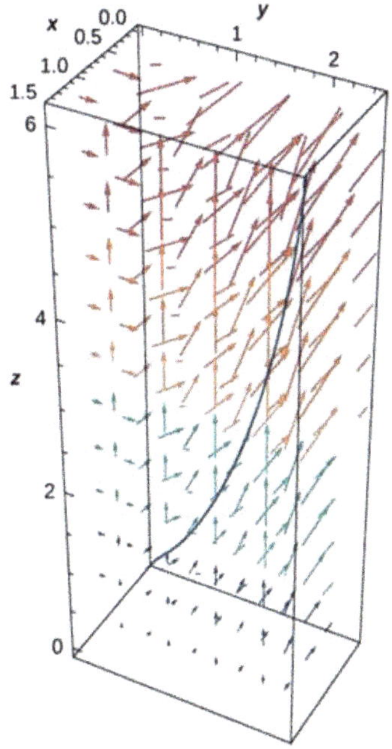

Figure 4.10: The curve and vector field in Example 4.10.

$$r'(t) = (2t, 1, 4t^3), \quad t \in [0, 1],$$

and

$$\begin{aligned}
W &= \int_0^1 \langle (t^5, t^3, t^6), (2t, 1, 4t^3) \rangle dt \\
&= \int_0^1 (2t^6 + t^3 + 4t^9) dt \\
&= 2\int_0^1 t^6 dt + \int_0^1 t^3 dt + 4\int_0^1 t^9 dt \\
&= \frac{2}{7}t^7\Big|_{t=0}^{t=1} + \frac{1}{4}t^4\Big|_{t=0}^{t=1} + \frac{2}{5}t^{10}\Big|_{t=0}^{t=1} \\
&= \frac{2}{7} + \frac{1}{4} + \frac{2}{5} \\
&= \frac{40 + 35 + 56}{140} \\
&= \frac{131}{140}.
\end{aligned}$$

Let Γ and G be as in Theorem 4.1. Suppose that

$$P(x,y) = 0,$$
$$Q(x,y) = x, \quad (x,y) \in G.$$

Then applying the Green formula, we find the area of G as follows:

$$\int\int_G dxdy = \int_{\Gamma^+} xdy,$$

whereupon

$$\mu G = \int_{\Gamma^+} xdy. \tag{4.8}$$

If we set

$$P(x,y) = -y,$$
$$Q(x,y) = 0, \quad (x,y) \in G,$$

then applying the Green formula, we arrive at

$$\int\int_G dxdy = -\int_{\Gamma^+} ydx,$$

or

$$\mu G = -\int_{\Gamma^+} ydx.$$

Hence by (4.8) we find

$$\mu G = \frac{1}{2}\int_{\Gamma^+} (xdy - ydx). \tag{4.9}$$

Example 4.11. We will find the area S bounded by the ellipse

$$\frac{x^2}{a^2} + \frac{y^2}{b^2} = 1, \quad a, b > 0,$$

using formula (4.9). We will employ the polar coordinates

$$x = a\cos t,$$
$$y = b\sin t, \quad t \in [0, 2\pi].$$

Then

$$\begin{aligned}
S &= \frac{1}{2}\int_0^{2\pi}(a\cos tb\cos tdt - a\sin t(-b\sin t)dt) \\
&= \frac{ab}{2}\int_0^{2\pi}((\sin t)^2 + (\cos t)^2)dt \\
&= \frac{ab}{2}(2\pi) \\
&= \pi ab.
\end{aligned}$$

Example 4.12. We will find the area S of the domain bounded by the astroid

$$\begin{aligned}
x &= a(\cos t)^3, \\
y &= b(\sin t)^3, \quad t \in [0, 2\pi], \quad a, b > 0,
\end{aligned}$$

using formula (4.9). We have

$$\begin{aligned}
S &= \frac{1}{2}\int_0^{2\pi}(a(\cos t)^3(3b(\sin t)^2\cos t)dt + b(\sin t)^3(3a(\cos t)^3\sin t)dt) \\
&= \frac{3ab}{2}\int_0^{2\pi}((\cos t)^4(\sin t)^2 + (\cos t)^2(\sin t)^4)dt \\
&= \frac{3ab}{2}\int_0^{2\pi}(\cos t)^2(\sin t)^2((\cos t)^2 + (\sin t)^2)dt \\
&= \frac{3ab}{8}\int_0^{2\pi}(\sin(2t))^2 dt \\
&= \frac{3ab}{16}\int_0^{2\pi}(1 - \cos(4t))dt \\
&= \frac{3ab}{16}\int_0^{2\pi}dt - \frac{3ab}{16}\int_0^{2\pi}\cos(4t)dt \\
&= \frac{3ab}{16}(2\pi) - \frac{3ab}{64}\sin(4t)\Big|_{t=0}^{t=2\pi} \\
&= \frac{3ab\pi}{8}.
\end{aligned}$$

Exercise 4.5. Find the area of the domain G, where

1. G is bounded by

$$\begin{aligned} y^2 &= 4 - x, \\ x &= 4, \\ y &= 1. \end{aligned}$$

2. G is bounded by

$$\begin{aligned} y &= 2x^2, \\ x - y + 1 &= 0. \end{aligned}$$

3. G is bounded by

$$\begin{aligned} y &= 1 - x^2, \\ x - y - 1 &= 0. \end{aligned}$$

4. G is bounded by

$$\begin{aligned} x &= t^2, \\ y &= t^3, \\ x &= 1, \quad t \in \mathbb{R}. \end{aligned}$$

5. G is bounded by

$$\begin{aligned} x &= a \cos t, \\ y &= b \sin t, \quad t \in \mathbb{R},\ a, b > 0. \end{aligned}$$

6. G is bounded by

$$\begin{aligned} x &= 12(\sin t)^3, \\ y &= 3(\cos t)^3, \quad t \in \mathbb{R}. \end{aligned}$$

7. G is bounded by

$$\begin{aligned} x &= a \sin(2\phi)(\cos\phi)^2, \\ y &= a \cos(2\phi)(\cos\phi)^2, \quad \phi \in \left[-\frac{\pi}{2}, \frac{\pi}{2}\right]. \end{aligned}$$

4.5 Independence of the line integrals on the path of integration

Suppose that G is a domain in $\mathbb{R}^2$. All curves that will be considered in this section will be supposed to be partially smooth. Assume that $P, Q \in \mathscr{C}(G)$. In this section, we will search for conditions under which the line integral of the second kind

$$\int_{\widehat{AB}} Pdx + Qdy, \tag{4.10}$$

where $A, B \in G$ are fixed points, does not depend on the curve that connects the points A and B. We will start with the following result.

Theorem 4.2. *The line integral* (4.10) *is independent of the curve that connects the points A and B if and only if*

$$\int_{\Gamma} Pdx + Qdy = 0 \tag{4.11}$$

for any closed curve Γ lying in G.

Proof. 1. Suppose that (4.11) holds for any closed curve Γ lying in G. In addition, assume that $\widehat{AB}_1$ and $\widehat{AB}_2$ are two curves that connect the points A and B (see Figure 4.11). Then the curve

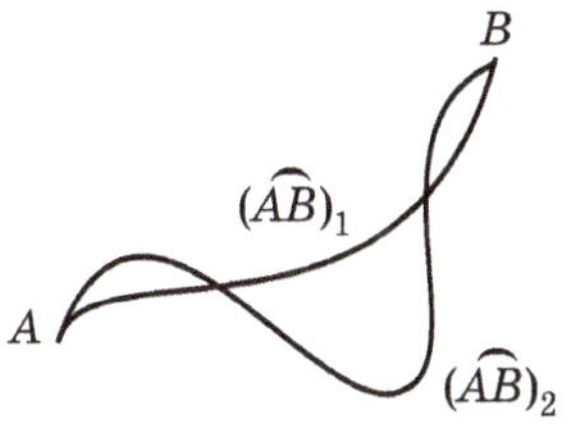

Figure 4.11: Two curves that connect the points A and B.

$$\widehat{AB}_1 \cup \widehat{BA}_2$$

is a closed curve lying in G. Now applying (4.11), we arrive at

$$\begin{aligned} 0 &= \int_{\widehat{AB}_1 \bigcup \widehat{BA}_2} Pdx + Qdy \\ &= \int_{\widehat{AB}_1} Pdx + Qdy + \int_{\widehat{BA}_2} Pdx + Qdy \\ &= \int_{\widehat{AB}_1} Pdx + Qdy - \int_{\widehat{AB}_2} Pdx + Qdy, \end{aligned}$$

whereupon

$$\int\limits_{\widehat{AB_1}} Pdx + Qdy = \int\limits_{\widehat{AB_2}} Pdx + Qdy.$$

2. Suppose that the line integral of the second kind (4.10) does not depend on the curve that connects the points A and B. Let Γ be an arbitrary closed curve lying in G. Take arbitrary $C, D \in \Gamma$. Then

$$\Gamma = \widehat{CD} \cup \widehat{DC},$$

and

$$\begin{aligned}\int\limits_{\Gamma} Pdx + Qdy &= \int\limits_{\widehat{CD}\bigcup\widehat{DC}} Pdx + Qdy \\ &= \int\limits_{\widehat{CD}} Pdx + Qdy + \int\limits_{\widehat{DC}} Pdx + Qdy \\ &= \int\limits_{\widehat{CD}} Pdx + Qdy - \int\limits_{\widehat{CD_1}} Pdx + Qdy \\ &= \int\limits_{\widehat{CD}} Pdx + Qdy - \int\limits_{\widehat{CD}} Pdx + Qdy \\ &= 0,\end{aligned}$$

i. e.,

$$\int\limits_{\Gamma} Pdx + Qdy = 0.$$

Here by $\widehat{CD_1}$ we have denoted the curve obtained from the curve $\widehat{DC}$ by the change of orientation. This completes the proof. □

Clearly, the theorem does not give a practical way to determine path independence, since it is impossible to check the line integrals around all possible closed curves in a region. What it mostly does is give an idea how line integrals behave and how seemingly unrelated line integrals can be related (in this case, a specific line integral between two points and all line integrals around closed curves).

Theorem 4.3. *The line integral* (4.10) *does not depend on the choice of the curve connecting the points A and B if and only if*

$$Pdx + Qdy$$

is a total differential of a function $u \in \mathscr{C}^1(G)$,

$$du = Pdx + Qdy. \tag{4.12}$$

If (4.12) *holds, then we have the equality*

$$\int_{\widehat{AB}} Pdx + Qdy = u(B) - u(A). \tag{4.13}$$

Proof. 1. Firstly, suppose that (4.12) holds. We will prove that the line integral (4.10) does not depend on the choice of the curve connecting the points A and B. Take arbitrary $M_0(z_1^0, z_2^0), M(z_1, z_2) \in G$ and suppose that $\widehat{M_0M}$ is a partially smooth curve that connects the points M_0 and M and lies in G (see Figure 4.12). Let

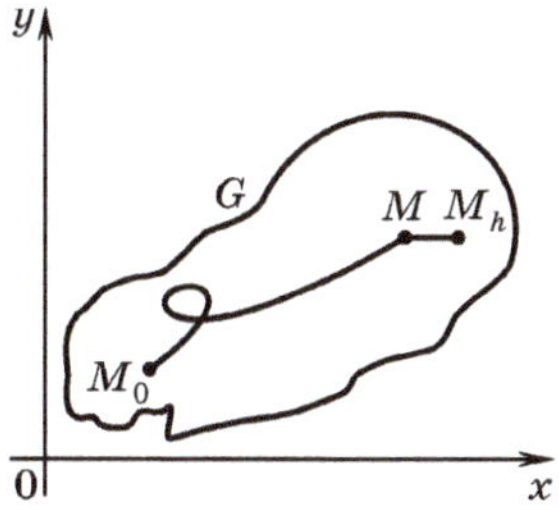

Figure 4.12: The curve M_0MM_*.

$$u(z_1, z_2) = \int_{\widehat{M_0M}} Pdx + Qdy.$$

Note that the function u is well defined since it does not depend on the choice of the curve that connects the points M_0 and M. Let $M_h = (z_1 + h, z_2) \in G$ for some $h \neq 0$ be such that the segment MM_h lies in G. Then

$$\begin{aligned} u(z_1 + h, z_2) - u(z_1, z_2) &= \int_{\widehat{M_0M_h}} Pdx + Qdy - \int_{\widehat{M_0M}} Pdx + Qdy \\ &= \int_{\widehat{MM_h}} Pdx + Qdy \\ &= \int_{z_1}^{z_1+h} P(t, z_2)dt. \end{aligned}$$

Now applying the mean value theorem, we find that there is $\theta \in (0, 1)$ such that

$$u(z_1 + h, z_2) - u(z_1, z_2) = P(z_1 + \theta h, z_2)h.$$

Hence

$$\frac{u(z_1 + h, z_2) - u(z_1, z_2)}{h} = P(z_1 + \theta h, z_2),$$

and

$$\begin{aligned}\frac{\partial u}{\partial z_1}(z_1, z_2) &= \lim_{h\to 0} \frac{u(z_1 + h, z_2) - u(z_1, z_2)}{h}\\ &= \lim_{h\to 0} P(z_1 + \theta h, z_2)\\ &= P(z_1, z_2).\end{aligned}$$

As above,

$$\frac{\partial u}{\partial z_2}(z_1, z_2) = Q(z_1, z_2).$$

Therefore there exists a function $u \in \mathscr{C}^1(G)$ such that (4.12) holds. Let now $\widehat{AB}$ be a curve given by

$$\begin{aligned}x &= x(t),\\ y &= y(t), \quad t \in [a, b].\end{aligned}$$

Then

$$\begin{aligned}A &= (x(a), y(a)),\\ B &= (x(b), y(b)),\end{aligned}$$

and

$$\begin{aligned}\int_{\widehat{AB}} Pdx + Qdy &= \int_a^b (P(x(t), y(t))x'(t) + Q(x(t), y(t))y'(t))dt\\ &= \int_a^b \left(\frac{\partial u}{\partial x}(x(t), y(t))x'(t) + \frac{\partial u}{\partial y}(x(t), y(t))y'(t)\right)dt\\ &= \int_a^b \frac{du}{dt}(x(t), y(t))dt\\ &= u(x(b), y(b)) - u(x(a), y(a))\\ &= u(B) - u(A),\end{aligned}$$

i. e., (4.13) holds.

2. Now we suppose that (4.12) and (4.13) hold. Let Γ be a closed curve with initial and end point A. Then

$$\begin{aligned}\int_{\Gamma} Pdx + Qdy &= u(A) - u(A)\\ &= 0.\end{aligned}$$

Hence, applying Theorem 4.2, we conclude that the integral (4.10) does not depend on the curve connecting the points A and B. This completes the proof. □

Before formulating the next criterion for path independence, we will prove the following auxiliary result, whose proof is similar to the proof of this criterion.

Theorem 4.4. *Let Γ be a smooth curve in G given by*

$$\begin{aligned}x &= x(t),\\ y &= y(t), \quad t \in [a,b].\end{aligned}$$

Let also, $\tau = \{t_j\}_{j=1}^{j_\tau}$ be a partition of the interval $[a,b]$, and let Λ_τ be an open polygon with edges $(x(t_j), y(t_j))$, $j \in \{0,1,\ldots,j_\tau\}$. Then

$$\lim_{|\tau|\to 0} \int_{\Lambda_\tau} Pdx + Qdy = \int_{\Gamma} Pdx + Qdy.$$

Proof. Firstly, observe that Γ is a compact set. Therefore there is a number $\eta > 0$ such that

$$d(\Gamma, \mathbb{R}^2 \backslash G) > \eta > 0.$$

Denote by Γ_η the set of all points $C \in \mathbb{R}^2$ such that

$$d(C, \Gamma) \le \eta.$$

We have that $\Gamma_\eta \subset G$ and Γ_η is bounded and closed. Since $x, y \in \mathscr{C}([a,b])$, we have that they are uniformly continuous on $[a,b]$. Then there is $\varepsilon > 0$ such that if $t^1, t^2 \in [a,b]$ satisfy the inequality

$$|t^1 - t^2| < \varepsilon,$$

then

$$d(M^1, M^2) < \eta,$$

where

$$M^1 = (x(t^1), y(t^1)),$$
$$M^2 = (x(t^2), y(t^2)).$$

Note that any points belonging to the segment M^1M^2 belong to Γ_η and G. If $|\tau| < \varepsilon$, then any point of Λ_τ belongs to G. Consider the integrals

$$\int_\Gamma P dx \quad \text{and} \quad \int_{\Lambda_\tau} P dx.$$

Set

$$\begin{aligned} x_1^j &= x(t_j), \\ x_2^j &= y(t_j), \\ P^j &= P(x_1^j, x_2^j), \\ \Delta x_1^j &= x_1^j - x_1^{j-1}, \quad j \in \{1, \ldots, j_\tau\}, \\ \sigma_\tau &= \sum_{j=1}^{j_\tau} P^j \Delta x_1^j. \end{aligned}$$

By the properties of the line integrals we have

$$\lim_{|\tau| \to 0} \sigma_\tau = \int_\Gamma P dx. \tag{4.14}$$

Denote

$$M_j = (x_1^j, x_2^j), \quad j \in \{0, 1, \ldots, j_\tau\}.$$

Then

$$\int_{\Lambda_\tau} P dx = \sum_{j=1}^{j_\tau} \int_{\widehat{M_{j-1}M_j}} P dx.$$

Next,

$$\begin{aligned} \int_{\widehat{M_{j-1}M_j}} dx &= \int_{\widehat{M_{j-1}M_j}} \cos a ds \\ &= |M_{j-1}M_j| \cos a \\ &= \Delta x_1^j, \quad j \in \{1, \ldots, j_\tau\}. \end{aligned}$$

Thus

$$\begin{aligned}\sigma_\tau &= \sum_{j=1}^{j_\tau} P^j \Delta x_1^j \\ &= \sum_{j=1}^{j_\tau} \int_{\widehat{M_{j-1}M_j}} P^j dx.\end{aligned}$$

Let L_τ be the length of the polygon Λ_τ, let S be the length of Γ, and let

$$\delta_\tau = \max_{j\in\{1,\dots,j_\tau\}} |M_{j-1}M_j|.$$

We have

$$\begin{aligned}\lim_{|\tau|\to 0} \delta_\tau &= 0, \\ L_\tau &\le S,\end{aligned}$$

and

$$\begin{aligned}\left|\int_{\Lambda_\tau} Pdx - \sigma_\tau\right| &= \left|\int_{\Lambda_\tau} Pdx - \sum_{j=1}^{j_\tau} \int_{\widehat{M_{j-1}M_j}} P^j dx\right| \\ &= \left|\sum_{j=1}^{j_\tau} \int_{\widehat{M_{j-1}M_j}} (P - P^j) dx\right| \\ &\le \sum_{j=1}^{j_\tau} \int_{\widehat{M_{j-1}M_j}} |P - P^j| dx \\ &\le \omega(|\tau|, P) \sum_{j=1}^{j_\tau} |\Delta x_1^j| \\ &\le \omega(|\tau|, P) L_\tau \\ &\le \omega(|\tau|, P) S.\end{aligned}$$

Since

$$\lim_{|\tau|\to 0} \omega(|\tau|, P) = 0,$$

applying (4.14), we get

$$\begin{aligned}\lim_{|\tau|\to 0} \int_{\Lambda_\tau} Pdx &= \lim_{|\tau|\to 0} \sigma_\tau \\ &= \int_{\Gamma} Pdx.\end{aligned}$$

As above, we get

$$\lim_{|\tau|\to 0} \int_{\Lambda_\tau} Qdy = \int_{\Gamma} Qdy.$$

This completes the proof. □

Definition 4.5. The domain G is said to be simply connected if any simple closed curve in the domain encloses only points in G.

The next criterion gives a practical way to determine path independence.

Theorem 4.5. *Suppose that $\frac{\partial P}{\partial y}$ and $\frac{\partial Q}{\partial x}$ exist and are continuous in G. Then, for the independence of the line integral* (4.10) *from the choice of the curve connecting the points A and B, it is necessary, and in the case where G is a simply connected domain, it is sufficient that*

$$\frac{\partial P}{\partial y} = \frac{\partial Q}{\partial x} \quad \text{in } G.$$

Proof. 1. Suppose that the line integral (4.10) does not depend on the curve connecting the points A and B. Then applying Theorem 4.3, we conclude that there is a function $u \in \mathscr{C}^1(G)$ such that

$$du = Pdx + Qdy$$

and

$$\frac{\partial u}{\partial x} = P,$$
$$\frac{\partial u}{\partial y} = Q \quad \text{in } G.$$

Since

$$\frac{\partial P}{\partial y} = \frac{\partial^2 u}{\partial y \partial x},$$
$$\frac{\partial Q}{\partial x} = \frac{\partial^2 u}{\partial x \partial y},$$

and $\frac{\partial P}{\partial y}, \frac{\partial Q}{\partial x} \in \mathscr{C}(G)$, we conclude that

$$\frac{\partial^2 u}{\partial y \partial x}, \frac{\partial^2 u}{\partial x \partial y} \in \mathscr{C}(G).$$

Therefore

$$\frac{\partial^2 u}{\partial y \partial x} = \frac{\partial^2 u}{\partial x \partial y} \quad \text{in } G,$$

and

$$\frac{\partial P}{\partial y} = \frac{\partial Q}{\partial x} \quad \text{in } G.$$

2. Suppose that G is simply connected and

$$\frac{\partial P}{\partial y} = \frac{\partial Q}{\partial x} \quad \text{in } G.$$

Let Γ be a closed polygon, and let D be a domain bounded by Γ. Then $D \subset G$. The functions P and Q are defined in

$$\overline{D} = D \cup \Gamma \subset G.$$

We represent D as a union of triangles. Then applying the Green formula, we get

$$\begin{aligned}\int_{\Gamma^+} Pdx + Qdy &= \int_D \left(\frac{\partial Q}{\partial x} - \frac{\partial P}{\partial y} \right) dxdy \\ &= 0.\end{aligned}$$

Therefore, for any closed polygon Γ, we have

$$\int_{\Gamma^+} Pdx + Qdy = 0. \tag{4.15}$$

Note that equality (4.15) holds for every polygon Γ, since it can be represented as a union of a finite number of closed polygons. Let now

$$\Gamma = \{r(t) : t \in [a, b]\}$$

be a partially smooth curve. Take a partition $\tau = \{t_j\}_{j=0}^{j_\tau}$ of the interval $[a, b]$, and let Λ_τ be a polygon with edges $r(t_j)$, $j \in \{0, \dots, j_\tau\}$, such that $\Lambda_\tau \subset G$. By the above we have

$$\int_{\Lambda_\tau} Pdx + Qdy = 0.$$

Hence by Theorem 4.4 we obtain

$$\begin{aligned}\int_{\Gamma} Pdx + Qdy &= \lim_{|\tau| \to 0} \int_{\Lambda_\tau} Pdx + Qdy \\ &= 0.\end{aligned}$$

Now apply Theorem 4.2, we conclude that the line integral (4.10) does not depend on the choice of the curve connecting the points A and B. This completes the proof. □

Example 4.13. Consider the integral

$$I = \int_{\widehat{AB}} xdy + ydx,$$

where $A(-1,3)$, $B(2,2)$. Let

$$u(x,y) = xy, \quad (x,y) \in \mathbb{R}^2.$$

Then

$$\begin{aligned} \frac{\partial u}{\partial x}(x,y) &= y, \\ \frac{\partial u}{\partial y}(x,y) &= x, \quad (x,y) \in \mathbb{R}^2. \end{aligned}$$

Thus

$$du(x,y) = xdy + ydx, \quad (x,y) \in \mathbb{R}^2,$$

and

$$\begin{aligned} I &= u(B) - u(A) \\ &= 2 \cdot 2 - (-1)3 \\ &= 7. \end{aligned}$$

Exercise 4.6. Prove that the integrand of the following integral is a total differential and compute it:

$$\int_{\widehat{AB}} xdx + ydy,$$

where $A(-1,0)$, $B(-3,4)$.

4.6 Advanced practical problems

Problem 4.1. Compute the following line integrals of the first kind:

1.

$$\int_{\Gamma} (x+y)ds,$$

where Γ is the triangle with edges $(0,0)$, $(1,0)$, and $(0,1)$.

2.
$$\int_\Gamma \frac{1}{y-x} ds,$$

where Γ is the segment with end points $(0,-2)$ and $(4,0)$.

3.
$$\int_\Gamma xyds,$$

where Γ is the one fourth of the ellipse

$$\frac{x^2}{a^2} + \frac{y^2}{b^2} = 1$$

lying in the first quadrant.

4.
$$\int_\Gamma xyds,$$

where Γ is the boundary of the rectangle $(0,0)$, $(4,0)$, $(4,2)$, and $(0,2)$.

5.
$$\int_\Gamma x^2 ds,$$

where Γ is the arc of the circle

$$x^2 + y^2 = a^2, \quad y \geq 0.$$

6.
$$\int_\Gamma (x^2 + y^2)^n ds,$$

where Γ is the circle

$$x^2 + y^2 = a^2, \quad a \in \mathbb{R},\ a > 0.$$

7.
$$\int_\Gamma (x-y) ds,$$

where Γ is the circle

$$x^2 + y^2 = ax.$$

8.

$$\int_\Gamma \sqrt{x^2 + y^2}\,ds,$$

where Γ is the circle

$$x^2 + y^2 = ax.$$

9.

$$\int_\Gamma (x + y)\,ds,$$

where Γ is the four-petal lemniscate given in polar coordinates

$$r^2 = a^2 \cos(2\phi).$$

10.

$$\int_\Gamma x\sqrt{x^2 - y^2}\,ds,$$

where Γ is the four-petal lemniscate given in polar coordinates

$$r^2 = a^2 \cos(2\phi).$$

11.

$$\int_\Gamma |y|\,ds,$$

where Γ is the lemniscate

$$r^2 = a^2 \cos(2\phi).$$

12.

$$\int_\Gamma (x^{\frac{4}{3}} + y^{\frac{4}{3}})\,ds,$$

where Γ is the astroid

$$x^{\frac{2}{3}} + y^{\frac{2}{3}} = a^{\frac{2}{3}}.$$

13.

$$\int_{\Gamma} y ds,$$

where Γ is the arc of the cycloid

$$x = a(t - \sin t),$$
$$y = a(1 - \cos t), \quad t \in [0, 2\pi].$$

14.

$$\int_{\Gamma} y^2 ds,$$

where Γ is the arc of the cycloid

$$x = a(t - \sin t),$$
$$y = a(1 - \cos t), \quad t \in [0, 2\pi].$$

15.

$$\int_{\Gamma} (x^2 + y^2) ds,$$

where Γ is the curve

$$x = a(\cos t + t \sin t),$$
$$y = a(\sin t - t \cos t), \quad t \in [0, 2\pi].$$

16.

$$\int_{\Gamma} \sqrt{x^2 + y^2} ds,$$

where Γ is the curve

$$x = a(\cos t + t \sin t),$$
$$y = a(\sin t - t \cos t), \quad t \in [0, 2\pi].$$

Problem 4.2. Compute the following line integrals of the first kind:

1.

$$\int_{\Gamma} \frac{1}{x^2 + y^2 + z^2} ds,$$

where Γ is the first loop of the helix

$$\begin{aligned} x &= a\cos t, \\ y &= a\sin t, \\ z &= bt. \end{aligned}$$

2.

$$\int_\Gamma (x^2 + y^2 + z^2)ds,$$

where Γ is the first loop of the helix

$$\begin{aligned} x &= a\cos t, \\ y &= a\sin t, \\ z &= bt. \end{aligned}$$

3.

$$\int_\Gamma \sqrt{2y^2 + z^2}ds,$$

where Γ is the circle

$$\begin{aligned} x^2 + y^2 + z^2 &= a^2, \\ x &= y. \end{aligned}$$

4.

$$\int_\Gamma xyzds,$$

where Γ is the fourth of the circle

$$\begin{aligned} x^2 + y^2 + z^2 &= a^2, \\ x &= y, \end{aligned}$$

lying in the first octant.

5.

$$\int_\Gamma (x + y)ds,$$

where Γ is the fourth of the circle

$$x^2 + y^2 + z^2 = a^2,$$
$$x = y,$$

lying in the first octant.

6.

$$\int_{\Gamma} x^2 ds,$$

where Γ is the fourth of the circle

$$x^2 + y^2 + z^2 = a^2,$$
$$x + y + z = 0,$$

lying in the first octant.

Problem 4.3. Compute the following line integrals of the second kind:

1.

$$\int_{\Gamma} xdy - ydx,$$

where Γ is the curve running from the point $A(0,0)$ to the point $B(1,2)$.

2.

$$\int_{\Gamma} xdy - ydx,$$

where Γ is the arc of the parabola running from the point $A(0,0)$ to the point $B(1,2)$.

3.

$$\int_{\Gamma} xdy - ydx,$$

where Γ is the curve ACB with $A(0,0)$, $B(1,2)$, and $C(0,1)$.

4.

$$\int_{\Gamma} xydx - y^2 dy,$$

where Γ is the arc of the parabola $y^2 = 2x$ running from the point $A(0,0)$ to the point $B(2,2)$.

5.

$$\int_{\Gamma} \frac{3x}{y}dx - \frac{2y^3}{x}dy,$$

where Γ is the arc of the parabola $x = y^2$ running from the point $A(4, 2)$ to the point $B(1, 1)$.

6.

$$\int_{\Gamma} \frac{x}{y}dx - \frac{y-x}{x}dy,$$

where Γ is the arc of the parabola $y = x^2$ running from the point $A(2, 4)$ to the point $B(1, 1)$.

7.

$$\int_{\Gamma} xdy,$$

where Γ is the arc of the circle

$$x^2 + y^2 = a^2, \quad x \geq 0,$$

running from the point $A(0, -a)$ to the point $B(0, a)$.

8.

$$\int_{\Gamma} x^3dy - xydx,$$

where Γ is the segment AB running from the point $A(0, -2)$ to the point $B(1, 3)$.

9.

$$\int_{\Gamma} -3x^2dx + y^3dy,$$

where Γ is the segment AB running from the point $A(0, 0)$ to the point $B(2, 4)$.

10.

$$\int_{\Gamma} (2x - y)dx + (4x + 5y)dy,$$

where Γ is the segment AB running from the point $A(3, -4)$ to the point $B(1, 2)$.

11.

$$\int_{\Gamma} (4x + 5y)dx + (2x - y)dy,$$

where Γ is the segment AB running from the point $A(1, -9)$ to the point $B(4, -3)$.

12.

$$\int_{\Gamma} \left(\frac{x}{x^2+y^2} + y \right) dx + \left(\frac{y}{x^2+y^2} + x \right) dy,$$

where Γ is the segment AB running from the point $A(1,0)$ to the point $B(3,4)$.

13.

$$\int_{\Gamma} (x+y)dx + (x-y)dy,$$

where Γ is the segment AB running from the point $A(0,1)$ to the point $B(2,3)$.

14.

$$\int_{\Gamma} xy^2 dx,$$

where Γ is the curve

$$x = \cos t,$$
$$y = \sin t, \quad t \in \left[0, \frac{\pi}{2}\right].$$

15.

$$\int_{\Gamma} xdy + ydx,$$

where Γ is the curve

$$x = R\cos t,$$
$$y = R\sin t, \quad t \in \left[0, \frac{\pi}{2}\right].$$

16.

$$\int_{\Gamma} ydx - xdy,$$

where Γ is the curve

$$x = a\cos t,$$
$$y = b\sin t, \quad t \in [0, 2\pi].$$

17.

$$\int_{\Gamma} y^2 dx + x^2 dy,$$

where Γ is the upper half of the ellipse

$$x = a\cos t,$$
$$y = b\sin t, \quad a, b > 0,\ t \in [0, 2\pi].$$

18.

$$\int_\Gamma (2a - y)dx + (y - a)dy,$$

where Γ is the curve

$$x = a(t - \sin t),$$
$$y = a(1 - \cos t), \quad t \in [0, 2\pi].$$

19.

$$\int_\Gamma \frac{x^2 dy - y^2 dx}{x^{\frac{5}{3}} + y^{\frac{5}{3}}},$$

where Γ is the curve

$$x = a(\cos t)^3$$
$$y = a(\sin t)^3, \quad t \in \left[0, \frac{\pi}{2}\right].$$

20.

$$\int_\Gamma (x^2 + y^2)dx,$$

where Γ is the boundary of the rectangle obtained by the lines

$$x = 1, \quad x = 3, \quad y = 1, \quad y = 5.$$

21.

$$\int_\Gamma (x^2 - 2xy)dx + (x - 2y)^2 dy,$$

where Γ is the boundary of the rectangle obtained by the lines

$$x = 0, \quad x = 2, \quad y = 0, \quad y = 1.$$

22.

$$\int_{\Gamma} (3x^2 - y)dx + (1 - 2x)dy,$$

where Γ is the boundary of the rectangle with edges $(0,0)$, $(1,0)$, $(1,1)$.

23.

$$\int_{\Gamma} (x^2 + y^2)dx + (x^2 - y^2)dy,$$

where Γ is the boundary of the rectangle with edges $(0,0)$, $(1,0)$, $(0,1)$.

24.

$$\int_{\Gamma} 2(x^2 + y^2)dx + (x + y)^2 dy,$$

where Γ is the boundary of the rectangle with edges $(1,1)$, $(1,3)$, $(2,2)$.

25.

$$\int_{\Gamma} \frac{dx + dy}{|x| + |y|},$$

where Γ is the boundary of the square with edges $(1,0)$, $(0,1)$, $(-1,0)$, $(0,-1)$.

26.

$$\int_{\Gamma} \frac{(x + y)dx + (y - x)dy}{x^2 + y^2},$$

where Γ is the circle

$$x^2 + y^2 = R^2, \quad R > 0.$$

27.

$$\int_{\Gamma} \frac{xy^2 dx - x^2 y dy}{x^2 + y^2},$$

where Γ is the first petal of the lemniscate

$$r^2 = a^2 \cos(2\phi), \quad a > 0.$$

28.

$$\int_{\Gamma} yzdx + z\sqrt{a^2 - y^2}dy + xydz,$$

where Γ is the curve

$$x = a\cos t,$$
$$y = a\sin t,$$
$$z = \frac{at}{2\pi}, \quad t \in [0, 2\pi],\ a > 0.$$

29.

$$\int_\Gamma (y+z)dx + (z+x)dy + (x+z)dz,$$

where Γ is the curve

$$x = a(\sin t)^2,$$
$$y = 2a\sin t\cos t,$$
$$z = a(\cos t)^2, \quad t \in [0, 2\pi],\ a > 0.$$

30.

$$\int_\Gamma ydx + zdy + xdz,$$

where Γ is the curve

$$x = a\cos\alpha\cos t,$$
$$y = a\cos\alpha\sin t,$$
$$z = a\sin\alpha, \quad a, \alpha > 0.$$

31.

$$\int_\Gamma xdx + ydy + (x+y-1)dz,$$

where Γ is the segment AB running from the point $A(1,1,1)$ to the point $B(2,3,4)$.

32.

$$\int_\Gamma \frac{xdx + ydy + zdz}{\sqrt{x^2+y^2+z^2-x-y+2z}},$$

where Γ is the segment AB running from the point $A(1,1,1)$ to the point $B(4,4,4)$.

33.

$$\int_\Gamma x(z-y)dx + y(x-z)dy + z(y-x)dz,$$

where Γ is the curve $ABCA$ with $A(a,0,0)$, $B(0,a,0)$, $C(0,0,a)$, $a > 0$.

Problem 4.4. Using the Green formula, compute the following line integrals:

1.

$$\int_{\Gamma} (xy + x + y)dx + (xy + x - y)dy,$$

where Γ is the closed curve

$$x^2 + y^2 = ax, \quad a > 0,$$

oriented in positive direction,

2.

$$\int_{\Gamma} e^x((1 - \cos y)dx + (\sin y - y)dy),$$

where Γ is the closed curve oriented in positive direction that is the boundary of

$$\{(x, y) \in \mathbb{R}^2 : 0 < x < \pi,\ 0 < y < \sin x\}.$$

3.

$$\int_{\Gamma} e^{y^2 - x^2}(\cos(2xy)dx + \sin(2xy)dy),$$

where Γ is the closed curve

$$x^2 + y^2 = R^2, \quad R > 0,$$

oriented in positive direction.

4.

$$\int_{\Gamma} (e^x \sin y - y)dx + (e^x \cos y - 1)dy,$$

where Γ is the closed curve oriented in positive direction that is the boundary of

$$\{(x, y) \in \mathbb{R}^2 : x^2 + y^2 < ax,\ y > 0\}, \quad a > 0.$$

5.

$$\int_{\Gamma} \frac{dx - dy}{x + y},$$

where Γ is the closed curve oriented in positive direction that is the boundary of the square with edges

$$(1,0), \quad (0,1), \quad (-1,0), \quad (0,-1).$$

6.

$$\int_{\Gamma} \sqrt{x^2+y^2}dx + y(xy + \log(x + \sqrt{x^2+y^2}))dy,$$

where Γ is the closed curve

$$x^2+y^2=R^2, \quad R>0,$$

oriented in positive direction.

7.

$$\int_{\Gamma} (x+y)^2 dx - (x-y)^2 dy,$$

where Γ is the closed curve oriented in positive direction that is the boundary of the domain obtained by the segment AB with $A(1,1)$ and $B(2,6)$ and the arc of the parabola

$$y = ax^2 + bx + c, \quad a,b,c \in \mathbb{R},$$

through the points A, B, and $O(0,0)$.

Problem 4.5. Find the area of the domain G, where

1. G is given by

$$\frac{x^2}{a^2} + \frac{y^2}{b^2} < 1,$$
$$\frac{x}{a} - \frac{y}{b} < \frac{\sqrt{3}-1}{2}, \quad a,b>0.$$

2. G is bounded by

$$(y-x)^2 + x^2 = 1.$$

3. G is bounded by

$$(x+y)^2 = ax,$$
$$y = 0.$$

4. G is bounded by

$$y^2 = x^2 - x^4.$$

5. G is bounded by

$$9y^2 = 4x^3 - x^4.$$

6. G is bounded by

$$(x^2 + y^2)^2 = a^2(x^2 - y^2), \quad x \geq 0, \ a > 0.$$

7. G is bounded by

$$(x^2 + y^2)^2 = 2ax^3, \quad a > 0.$$

8. G is bounded by

$$x^3 + y^3 = x^2 + y^2,$$
$$x = 0,$$
$$y = 0.$$

9. G is bounded by

$$x = \frac{3t}{1 + t^3},$$
$$y = \frac{3t^2}{1 + t^3}.$$

10. G is bounded by

$$x = a\cos\phi,$$
$$y = a\sin(2\phi), \quad x \geq 0.$$

11. G is bounded by

$$(\sqrt{x} + \sqrt{y})^{12} = xy.$$

Problem 4.6. Prove that the integrands of the following line integrals are total differentials and compute them

1.

$$\int_{\widehat{AB}} (x + y)dx + (x - y)dy,$$

where $A(2, -1)$ and $B(1, 0)$.

2.

$$\int_{\widehat{AB}} 2xydx + x^2dy,$$

where $A(0,0)$ and $B(-2,-1)$.

3.

$$\int_{\widehat{AB}} (x^4 + 4xy^3)dx + (6x^2y^2 - 5y^4)dy,$$

where $A(-2,-1)$ and $B(0,3)$.

4.

$$\int_{\widehat{AB}} (x^2 + 2xy - y^2)dx + (x^2 - 2xy - y^2)dy,$$

where $A(3,0)$ and $B(0,-3)$.

5.

$$\int_{\widehat{AB}} (3x^2 - 2xy + y^2)dx + (2xy - x^2 - 3y^2)dy,$$

where $A(-1,2)$ and $B(1,-2)$.

5 Surface integrals

In mathematics, particularly multivariable calculus, a surface integral is a generalization of multiple integrals to integration over surfaces. It can be thought of as the double integral analogue of the line integral. Given a surface, we may integrate over this surface a scalar field (that is, a function of position that returns a scalar as a value) or a vector field (that is, a function that returns a vector as a value). If a region R is not flat, then it is called a surface as shown in the illustration. Surface integrals have applications in physics, particularly with the theories of classical electromagnetism.

In this chapter, surface integrals of the first and second kinds are defined, and some of their properties are listed. The Gauss–Ostrogradsky and Stokes formulas are formulated and proved.

5.1 Surface integrals of the first kind

Surface integrals of the first kind are an analogue of line integrals of the first kind in one higher dimension. The domain of integration of a line integral of the first kind is a parameterized curve, whereas the domain of integration of a surface integral of the first kind is a parameterized surface. Therefore the definition of a surface integral of the first kind quite closely follows the definition of a line integral of the first kind.

Let S be a smooth surface with parameterization

$$r(u,v) = \big(x(u,v), y(u,v), z(u,v)\big), \quad (u,v) \in D,$$

and let $f(x,y,z)$ be a function with a domain that contains S. Without loss of generality, assume that D is a rectangle. We can extend the basic logic of how we proceed to any parameter domain. Divide the rectangle D into subrectangles D_{ij}, $i \in \{1,\dots,m\}$, $j \in \{1,\dots,l\}$, with horizontal width Δu and vertical high Δv. Thus D is subdivided into mn rectangles. The division of D into subrectangles gives the corresponding division of S into pieces S_{ij}, $i \in \{1,\dots,m\}$, $j \in \{1,\dots,l\}$. Choose points P_{ij}, $i \in \{1,\dots,m\}$, $j \in \{1,\dots,l\}$, and then multiply by area S_{ij} to form the Riemann sum

$$\sum_{i=1}^{m}\sum_{j=1}^{l} f(P_{ij})\Delta S_{ij}.$$

Definition 5.1. The surface integral of the first kind of a function f over S is defined by

$$\int_S f(x,y,z)dS = \lim_{m,l\to\infty} \sum_{i=1}^{m}\sum_{j=1}^{l} f(P_{ij})\Delta S_{ij},$$

provided that the limit exists.

https://doi.org/10.1515/9783112219607-005

To develop a method that makes surface integrals easier to compute, we approximate surface areas ΔS_{ij} with small pieces of the tangent plane. Recall the definition of r_u and r_v:

$$r_u(u,v) = (x_u(u,v), y_u(u,v), z_u(u,v)),$$
$$r_v(u,v) = (x_v(u,v), y_v(u,v), z_v(u,v)).$$

Then

$$\Delta S_{ij} \approx \|r_u(P_{ij}) \times r_v(P_{ij})\| \Delta u \Delta v.$$

We introduce the quantities

$$E = \left(\frac{\partial x}{\partial u}\right)^2 + \left(\frac{\partial y}{\partial u}\right)^2 + \left(\frac{\partial z}{\partial u}\right)^2,$$
$$F = \frac{\partial x}{\partial u}\frac{\partial x}{\partial v} + \frac{\partial y}{\partial u}\frac{\partial y}{\partial v} + \frac{\partial z}{\partial u}\frac{\partial z}{\partial v},$$
$$G = \left(\frac{\partial x}{\partial v}\right)^2 + \left(\frac{\partial y}{\partial v}\right)^2 + \left(\frac{\partial z}{\partial v}\right)^2.$$

Then the surface integral of the first kind can be rewritten in the following form:

$$\int_S f(x,y,z)dS = \int\int_D f(x(u,v), y(u,v), z(u,v)) \sqrt{E(u,v)G(u,v) - (F(u,v))^2} dudv. \tag{5.1}$$

If f is a continuous function along the surface S, then using that E, F, and G are continuous along the surface S, we conclude that the surface integral of the first kind (5.1) exists. Conditions for the existence of surface integrals of the first kind can be obtained by the corresponding conditions for the existence of double integrals. Since the surface integrals are reduced to double integrals, many properties of the double integrals are valid for the surface integrals of the first kind. If $f \equiv 1$, then

$$\mu S = \int\int_D \sqrt{EG - F^2} dudv$$
$$= \int_S dS.$$

Example 5.1. We will compute the integral

$$I = \int_S (x^2 + y)dS,$$

where S is the cylinder

$$x^2 + y^2 = 1, \quad 0 \le z \le 3$$

(see Figure 5.1). A parameterization of the given cylinder is

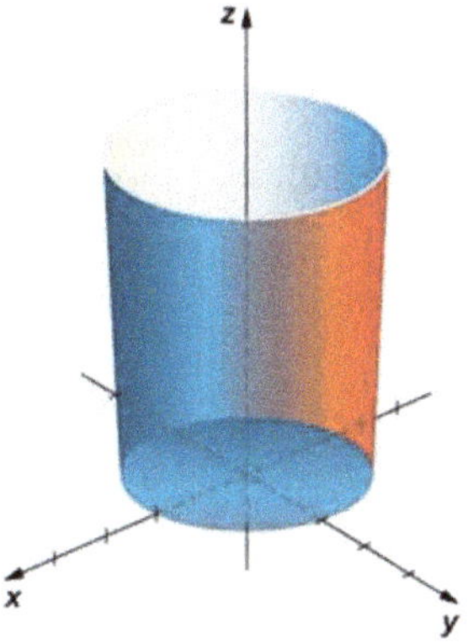

Figure 5.1: A cylinder.

$$r(u, v) = (\cos u, \sin u, v), \quad 0 \le u \le 2\pi,\ 0 \le v \le 3.$$

We have

$$\begin{aligned} r_u(u, v) &= (-\sin u, \cos u, 0), \\ r_v(u, v) &= (0, 0, 1), \end{aligned}$$

and

$$\begin{aligned} E &= (-\sin u)^2 + (\cos u)^2 \\ &= 1, \\ F &= 0, \\ G &= 1. \end{aligned}$$

Then

$$\begin{aligned} I &= \int_0^3 \int_0^{2\pi} ((\cos u)^2 + \sin u) du dv \\ &= 3 \int_0^{2\pi} \frac{1 + \cos(2u)}{2} du + 3 \int_0^{2\pi} \sin u du \\ &= \frac{3}{2}(2\pi) + \frac{3}{4} \sin(2u)\Big|_{u=0}^{u=2\pi} - 3 \cos u\Big|_{u=0}^{u=2\pi} \\ &= 3\pi. \end{aligned}$$

Example 5.2. Consider the integral

$$\int_S (x + y + z)dS,$$

where

$$S = \{(x, y, z) \in \mathbb{R}^3 : x + 2y + 4z = 4,\ x \geq 0,\ y \geq 0,\ z \geq 0\}.$$

By the definition of S we find

$$\begin{aligned} x &= 4 - 2y - 4z \\ &\geq 0 \end{aligned}$$

and

$$\begin{aligned} &0 \leq y \leq 2 - 2z, \\ &0 \leq z \leq 1. \end{aligned}$$

Thus, if we set

$$\begin{aligned} y &= u, \\ z &= v, \end{aligned}$$

then we get that S is given by

$$r(u, v) = (4 - 2u - 4v, u, v), \quad (u, v) \in D,$$

where

$$D = \{(u, v) \in \mathbb{R}^2 : 0 \leq u \leq 2 - 2v,\ 0 \leq v \leq 1\}.$$

Here

$$\begin{aligned} x(u, v) &= 4 - 2u - 2v, \\ y(u, v) &= u, \\ z(u, v) &= v, \quad (u, v) \in D. \end{aligned}$$

Then

$$\begin{aligned} \frac{\partial x}{\partial u}(u, v) &= -2, \\ \frac{\partial x}{\partial v}(u, v) &= -4, \end{aligned}$$

$$\frac{\partial y}{\partial u}(u,v) = 1,$$
$$\frac{\partial y}{\partial v}(u,v) = 0,$$
$$\frac{\partial z}{\partial u}(u,v) = 0,$$
$$\frac{\partial z}{\partial v}(u,v) = 1,$$

and

$$\begin{aligned}
E(u,v) &= \left(\frac{\partial x}{\partial u}(u,v)\right)^2 + \left(\frac{\partial y}{\partial u}(u,v)\right)^2 + \left(\frac{\partial z}{\partial u}(u,v)\right)^2 \\
&= (-2)^2 + 1^2 \\
&= 4 + 1 \\
&= 5, \\
F(u,v) &= \frac{\partial x}{\partial u}(u,v)\frac{\partial x}{\partial v}(u,v) + \frac{\partial y}{\partial u}(u,v)\frac{\partial y}{\partial v}(u,v) \\
&\quad + \frac{\partial z}{\partial u}(u,v)\frac{\partial z}{\partial v}(u,v) \\
&= -2(-4) + 1\cdot 0 + 1\cdot 0 \\
&= 8, \\
G(u,v) &= \left(\frac{\partial x}{\partial v}(u,v)\right)^2 + \left(\frac{\partial y}{\partial v}(u,v)\right)^2 + \left(\frac{\partial z}{\partial v}(u,v)\right)^2 \\
&= (-4)^2 + 1^2 \\
&= 16 + 1 \\
&= 17, \quad (u,v) \in D,
\end{aligned}$$

and

$$\begin{aligned}
E(u,v)G(u,v) - (F(u,v))^2 &= 5\cdot 17 - 8^2 \\
&= 85 - 64 \\
&= 21.
\end{aligned}$$

Therefore

$$\begin{aligned}
I &= \int_0^1 \int_0^{2-2v} (4 - 2u - 4v + u + v)\sqrt{21}dudv \\
&= \sqrt{21}\int_0^1 \int_0^{2-2v} (4 - u - 3v)dudv
\end{aligned}$$

$$
\begin{aligned}
&= \sqrt{21}\int_0^1 \left(4(2-2v) - \frac{1}{2}(2-2v)^2 - 3v(2-2v)\right)dv \\
&= \sqrt{21}\int_0^1 (8-8v-2(1-v)^2-6v+6v^2)dv \\
&= \sqrt{21}\int_0^1 (8-8v-2+4v-2v^2-6v+6v^2)dv \\
&= \sqrt{21}\int_0^1 (6-10v+4v^2)dv \\
&= \sqrt{21}\left(6v\Big|_{v=0}^{v=1} - 5v^2\Big|_{v=0}^{v=1} + \frac{4}{3}v^3\Big|_{v=0}^{v=1}\right) \\
&= \sqrt{21}\left(6-5+\frac{4}{3}\right) \\
&= \sqrt{21}\left(1+\frac{4}{3}\right) \\
&= \frac{7\sqrt{21}}{3}.
\end{aligned}
$$

Example 5.3. Consider the integral

$$I = \int_S (x^2+y^2)dS,$$

where S is the sphere

$$x^2+y^2+z^2 = R^2, \quad R > 0.$$

Consider the spherical coordinates

$$
\begin{aligned}
x(u,v) &= R\cos u \sin v, \\
y(u,v) &= R\sin u \sin v, \\
z(u,v) &= R\cos v, \quad u \in [0,2\pi],\ v \in [0,\pi].
\end{aligned}
$$

Then

$$
\begin{aligned}
\frac{\partial x}{\partial u}(u,v) &= -R\sin u \sin v, \\
\frac{\partial x}{\partial v}(u,v) &= R\cos u \cos v, \\
\frac{\partial y}{\partial u}(u,v) &= R\cos u \sin v,
\end{aligned}
$$

$$\frac{\partial y}{\partial v}(u,v) = R\sin u\cos v,$$
$$\frac{\partial z}{\partial u}(u,v) = 0,$$
$$\frac{\partial z}{\partial v}(u,v) = -R\sin v, \quad u \in [0,2\pi],\ v \in [0,\pi].$$

Hence

$$\begin{aligned}
E(u,v) &= \left(\frac{\partial x}{\partial u}(u,v)\right)^2 + \left(\frac{\partial y}{\partial u}(u,v)\right)^2 + \left(\frac{\partial z}{\partial u}(u,v)\right)^2\\
&= (-R\sin u\sin v)^2 + (\cos u\sin v)^2\\
&= R^2(\sin u)^2(\sin v)^2 + R^2(\cos v)^2(\sin v)^2\\
&= R^2(\sin v)^2,\\
F(u,v) &= \frac{\partial x}{\partial u}(u,v)\frac{\partial x}{\partial v}(u,v) + \frac{\partial y}{\partial u}(u,v)\frac{\partial y}{\partial v}(u,v)\\
&\quad + \frac{\partial z}{\partial u}(u,v)\frac{\partial z}{\partial v}(u,v)\\
&= -R\sin u\cos u\sin v\cos v + R\sin u\cos u\sin v\cos v\\
&= 0,\\
G(u,v) &= \left(\frac{\partial x}{\partial v}(u,v)\right)^2 + \left(\frac{\partial y}{\partial v}(u,v)\right)^2 + \left(\frac{\partial z}{\partial v}(u,v)\right)^2\\
&= (R\cos u\cos v)^2 + (R\sin u\cos v)^2 + (-R\sin v)^2\\
&= R^2(\cos u)^2(\cos v)^2 + R^2(\sin u)^2(\cos v)^2 + R^2(\sin v)^2\\
&= R^2(\cos v)^2 + R^2(\sin v)^2\\
&= R^2, \quad u \in [0,2\pi[,\ v \in [0,\pi],
\end{aligned}$$

and

$$\begin{aligned}
(x(u,v))^2 + (y(u,v))^2 &= (R\cos u\sin v)^2 + (R\sin u\sin v)^2\\
&= R^2(\cos u)^2(\sin v)^2 + R^2(\sin u)^2(\sin v)^2\\
&= R^2(\sin v)^2, \quad u \in [0,2\pi],\ v \in [0,\pi].
\end{aligned}$$

Consequently,

$$\begin{aligned}
I &= \int_0^{2\pi}\int_0^{\pi} R^2(\sin v)^2\sqrt{R^4(\sin v)^2}\,dv\,du\\
&= R^4\int_0^{2\pi}\int_0^{\pi}(\sin v)^3\,dv\,du
\end{aligned}$$

$$
\begin{aligned}
&= 2\pi R^4 \int_0^{\pi} (\sin v)^3 dv \\
&= -2\pi R^4 \int_0^{\pi} (\sin v)^2 d\cos v \\
&= -2\pi R^4 \int_0^{\pi} (1 - (\cos v)^2) d\cos v \\
&= -2\pi R^4 \left(\int_0^{\pi} d\cos v - \int_0^{\pi} (\cos v)^2 d\cos v \right) \\
&= -2\pi R^4 \left(\cos v\Big|_{v=0}^{v=\pi} - \frac{1}{3}(\cos v)^3 \Big|_{v=0}^{v=\pi} \right) \\
&= -2\pi R^4 \left(-1 - 1 - \frac{1}{3}(-1 - 1) \right) \\
&= -2\pi R^4 \left(-2 + \frac{2}{3} \right) \\
&= \frac{8\pi R^4}{3}.
\end{aligned}
$$

Example 5.4. We will compute the integral

$$I = \int_S (x^2 + y^2 + z^2) dS,$$

where S is the sphere

$$x^2 + y^2 + z^2 = R^2, \quad R > 0.$$

Consider the spherical coordinates given in Example 5.3. By the computations in Example 5.3 we find

$$
\begin{aligned}
I &= \int_0^{2\pi} \int_0^{\pi} R^2 \sqrt{R^4 (\sin v)^2} dv \\
&= 2\pi R^4 \int_0^{\pi} \sin v dv \\
&= -2\pi R^4 \cos v\Big|_{v=0}^{v=\pi} \\
&= 4\pi R^4.
\end{aligned}
$$

Exercise 5.1. Compute the following surface integrals of the first kind:

1.

$$\int_S (x + y + z)dS,$$

where S is the surface given by

$$x^2 + y^2 + z^2 = 1, \quad z \geq 0.$$

2.

$$\int_S (x^2 + y^2)dS,$$

where S is the surface given by

$$\sqrt{x^2 + y^2} \leq z \leq 1.$$

3.

$$\int_S (x^2 + y^2 + z^2)dS,$$

where S is the surface given by

$$|x| \leq a, \quad |y| \leq a, \quad |z| \leq a, \quad a > 0.$$

4.

$$\int_S \frac{1}{(1 + x + y)^2}dS,$$

where S is the surface given by

$$x + y + z \leq 1, \quad x \geq 0, \quad y \geq 0, \quad z \geq 0.$$

5.

$$\int_S xyzdS,$$

where S is the surface given by

$$z = x^2 + y^2, \quad z \leq 1.$$

5.2 Surface integrals of the second kind

Let S be an oriented surface with unit normal vector N, and let v be a velocity field of a fluid flowing through S. Suppose that the fluid has a density $\rho(x,y,z)$. Imagine the fluid flows through S and S is completely permeable, so that it does not impede the fluid flow (see Figure 5.2). The mass flux of the fluid is the rate of mass flow per unit area. The mass flux is measured in mass per unit time per unit area. The rate of flow, measured in mass per unit time per unit area, is ρN. To compute the mass flux across S, chop S into small pieces S_{ij}, and if S_{ij} is small enough, then it can be approximated by a tangent plane at some point P in S_{ij}. Consequently, the unit normal vector at P can be used to approximate $N(x,y,z)$ across the entire piece S_{ij}, because the normal vector to a plane does not change as we move across the plane. The component of the vector ρv at P in the direction N is $\langle \rho v, N\rangle$ at P. Since S_{ij} is small, the inner product $\langle \rho v, N\rangle$ changes very little as we vary across S_{ij}, and so $\langle \rho v, N\rangle$ can be taken as approximately constant across S_{ij}. The mass of the fluid per unit time flowing across S_{ij} in the direction N can be approximated by

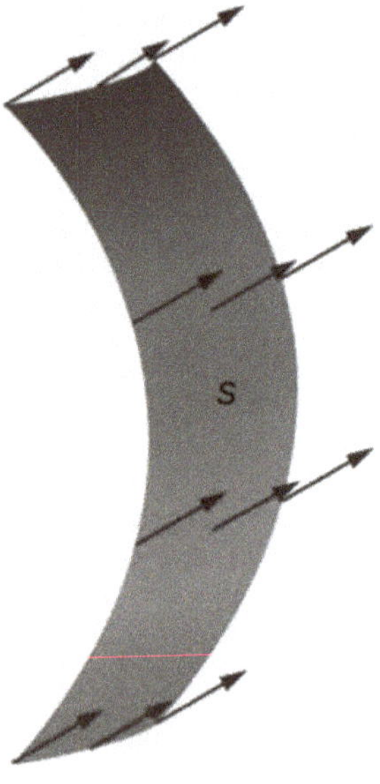

Figure 5.2: The fluid flow across a completely permeable surface S.

$$\langle \rho v, N\rangle \Delta S_{ij},$$

where N, ρ, and v are evaluated at P (see Figure 5.3). To approximate the mass flux across S, form the sum

$$\sum_{i=1}^{m}\sum_{j=1}^{l}\langle \rho v, N\rangle \Delta S_{ij}.$$

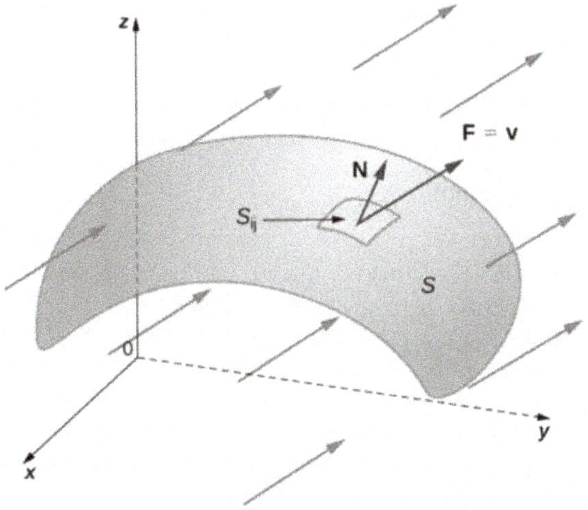

Figure 5.3: The mass of fluid per unit time flowing across S_{ij} in the direction N can be approximated by $\langle \rho v, N\rangle \Delta S_{ij}$.

As pieces S_{ij} get smaller, the last sum gets arbitrarily close to the mass flux. Thus the mass flux is

$$\lim_{m,l\to\infty} \sum_{i=1}^{m}\sum_{j=1}^{l} \langle \rho v, N\rangle \Delta S_{ij}.$$

Letting the vector field ρv to be an arbitrary vector field F leads to the following definition.

Definition 5.2. Let F be a continuous vector field with a domain that contains the oriented surface S with unit normal vector N. The surface integral of the second kind of F over S is defined as follows:

$$\int_S \langle F, N\rangle dS = \lim_{m,l\to\infty} \sum_{i=1}^{m}\sum_{j=1}^{l} \langle \rho v, N\rangle \Delta S_{ij},$$

provided that the limit exists.

Surface integrals of the second type are easier to compute after the surface S is parameterized. Let

$$r(u,v) = (x(u,v), y(u,v), z(u,v))$$

be a parameterization of S. Then

$$N = \frac{r_u \times r_v}{\|r_u \times r_v\|},$$

and

$$\int_S \langle F, N\rangle dS = \int\int_D \left\langle F, \frac{r_u \times r_v}{\|r_u \times r_v\|}\right\rangle \|r_u \times r_v\| dudv$$
$$= \int\int_D \langle F, r_u \times r_v\rangle dudv$$
$$= \int\int_D \det\begin{pmatrix} P(u,v) & Q(u,v) & R(u,v) \\ x_u(u,v) & y_u(u,v) & z_u(u,v) \\ x_v(u,v) & y_v(u,v) & z_v(u,v) \end{pmatrix} dudv.$$

Example 5.5. Consider the integral

$$I = \int_S (2z - x)dydz + (x + 2z)dzdx + 3zdxdy,$$

where S is given by

$$x + 4y + z = 4, \quad x \geq 0, \quad y \geq 0, \quad z \geq 0.$$

Introduce

$$x(u,v) = 4 - 4u - v,$$
$$y(u,v) = u,$$
$$z(u,v) = v, \quad (u,v) \in \mathbb{R}^2.$$

By the definition of the surface S we have

$$4 - 4u - v \geq 0,$$
$$u \geq 0,$$
$$v \geq 0,$$

or

$$0 \leq v \leq 4 - 4u,$$
$$0 \leq u \leq 1.$$

Thus S is given by

$$r(u,v) = (4 - 4u - v, u, v), \quad (u,v) \in D,$$

where

$$D = \{(u,v) \in \mathbb{R}^2 : u \in [0,1],\ v \in [0, 4 - 4u]\}.$$

Here

$$\begin{aligned}P(u,v) &= 2z(u,v) - x(u,v)\\ &= 2v - 4 + 4u + v\\ &= 4u + 3v - 4,\\ Q(u,v) &= x(u,v) + 2z(u,v)\\ &= 4 - 4u - v + 2v\\ &= 4 - 4u + v,\\ R(u,v) &= 3z(u,v)\\ &= 3v, \quad (u,v) \in D.\end{aligned}$$

Then

$$\begin{aligned}x_u(u,v) &= -4,\\ x_v(u,v) &= -1,\\ y_u(u,v) &= 1,\\ y_v(u,v) &= 0,\\ z_u(u,v) &= 0,\\ z_v(u,v) &= 1, \quad (u,v) \in D,\end{aligned}$$

and

$$\begin{aligned}\det\begin{pmatrix} P(u,v) & Q(u,v) & R(u,v)\\ x_u(u,v) & y_u(u,v) & z_u(u,v)\\ x_v(u,v) & y_v(u,v) & z_v(u,v)\end{pmatrix} &= \det\begin{pmatrix} 4u+3v-4 & 4-4u+v & 3v\\ -4 & 1 & 0\\ -1 & 0 & 1\end{pmatrix}\\ &= 4u + 3v - 4 + 3v + 4(4 - 4u + v)\\ &= 4u + 3v - 4 + 3v + 16 - 16u + 4v\\ &= 12 - 12u + 10v, \quad (u,v) \in D.\end{aligned}$$

Consequently,

$$\begin{aligned}I &= \int_0^1 \int_0^{4-4u} (12 - 12u + 10v)\,dv\,du\\ &= \int_0^1 \left(12(4-4u) - 12u(4-4u) + 5v^2\Big|_{v=0}^{v=4-4u}\right)du\\ &= \int_0^1 \left(48 - 48u - 48u + 48u^2 + 80(1-u)^2\right)du\end{aligned}$$

$$
\begin{aligned}
&= \int_0^1 (48 - 96u + 48u^2 + 80 - 160u + 80u^2)du \\
&= \int_0^1 (128 - 256u + 128u^2)du \\
&= 128 - 128u^2\Big|_{u=0}^{u=1} + \frac{128}{3}u^3\Big|_{u=0}^{u=1} \\
&= 128 - 128 + \frac{128}{3} \\
&= \frac{128}{3}.
\end{aligned}
$$

Example 5.6. Now we will compute the integral

$$\int_S (x^5 + z)dydz,$$

where S is the interior side of the semi-sphere

$$x^2 + y^2 + z^2 = R^2, \quad z \le 0.$$

Introduce the spherical coordinates

$$
\begin{aligned}
x(u,v) &= R\cos u \sin v, \\
y(u,v) &= R\sin u \sin v, \\
z(u,v) &= R\cos v, \quad u \in [0,2\pi], \ v \in [0,\pi].
\end{aligned}
$$

By the definition of S we have that $z \le 0$. Therefore $v \in [\frac{\pi}{2},\pi]$, and S can be represented in the form

$$r(u,v) = (R\cos u \sin v, R\sin u \sin v, R\cos v), \quad u \in [0,2\pi], \ v \in \left[\frac{\pi}{2},\pi\right].$$

Here

$$
\begin{aligned}
P(u,v) &= (x(u,v))^5 + z(u,v) \\
&= R^5(\cos u)^5(\sin v)^5 + R\cos v, \\
Q(u,v) &= 0, \\
R(u,v) &= 0, \quad u \in [0,2\pi], \ v \in \left[\frac{\pi}{2},\pi\right].
\end{aligned}
$$

We have

$$\begin{aligned}
x_u(u,v) &= -R\cos u\sin v,\\
x_v(u,v) &= R\cos u\cos v,\\
y_u(u,v) &= R\cos u\sin v,\\
y_v(u,v) &= R\sin u\cos v,\\
z_u(u,v) &= 0,\\
z_v(u,v) &= -R\sin v, \quad u\in[0,2\pi],\ v\in\left[\frac{\pi}{2},\pi\right],
\end{aligned}$$

and

$$\begin{aligned}
&\det\begin{pmatrix} P(u,v) & Q(u,v) & R(u,v)\\ x_u(u,v) & y_u(u,v) & z_u(u,v)\\ x_v(u,v) & y_v(u,v) & z_v(u,v)\end{pmatrix}\\
&= \det\begin{pmatrix} R^5(\cos u)^5(\sin v)^5 + R\cos v & 0 & 0\\ -R\sin u\sin v & R\cos u\sin v & 0\\ R\cos u\cos v & R\sin u\cos v & -R\sin v\end{pmatrix}\\
&= (R^5(\cos u)^5(\sin v)^5 + R\cos v)\det\begin{pmatrix} R\cos u\sin v & 0\\ R\sin u\cos v & -R\sin v\end{pmatrix}\\
&= (R^5(\cos u)^5(\sin v)^5 + R\cos v)(-R^2\cos u(\sin v)^2)\\
&= -R^7(\cos u)^6(\sin v)^7 - R^3\cos u\cos v(\sin v)^2, \quad u\in[0,2\pi],\ v\in\left[\frac{\pi}{2},\pi\right].
\end{aligned}$$

Hence

$$\begin{aligned}
I &= \int_0^{2\pi}\int_{\frac{\pi}{2}}^{\pi}(-R^7(\cos u)^6(\sin v)^7 - R^3\cos u\cos v(\sin v)^2)dvdu\\
&= -R^7\left(\int_0^{2\pi}(\cos u)^6du\right)\left(\int_{\frac{\pi}{2}}^{\pi}(\sin v)^7dv\right) - R^3\left(\int_0^{2\pi}\cos udu\right)\left(\int_{\frac{\pi}{2}}^{\pi}\cos v(\sin v)^2dv\right)\\
&= -R^7\left(\int_0^{2\pi}(\cos u)^6du\right)\left(\int_{\frac{\pi}{2}}^{\pi}(\sin v)^7dv\right) - R^3\left(\sin u\Big|_{u=0}^{u=2\pi}\right)\left(\int_{\frac{\pi}{2}}^{\pi}\cos v(\sin v)^2dv\right)\\
&= -R^7\left(\int_0^{2\pi}(\cos u)^6du\right)\left(\int_{\frac{\pi}{2}}^{\pi}(\sin v)^7dv\right).
\end{aligned}$$

Let

$$I_1 = \int_0^{2\pi}(\cos u)^6du,$$

$$I_2 = \int_{\frac{\pi}{2}}^{\pi} (\sin v)^7 dv.$$

We have

$$\begin{aligned}
I_1 &= \int_0^{2\pi} (\cos u)^5 d \sin u \\
&= (\cos u)^5 \sin u \Big|_{u=0}^{u=2\pi} + 5 \int_0^{2\pi} (\cos u)^4 (\sin u)^2 du \\
&= 5 \int_0^{2\pi} (\cos u)^4 (1 - (\cos u)^2) du \\
&= 5 \int_0^{2\pi} (\cos u)^4 du - 5 \int_0^{2\pi} (\cos u)^6 du \\
&= -5I_1 + 5 \int_0^{2\pi} (\cos u)^4 du \\
&= -5I_1 + 5 \int_0^{2\pi} (\cos u)^3 d \sin u \\
&= -5I_1 + 5(\cos u)^3 \sin u \Big|_{u=0}^{u=2\pi} + 15 \int_0^{2\pi} (\cos u)^2 (\sin u)^2 du \\
&= -5I_1 + \frac{15}{4} \int_0^{2\pi} (\sin(2u))^2 du \\
&= -5I_1 + \frac{15}{4} \int_0^{2\pi} \frac{1 - \cos(4u)}{2} du \\
&= -5I_1 + \frac{15}{8} \int_0^{2\pi} du - \frac{15}{8} \int_0^{2\pi} \cos(4u) du \\
&= -5I_1 + \frac{15}{8}(2\pi) - \frac{15}{32} \sin(4u) \Big|_{u=0}^{u=2\pi} \\
&= -5I_1 + \frac{15}{4}\pi,
\end{aligned}$$

whereupon

$$6I_1 = \frac{15}{4}\pi,$$

or

$$I_1 = \frac{5}{8}\pi.$$

Next,

$$\begin{aligned}
I_2 &= -\int_{\frac{\pi}{2}}^{\pi} (\sin v)^6 d\cos v \\
&= -\int_{\frac{\pi}{2}}^{\pi} \left(1 - (\cos v)^2\right)^3 d\cos v \\
&= -\int_{\frac{\pi}{2}}^{\pi} \left(1 - 3(\cos v)^2 + 3(\cos v)^4 - (\cos v)^6\right) d\cos v \\
&= -\int_{\frac{\pi}{2}}^{\pi} d\cos v + 3\int_{\frac{\pi}{2}}^{\pi} (\cos v)^2 d\cos v - 3\int_{\frac{\pi}{2}}^{\pi} (\cos v)^4 d\cos v \\
&\quad + \int_{\frac{\pi}{2}}^{\pi} (\cos v)^6 d\cos v \\
&= -\cos v\Big|_{\frac{\pi}{2}}^{\pi} + (\cos v)^3\Big|_{\frac{\pi}{2}}^{\pi} - \frac{3}{5}(\cos v)^5\Big|_{\frac{\pi}{2}}^{\pi} \\
&\quad + \frac{1}{7}(\cos v)^7\Big|_{\frac{\pi}{2}}^{\pi} \\
&= 1 - 1 + \frac{3}{5} - \frac{1}{7} \\
&= \frac{16}{35}.
\end{aligned}$$

Consequently,

$$\begin{aligned}
I &= -R^7 \frac{5}{8}\pi \frac{16}{35} \\
&= -\frac{2\pi R^7}{7}.
\end{aligned}$$

Exercise 5.2. Compute the following surface integrals of the second kind:

1.

$$\int_S (x^2 + y^2) dxdy,$$

where S is the lower half of the circle

$$x^2 + y^2 \leq 4, \quad z = 0.$$

2.

$$\int_S xzdxdy,$$

where S is the interior side of

$$x + y + z \leq 1, \quad x \geq 0, \quad y \geq 0, \quad z \geq 0.$$

3.

$$\int_S yzdydz + xzdzdx + xydxdy,$$

where S is the interior side of

$$x + y + z \leq 1, \quad x \geq 0, \quad y \geq 0, \quad z \geq 0.$$

4.

$$\int_S ydzdx,$$

where S is the exterior side of the sphere

$$x^2 + y^2 + z^2 = R^2.$$

5.

$$\int_S x^2 dydz,$$

where S is the exterior side of the sphere

$$x^2 + y^2 + z^2 = R^2.$$

5.3 The Gauss–Ostrogradsky formula

In differential and integral calculus, the Gauss–Ostrogradsky theorem, also known as the divergence theorem, is a theorem relating the flux of a vector field through a closed surface to the divergence of the field in the volume enclosed. The Gauss–Ostrogradsky theorem is an important result for the mathematics of physics and engineering, particularly in electrostatics and fluid dynamics. In these fields, it is usually applied in three dimensions. However, it generalizes to any number of dimensions. In one dimension, it is equivalent to the fundamental theorem of calculus. In two dimensions, it is equivalent to the Green theorem.

Suppose that G is a domain in $\mathbb{R}^3$ and D is a domain in $\mathbb{R}^2$ such that the boundary S of G consists of two surfaces S_1 and S_2 given by

$$z = \phi_1(x,y), \quad (x,y) \in D,$$

and

$$z = \phi_2(x,y), \quad (x,y) \in D,$$

respectively, where $\phi_1, \phi_2 \in \mathscr{C}(\overline{D})$,

$$\phi_1(x,y) < \phi_2(x,y), \quad (x,y) \in D,$$

and the part S_0 of the cylinder with bases S_1 and S_2 and generatrix Oz, i. e.,

$$\begin{aligned} \partial G &= S \\ &= S_0 \cup S_1 \cup S_2, \end{aligned}$$

and

$$G = \{(x,y,z) \in \mathbb{R}^3 : (x,y) \in D, \phi_1(x,y) < z < \phi_2(x,y)\}.$$

Let f be a function on S. Then we have

$$\begin{aligned} \int\limits_{S^+} f(x,y,z)dxdy &= \int\limits_{S_1} f(x,y,z)dxdy + \int\limits_{S_2} f(x,y,z)dxdy \\ &\quad + \int\limits_{S_0} f(x,y,z)dxdy \\ &= \int\limits_{S_1} f(x,y,z)dxdy + \int\limits_{S_2} f(x,y,z)dxdy, \end{aligned}$$

where we have used that

$$\int\limits_{S_0} f(x,y,z)dxdy = 0.$$

Theorem 5.1 (The Gauss–Ostrogradsky theorem). *Suppose that $P, Q, R \in \mathscr{C}(G)$, $\frac{\partial P}{\partial x}, \frac{\partial Q}{\partial y}, \frac{\partial R}{\partial z}$ exist, and $\frac{\partial P}{\partial x}, \frac{\partial Q}{\partial y}, \frac{\partial R}{\partial z} \in \mathscr{C}^1(G)$. Then*

$$\int_G \left(\frac{\partial P}{\partial x} + \frac{\partial Q}{\partial y} + \frac{\partial R}{\partial z}\right) dxdydz = \int_{S^+} Pdydz + Qdzdx + Rdxdy. \tag{5.2}$$

Definition 5.3. Formula (5.2) is said to be the Gauss–Ostrogradsky[1,2] formula.

Proof. We have

$$\begin{aligned}
\int_G \frac{\partial R}{\partial z} dxdydz &= \int_D \left(\int_{\phi_1(x,y)}^{\phi_2(x,y)} \frac{\partial R}{\partial z} dz \right) dxdy \\
&= \int_D (R(x,y,\phi_2(x,y)) - R(x,y,\phi_1(x,y)))dxdy \\
&= \int_D R(x,y,\phi_2(x,y))dxdy - \int_D R(x,y,\phi_1(x,y))dxdy \\
&= \int_{S_2} R(x,y,z)dxdy + \int_{S_1} R(x,y,z)dxdy \\
&= \int_{S^+} R(x,y,z)dxdy.
\end{aligned}$$

As above, we have

$$\begin{aligned}
\int_G \frac{\partial P}{\partial x} dx &= \int_{S^+} P(x,y,z)dydz, \\
\int_G \frac{\partial Q}{\partial y} dx &= \int_{S^+} Q(x,y,z)dzdx,
\end{aligned}$$

whereupon we get the desired result. This completes the proof. □

Example 5.7. Using the Gauss–Ostrogradsky formula, we will compute the integral

$$I = \int_S (1 + 2x)dydz + (2x + 3y)dzdx + (3y + 4z)dxdy,$$

1 Johann Carl Friedrich Gauss (30 April 1777–23 February 1855) was a German mathematician, astronomer, geodesist, and physicist, who made contributions to many fields in mathematics and science. Gauss ranks among history's most influential mathematicians and has been referred to as the "Prince of Mathematicians". He was director of the Göttingen Observatory and professor at the university for nearly half a century, from 1807 until his death in 1855.

2 Mikhail Vasilyevich Ostrogradsky (24 September 1801–1 January 1862) was a Ukrainian mathematician, mechanician, and physicist of Ukrainian Cossack ancestry. Ostrogradsky was a student of Timofei Osipovsky and is considered to be a disciple of Leonhard Euler, who was known as one of the leading mathematicians of Imperial Russia.

where S is the exterior side of

$$\frac{x}{a}+\frac{y}{b}+\frac{z}{c}=1,\quad x\geq 0,\ y\geq 0,\ z\geq 0.$$

Then

$$\begin{aligned}0&\leq \frac{z}{c}\leq 1-\frac{x}{a}-\frac{y}{b},\\ 0&\leq \frac{y}{b}\leq 1-\frac{x}{a},\\ 0&\leq \frac{x}{a}\leq 1,\end{aligned}$$

or

$$\begin{aligned}&0\leq z\leq c\left(1-\frac{x}{a}-\frac{y}{b}\right),\\ &0\leq y\leq b\left(1-\frac{x}{a}\right),\\ &0\leq x\leq a.\end{aligned}$$

Thus

$$G=\left\{(x,y,z)\in\mathbb{R}^3:0\leq z\leq c\left(1-\frac{x}{a}-\frac{y}{b}\right),\ 0\leq y\leq b\left(1-\frac{x}{a}\right),\right.$$
$$\left.0\leq x\leq a\right\}.$$

Here

$$\begin{aligned}P(x,y,z)&=1+2x,\\ Q(x,y,z)&=2x+3y,\\ R(x,y,z)&=3y+4z,\quad (x,y,z)\in G.\end{aligned}$$

We have

$$\begin{aligned}\frac{\partial P}{\partial x}(x,y,z)&=2,\\ \frac{\partial Q}{\partial y}(x,y,z)&=3,\\ \frac{\partial R}{\partial z}(x,y,z)&=4,\quad (x,y,z)\in G,\end{aligned}$$

and

$$\begin{aligned}\frac{\partial P}{\partial x}(x,y,z)+\frac{\partial Q}{\partial y}(x,y,z)+\frac{\partial R}{\partial z}(x,y,z)&=2+3+4\\ &=9,\quad (x,y,z)\in G.\end{aligned}$$

Now applying the Gauss–Ostrogradsky formula, we find

$$
\begin{aligned}
I &= 9\int_0^a \int_0^{b(1-\frac{x}{a})} \int_0^{c(1-\frac{x}{a}-\frac{y}{b})} dzdydx \\
&= 9c\int_0^a \int_0^{b(1-\frac{x}{a})} \left(1-\frac{x}{a}-\frac{y}{b}\right)dydx \\
&= 9c\int_0^a \left(b\left(1-\frac{x}{a}\right)^2 - \frac{1}{2b}y^2\Big|_{y=0}^{y=b(1-\frac{x}{a})}\right)dx \\
&= 9c\int_0^a \left(b\left(1-\frac{x}{a}\right)^2 - \frac{1}{2b}b^2\left(1-\frac{x}{a}\right)^2\right)dx \\
&= \frac{9cb}{2}\int_0^a \left(1-\frac{x}{a}\right)^2 dx \\
&= \frac{9abc}{6}\left(1-\frac{x}{a}\right)^3\Big|_{x=a}^{x=0} \\
&= \frac{3abc}{2}(1-0) \\
&= \frac{3abc}{2}.
\end{aligned}
$$

Exercise 5.3. Using the Gauss–Ostrogradsky formula, compute the following surface integrals of the second kind:

1.
$$\int_S (1+2x)dydz + (2x+3y)dzdx + (3y+4z)dxdy,$$

   where S is the interior side of

$$|x-y+z| + |y-z+x| + |z-x+y| = a, \quad a > 0.$$

2.
$$\int_S zdxdy + (5x+y)dydz,$$

   where S is the exterior side of

$$x^2+y^2 \le z^2, \quad 0 \le z \le 4.$$

3.

$$\int_S zdxdy + (5x + y)dydz,$$

where S is the interior side of

$$\frac{x^2}{4} + \frac{y^2}{9} + z^2 = 1.$$

4.

$$\int_S zdxdy + (5x + y)dydz,$$

where S is the exterior side of the boundary of

$$1 < x^2 + y^2 + z^2 < 4.$$

5.

$$\int_S x^2dydz + y^2dzdx + z^2dxdy,$$

where S is the interior side of

$$0 \le x \le a, \quad 0 \le y \le b, \quad 0 \le z \le c, \quad a, b, c > 0.$$

5.4 The Stokes formula

The Stokes theorem, also known as the Kelvin–Stokes theorem after Lord Kelvin and George Stokes, the fundamental theorem for curls, or simply the curl theorem, is a theorem in differential and integral calculus on $\mathbb{R}^3$.

Suppose that S is a twice continuously differentiable surface without singular points. Let S be given by

$$r = r(u, v) = (x(u, v), y(u, v), z(u, v)), \quad (u, v) \in \overline{D},$$

where D is a bounded domain in R^2 for which the Green formula holds, represented by

$$\begin{aligned} u &= u(t), \\ v &= v(t), \quad t \in [a, b]. \end{aligned}$$

Let $\Gamma_0 = \partial D$ be positively oriented. Suppose that on S, there is a unit normal vector

$$N = (\cos\alpha, \cos\beta, \cos\gamma).$$

Denote by Γ a contour given by

$$r = r(u(t), v(t)), \quad t \in [a, b].$$

Theorem 5.2 (The Stokes theorem). *Suppose that $P, Q, R \in \mathscr{C}^1(G)$. Then*

$$\begin{aligned}&\int\limits_{\Gamma} Pdx + Qdy + Rdz \\ &= \int\limits_{S} \left(\left(\frac{\partial R}{\partial y} - \frac{\partial Q}{\partial z} \right) \cos\alpha + \left(\frac{\partial P}{\partial z} - \frac{\partial Q}{\partial x} \right) \cos\beta + \left(\frac{\partial Q}{\partial x} - \frac{\partial P}{\partial y} \right) \cos\gamma \right) dS.\end{aligned} \tag{5.3}$$

Definition 5.4. Formula (5.3) is called the Stokes[3] formula or Kelvin[4]–Stokes formula.

Proof. We have

$$\begin{aligned}&\int\limits_{\Gamma} P(x,y,z)dx \\ &= \int\limits_{a}^{b} P(x(u(t),v(t)), y(u(t),v(t)), z(u(t),v(t)))x'_{1t}(u(t),v(t))dt \\ &= \int\limits_{\Gamma_0} P(x(u,v), y(u,v), z(u,v)) \left(\frac{\partial x}{\partial u}(u,v)du + \frac{\partial x}{\partial v}(u,v)dv \right) \\ &= \int\limits_{\Gamma_0} P\frac{\partial x}{\partial u}du + P\frac{\partial x}{\partial v}dv.\end{aligned}$$

3 Sir George Gabriel Stokes, 1st Baronet, FRS (13 August 1819–1 February 1903) was an Irish physicist and mathematician. Born in County Sligo, Ireland, Stokes spent all of his career at the University of Cambridge, where he was the Lucasian Professor of Mathematics from 1849 until his death in 1903. As a physicist, Stokes made seminal contributions to fluid mechanics, including the Navier–Stokes equations, and to physical optics, with notable works on polarization and fluorescence. As a mathematician, he popularized "Stokes' theorem" in vector calculus and contributed to the theory of asymptotic expansions. Stokes, along with Felix Hoppe-Seyler, first demonstrated the oxygen transport function of haemoglobin and showed color changes produced by the aeration of haemoglobin solutions.

4 William Thomson, 1st Baron Kelvin (26 June 1824–17 December 1907), was a British mathematician, mathematical physicist and engineer. Born in Belfast, he was the professor of Natural Philosophy at the University of Glasgow for 53 years, where he undertook significant research on the mathematical analysis of electricity, was instrumental in the formulation of the first and second laws of thermodynamics, and contributed significantly to unifying physics, which was then in its infancy of development as an emerging academic discipline. He received the Royal Society's Copley Medal in 1883 and served as its president from 1890 to 1895. In 1892, he became the first scientist to be elevated to the House of Lords.

Now applying the Green formula, we get

$$\begin{aligned}
\int_{\Gamma} P dx &= \int_{D} \left(\frac{\partial}{\partial u}\left(P\frac{\partial x}{\partial v}\right) - \frac{\partial}{\partial v}\left(P\frac{\partial x}{\partial u}\right)\right) dudv \\
&= \int_{D} \left(\frac{\partial P}{\partial u}\frac{\partial x}{\partial v} + P\frac{\partial^2 x}{\partial u \partial v} - \frac{\partial P}{\partial v}\frac{\partial x}{\partial u} - P\frac{\partial^2 x}{\partial u \partial v}\right) dudv \\
&= \int_{D} \left(\left(\frac{\partial P}{\partial x}\frac{\partial x}{\partial u} + \frac{\partial P}{\partial y}\frac{\partial y}{\partial u} + \frac{\partial P}{\partial z}\frac{\partial z}{\partial u}\right)\frac{\partial x}{\partial v} \right. \\
&\quad \left. - \left(\frac{\partial P}{\partial x}\frac{\partial x}{\partial v} + \frac{\partial P}{\partial y}\frac{\partial y}{\partial v} + \frac{\partial P}{\partial z}\frac{\partial z}{\partial v}\right)\frac{\partial x}{\partial u}\right) dudv \\
&= \int_{S^+} \frac{\partial P}{\partial z} dzdx - \int_{S^-} \frac{\partial P}{\partial y} dxdy \\
&= \int_{S} \left(\frac{\partial P}{\partial z}\cos\beta - \frac{\partial P}{\partial y}\cos\gamma\right) dS.
\end{aligned}$$

As above, we get

$$\int_{\Gamma} Q dy = \int_{S} \left(\frac{\partial Q}{\partial x}\cos\gamma - \frac{\partial Q}{\partial z}\cos\alpha\right) dS,$$

$$\int_{\Gamma} R dz = \int_{S} \left(\frac{\partial R}{\partial y}\cos\alpha - \frac{\partial R}{\partial x}\cos\beta\right) dS,$$

whereupon we get the desired result. This completes the proof. □

Example 5.8. Let

$$(P(x,y,z), Q(x,y,z), R(x,y,z)) = (-z, x, 0),$$

and let S be the hemisphere, oriented outward, with parameterization

$$r(\phi,\theta) = (\sin\phi\cos\theta, \sin\phi\sin\theta, \cos\phi), \quad \phi,\theta \in [0,\pi]$$

(see Figure 5.4). Let Γ be the boundary of S. Note that Γ is a circle of radius 1, centered at the origin and sitting in the plane $y = 0$. Its parameterization is

$$(\cos t, 0, \sin t), \quad t \in [0, 2\pi].$$

Then

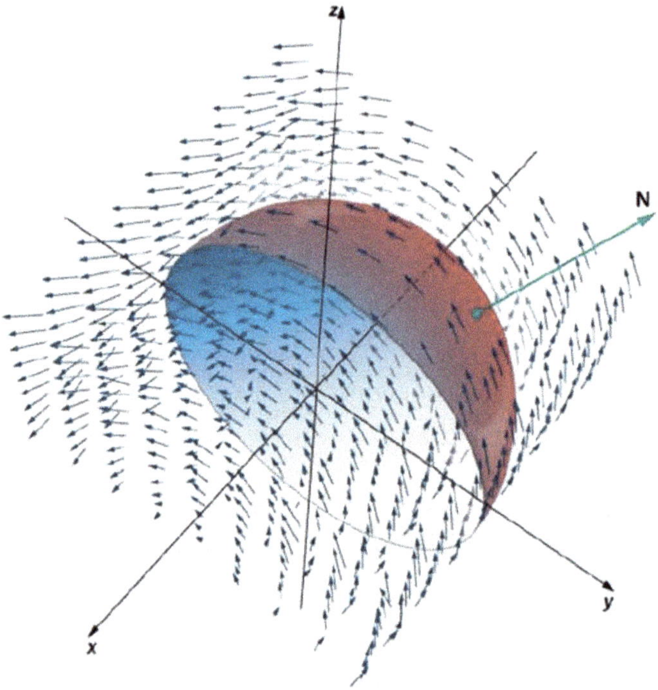

Figure 5.4: A hemisphere.

$$
\begin{aligned}
\int_{\Gamma} Pdx + Qdy + Rdz &= \int_0^{2\pi} (-\sin t(-\sin t))dt \\
&= \int_0^{2\pi} (\sin t)^2 dt \\
&= \frac{1}{2}\int_0^{2\pi} (1 - \cos(2t))dt \\
&= \frac{1}{2}\int_0^{2\pi} dt - \frac{1}{2}\int_0^{2\pi} \cos(2t)dt \\
&= \pi - \frac{1}{4}\sin(2t)\Big|_{t=0}^{t=2\pi} \\
&= \pi,
\end{aligned}
$$

and

$$\int_S \left(\left(\frac{\partial R}{\partial y} - \frac{\partial Q}{\partial z}\right)\cos\alpha + \left(\frac{\partial P}{\partial z} - \frac{\partial Q}{\partial x}\right)\cos\beta + \left(\frac{\partial Q}{\partial x} - \frac{\partial P}{\partial y}\right)\cos\gamma\right)dS$$

$$= \int_0^\pi \int_0^\pi \langle (0,1,1), (\cos\theta(\sin\phi)^2, \sin\theta(\sin\phi)^2, \sin\phi\cos\phi)\rangle d\phi d\theta$$

$$= \int_0^\pi \int_0^\pi (\sin\theta(\sin\phi)^2 + \sin\phi\cos\phi) d\phi d\theta$$

$$= \int_0^\pi \int_0^\pi \sin\theta(\sin\phi)^2 d\phi d\theta + \int_0^\pi \int_0^\pi \sin\phi\cos\phi d\phi d\theta$$

$$= \frac{1}{2}\left(\int_0^\pi \sin\theta d\theta\right)\int_0^\pi (1 - \cos(2\phi))d\phi + \pi \int_0^\pi \sin\phi d\sin\phi$$

$$= -\frac{1}{2}(\cos\theta\Big|_{\theta=0}^{\theta=\pi})\pi + \frac{\pi}{2}(\sin\phi)^2\Big|_{\phi=0}^{\phi=\pi}$$

$$= \pi.$$

Thus we have verified the Stokes theorem.

Exercise 5.4. Verify the Stokes theorem for

$$(P(x,y,z), Q(x,y,z), R(x,y,z)) = (y, x, -z)$$

and surface that is the upwardly oriented portion of the graph of $f(x,y) = x^2y$ over a triangle in the (x,y)-plane with vertices $(0,0)$, $(2,0)$, and $(0,2)$.

5.5 Advanced practical problems

Problem 5.1. Compute the following surface integrals of the first kind:

1.

$$\int_S (x^2 + y^2 + z^2)dS,$$

where S is the surface given by

$$|x| + |y| + |z| \leq a, \quad a > 0.$$

2.

$$\int_S (x^2 + y^2 + z^2)dS,$$

where S is the surface given by

$$x^2 + y^2 \leq r^2, \quad z \in [0, H],\ r, H > 0.$$

3.

$$\int_S |xy| z dS,$$

where S is the surface given by

$$z = x^2 + y^2, \quad z \leq 1.$$

4.

$$\int_S (x^2 + y^2) dS,$$

where S is the surface given by

$$z = \sqrt{x^2 + y^2}, \quad z \leq 1.$$

5.

$$\int_S \sqrt{x^2 + y^2} dS,$$

where S is the surface given by

$$z = \sqrt{x^2 + y^2}, \quad z \leq 1.$$

6.

$$\int_S (xy + yz + xz) dS,$$

where S is the surface given by

$$z = \sqrt{x^2 + y^2}, \quad x^2 + y^2 = 2x.$$

7.

$$\int_S (x^2y^2 + y^2z^3 + x^2z^2) dS,$$

where S is the surface given by

$$z = \sqrt{x^2 + y^2}$$

inside of

$$x^2 + y^2 = 2x.$$

8.

$$\int_S \sqrt{\frac{x^2}{a^4} + \frac{y^2}{b^4} + \frac{z^2}{c^4}} dS,$$

where S is the surface given by

$$\frac{x^2}{a^2} + \frac{y^2}{b^2} + \frac{z^2}{c^2} = 1.$$

9.

$$\int_S \frac{1}{\sqrt{\frac{x^2}{a^4} + \frac{y^2}{b^4} + \frac{z^2}{c^4}}} dS,$$

where S is the surface given by

$$\frac{x^2}{a^2} + \frac{y^2}{b^2} + \frac{z^2}{c^2} = 1.$$

10.

$$\int_S \frac{1}{(x^2 + y^2 + z^2)^{\frac{3}{2}} \sqrt{\frac{x^2}{a^4} + \frac{y^2}{b^4} + \frac{z^2}{c^4}}} dS,$$

where S is the surface given by

$$\frac{x^2}{a^2} + \frac{y^2}{b^2} + \frac{z^2}{c^2} = 1.$$

11.

$$\int_S \frac{1}{(x^2 + y^2 + (z - a)^2)^{\frac{n}{2}}} dS, \quad n \in \mathbb{N},$$

where S is the surface given by

$$x^2 + y^2 + z^2 = R^2, \quad R > 0.$$

12.

$$\int_S z^2 dS,$$

where S is the surface given by

$$x = u\cos v \sin\alpha,$$
$$y = u\sin v \sin\alpha,$$
$$z = u\cos\alpha, \quad \alpha = \text{const} \in \left(0, \frac{\pi}{2}\right), \; u \in [0,1], \; v \in [0,2\pi].$$

13.

$$\int_S z dS,$$

where S is the surface given by

$$x = u\cos v,$$
$$y = u\sin v,$$
$$z = v, \quad u \in [0,1], \; v \in [0,2\pi].$$

14.

$$\int_S f(r) dS,$$

where $r = \sqrt{x^2 + y^2 + z^2}$,

$$f(r) = \begin{cases} 1 - r^2 & \text{if } r \le 1, \\ 0 & \text{if } r \ge 1, \end{cases}$$

and S is the surface given by

$$x + y + z = a, \quad a > 0.$$

15.

$$\int_S f(r, z) dS,$$

where $r = \sqrt{x^2 + y^2}$,

$$f(r,z) = \begin{cases} r^2 & \text{if } r \le z, \\ 0 & \text{if } r \ge z, \end{cases}$$

and S is the surface given by

$$x^2 + y^2 + z^2 = R^2, \quad R > 0.$$

Problem 5.2. Compute the following surface integrals of the second kind:

1.

$$\int_S x^2 dydz,$$

where S is the interior side of the semi-sphere

$$x^2 + y^2 + z^2 = R^2, \quad z \le 0.$$

2.

$$\int_S x^2 dydz + z^2 dxdy,$$

where S is the interior side of the sphere

$$x^2 + y^2 + z^2 = R^2, \quad x \le 0,\ y \ge 0.$$

3.

$$\int_S x^2 dydz + y^2 dzdx + z^2 dxdy,$$

where S is the exterior side of the sphere

$$(x-a)^2 + (y-b)^2 + (z-c)^2 = R^2, \quad a, b, c > 0.$$

4.

$$\int_S z^2 dxdy,$$

where S is the interior side of the semi-sphere

$$(x-a)^2 + (y-b)^2 + z^2 = R^2, \quad z \ge 0.$$

5.

$$\int_S (x-1)^3 dydz,$$

where S is the exterior side of the semi-sphere

$$x^2 + y^2 + z^2 = 2x, \quad z \le 0.$$

6.

$$\int_S dzdx,$$

where S is the exterior side of

$$\frac{x^2}{a^2} + \frac{y^2}{b^2} + \frac{z^2}{c^2} = 1, \quad a, b, c > 0.$$

7.

$$\int_S xdydz,$$

where S is the exterior side of

$$\frac{x^2}{a^2} + \frac{y^2}{b^2} + \frac{z^2}{c^2} = 1, \quad a, b, c > 0.$$

8.

$$\int_S x^2 dydz,$$

where S is the exterior side of

$$\frac{x^2}{a^2} + \frac{y^2}{b^2} + \frac{z^2}{c^2} = 1, \quad a, b, c > 0.$$

9.

$$\int_S \frac{1}{z} dxdy,$$

where S is the exterior side of

$$\frac{x^2}{a^2} + \frac{y^2}{b^2} + \frac{z^2}{c^2} = 1, \quad a, b, c > 0.$$

10.

$$\int_S yzdzdx,$$

where S is the exterior side of

$$\frac{x^2}{a^2}+\frac{y^2}{b^2}+\frac{z^2}{c^2}=1, \quad z\ge 0, \quad a,b,c>0.$$

11.

$$\int_S x^3dydz+y^3dzdx,$$

where S is the exterior side of

$$\frac{x^2}{a^2}+\frac{y^2}{b^2}+\frac{z^2}{c^2}=1, \quad z\ge 0,\ a,b,c>0.$$

12.

$$\int_S (2x^2+y^2+z^2)dydz,$$

where S is the exterior side of

$$\sqrt{y^2+z^2}\le x\le H, \quad H>0.$$

13.

$$\int_S (y-z)dydz+(z-x)dzdx+(x-y)dxdy,$$

where S is the exterior side of

$$x^2+y^2=z^2, \quad 0<z\le H.$$

14.

$$\int_S yz^2dxdz,$$

where S is the interior side of

$$x^2+y^2=r^2, \quad y\le 0,\ 0\le z\le r.$$

15.

$$\int_S yzdxdy + zxdydz + xydzdx,$$

where S is the interior side of

$$x^2 + y^2 = r^2, \quad x \le 0,\ y \ge 0,\ 0 \le z \le H.$$

16.

$$\int_S x^6 dydz + y^4 dzdx + z^2 dxdy,$$

where S is the down side of

$$z = x^2 + y^2, \quad z \le 1.$$

17.

$$\int_S xdydz + ydzdx + zdxdy,$$

where S is the upper side of

$$z = x^2 - y^2, \quad |y| \le x \le a.$$

Problem 5.3. Using the Gauss–Ostrogradsky formula, compute the following surface integrals of the second kind:

1.

$$\int_S x^2 dydz + y^2 dzdx + z^2 dxdy,$$

where S is the exterior side of

$$\frac{x^2}{a^2} + \frac{y^2}{b^2} \le \frac{z^2}{c^2}, \quad 0 \le z \le c.$$

2.

$$\int_S x^3 dydz + y^3 dzdx + z^3 dxdy,$$

where S is the exterior side of

$$x + y + z \le a, \quad x \ge 0,\ y \ge 0,\ z \ge 0.$$

3.

$$\int_S x^3 dydz + y^3 dzdx + z^3 dxdy$$

where S is the interior side of

$$x^2 + y^2 + z^2 = R^2, \quad R > 0.$$

4.

$$\int_S x^4 dydz + y^4 dzdx + z^4 dxdy,$$

where S is

$$x^2 + y^2 + z^2 = R^2, \quad R > 0.$$

5.

$$\int_S x^4 dydz + y^4 dzdx + z^4 dxdy$$

where S is the exterior side of

$$x^2 + y^2 + z^2 \leq R^2, \quad z \geq 0,\ R > 0.$$

Problem 5.4. Verify the Stokes theorem for

$$(P(x,y,z), Q(x,y,z), R(x,y,z)) = (2y, x^2, xz)$$

and the surface given by

$$z = x^2 + y.$$

Solutions, hints, and answers to the exercises

Chapter 2

Exercise 2.1. Answer:

1. 0.
2. $-\frac{e-5\frac{1}{e}}{2}$.

Exercise 2.2. Answer:

1. $\frac{4}{27}$.
2. $\frac{2a^5}{15}$.
3. $\frac{4R^5}{15}$.
4. $\frac{76}{3}$.
5. $14a^4$.
6. $\frac{31}{30}$.
7. $-\frac{45\pi-10}{6\pi^2}$.
8. $2\cosh 1 - 2$.
9. 0.
10. $\frac{135}{4}$.

Exercise 2.3. Answer:

1. $$\frac{c^2(e^a - 1)(1 - \cos b)}{2}.$$
2. $$c(ae^a + be^{-b} - (a + b)e^{a-b}).$$
3. 0.
4. $\frac{3}{4}$.
5. −8.
6. 28.
7. 1.126.

Exercise 2.4. Answer:

1. $\pi \log 3$.
2. $\frac{15a^4}{2}$.
3. $\frac{2a^5}{15}$.
4. $\frac{\pi}{8}$.
5. $\pi(\log a)^2$.
6. 0.
7. $-\frac{\pi R^4}{8}$.

https://doi.org/10.1515/9783112219607-006

8. $\frac{\pi}{40}$.
9. $\frac{128}{525}$.
10. $\frac{8\pi}{5}$.

Chapter 3.2

Exercise 3.1. Answer:

3. $\sqrt{\pi}$.

Exercise 3.2. Answer:

1. Convergent for $\alpha > 2$ and divergent for $\alpha \le 2$.
2. Convergent for $\alpha > 1$ and divergent for $\alpha \le 1$.
3. Convergent for $\alpha > \frac{1}{2}$ and divergent for $\alpha \le \frac{1}{2}$.

Exercise 3.3. Answer:

1. Convergent.
2. Convergent for $\alpha < 3$. Divergent for $\alpha \ge 3$.
3. Divergent.

Exercise 3.4. Answer:

1. Divergent for all p.
2. Divergent.
3. Convergent for $\alpha < -1$. Divergent for $\alpha \ge -1$.
4. Convergent for $\alpha < 2$. Divergent for $\alpha \ge 2$.
5. Convergent for $\alpha < 1$. Divergent for $\alpha \ge 1$.
6. Convergent for $\alpha < 1$. Divergent for $\alpha \ge 1$.
7. Convergent for $\alpha < 1$ and $\beta < 1$ and divergent in the remaining cases.
8. Convergent for $\frac{1}{\alpha} + \frac{1}{\beta} > 1$. Divergent for $\frac{1}{\alpha} + \frac{1}{\beta} \le 1$.
9. Convergent for $\frac{1}{\alpha} + \frac{1}{\beta} > p$. Divergent for $\frac{1}{\alpha} + \frac{1}{\beta} \le p$.
10. Convergent for $p < 1$. Divergent for $p \ge 1$.

Chapter 4

Exercise 4.1. Answer:

1. $\frac{\sqrt{5}}{2}$.
2. $3 + 2\sqrt{5}$.
3. $\frac{8\pi b^2\sqrt{a^2+b^2}}{3a^2}$.
4. $\frac{2^{\frac{3}{2}}}{3}((1 + 2\pi^2)^{\frac{3}{2}} - 1)$.
5. $\frac{a^2\sqrt{2}}{512}(100\sqrt{38} - 72 - 17\log(\frac{25+4\sqrt{38}}{17}))$.

Exercise 4.2. Answer:

1. π.
2. $\frac{14-3\log 4}{3}$.
3. 8.
4. 4.
5. $\frac{12}{5}$.
6. $2\sin 2$.
7. $-\frac{8}{15}$.
8. $-\frac{14}{15}$.
9. $-\pi a^2$.
10. $\frac{1}{35}$.

Exercise 4.3. Answer:

1. 0.
2. πab.
3. 0.
4. $-\frac{140}{3}$.
5. 0.

Exercise 4.5. Answer:

1. $\frac{1}{3}$.
2. $\frac{9}{8}$.
3. $\frac{9}{2}$.
4. $\frac{4}{5}$.
5. πab.
6. $\frac{27\pi}{2}$.
7. $\frac{3\pi a^2}{8}$.

Exercise 4.6. Answer: 12.

Chapter 5

Exercise 5.1. Answer:

1. π.
2. $\frac{\pi(1+\sqrt{2})}{2}$.
3. $40a^4$.
4. $(\sqrt{3}-1)(\log 2+\frac{\sqrt{3}}{2})$.
5. 0.

Exercise 5.2. Answer:

1. -8π.
2. $-\frac{1}{24}$.
3. 0.

4. $\frac{4\pi R^3}{3}$.
5. 0.

Exercise 5.3. Answer:
1. $-3a^3$.
2. 128π.
3. -48π.
4. 56π.
5. $(a + b + c)abc$.

Solutions, hints, and answers to the problems

Chapter 1

Problem 1.1. Answer:

1. $b - a$.
2. $b - a$.

Problem 1.13.

1. *Solution.* We have

$$\begin{aligned}\mu X_{ij} &= \frac{2}{n} \cdot \frac{1}{n} \\ &= \frac{2}{n^2}, \quad i,j \in \{-n+1, -n, \ldots, n\},\end{aligned}$$

and

$$\begin{aligned}m_{ij} &= \inf_{(x,y)\in X_{ij}} f(x,y) \\ &= 2\frac{2(i-1)}{n} - \frac{j}{n} \\ &= \frac{4i-j-4}{n}, \\ M_{ij} &= \sup_{(x,y)\in X_{ij}} f(x,y) \\ &= \frac{4i}{n} - \frac{j-1}{n} \\ &= \frac{4i-j+1}{n}, \quad i,j \in \{-n+1, -n, \ldots, n\}.\end{aligned}$$

Then

$$\begin{aligned}s_{\tau_n} &= \sum_{i,j=-n+1}^{n} m_{ij}\mu X_{ij} \\ &= \sum_{i,j=-n+1}^{n} \frac{4i-j-4}{n} \cdot \frac{2}{n^2} \\ &= \frac{2}{n^3} \sum_{i,j=-n+1}^{n} (4i-j-4) \\ &= \frac{2}{n^2}\left(4\sum_{i,j=-n+1}^{n} i - \sum_{i,j=-n+1}^{n} j - 4\sum_{i,j=-n+1}^{n} 1\right) \\ &= \frac{2}{n^3}(4n(-n+1+n)2n - 2n(-n+1+n)n - 4 \cdot 2n \cdot 2n)\end{aligned}$$

https://doi.org/10.1515/9783112219607-007

$$= \frac{2}{n^3}(8n^2 - 2n^2 - 16n^2)$$
$$= -\frac{20}{n}, \quad n \in \mathbb{N},$$

and

$$S_{\tau_n} = \sum_{i,j=-n+1}^{n} M_{ij}\mu X_{ij}$$
$$= \sum_{i,j=-n+1}^{n} \frac{4i - j + 1}{n} \cdot \frac{2}{n^2}$$
$$= \frac{2}{n^3} \sum_{i,j=-n+1}^{n} (4i - j + 1)$$
$$= \frac{2}{n^2}\left(4 \sum_{i,j=-n+1}^{n} i - \sum_{i,j=-n+1}^{n} j + \sum_{i,j=-n+1}^{n} 1\right)$$
$$= \frac{2}{n^3}(4n(-n + 1 + n)2n - 2n(-n + 1 + n)n + 2n \cdot 2n)$$
$$= \frac{2}{n^3}(8n^2 - 2n^2 + 4n^2)$$
$$= \frac{20}{n}, \quad n \in \mathbb{N}.$$

Hence

$$\lim_{n\to\infty} s_{\tau_n} = \lim_{n\to\infty}\left(-\frac{20}{n}\right)$$
$$= 0,$$

and

$$\lim_{n\to\infty} S_{\tau_n} = \lim_{n\to\infty}\left(\frac{20}{n}\right)$$
$$= 0.$$

2. Answer:

$$s_{\tau_n} = \frac{2}{n^2} \cdot \frac{\sinh 2 \sinh 1}{\sinh(\frac{1}{n}) \sinh(\frac{1}{2n})} e^{-\frac{3}{2n}},$$
$$S_{\tau_n} = \frac{2}{n^2} \cdot \frac{\sinh 2 \sinh 1}{\sinh(\frac{1}{n}) \sinh(\frac{1}{2n})} e^{\frac{3}{2n}}, \quad n \in \mathbb{N},$$
$$I = 4 \sinh 2 \sinh 1.$$

3. Answer:

$$s_{\tau_n} = -\frac{8}{n},$$
$$S_{\tau_n} = \frac{8}{n},$$
$$I = 0.$$

4. Answer:

$$s_{\tau_n} = \frac{20(n-1)(2n-1)}{3n^2},$$
$$S_{\tau_n} = \frac{20(n+1)(2n+1)}{3n^2}, \qquad n \in \mathbb{N},$$
$$I = \frac{40}{3}.$$

Problem 1.14. Answer:

1.
$$\frac{\varepsilon}{\pi\sqrt{p^2+q^2}}.$$

2.
$$\frac{1}{\sqrt{p^2+q^2}}\log\left(1+\frac{\varepsilon}{\pi}e^{-\sqrt{p^2+q^2}}\right).$$

3.
$$\frac{2-\sqrt{p^2+q^2}}{\sqrt{p^2+q^2}}(1-e^{-\frac{\varepsilon}{\pi}}).$$

4.
$$\frac{\varepsilon}{2\pi e}.$$

5.
$$\frac{8\varepsilon}{9\pi}.$$

Chapter 2

Problem 2.1. Answer:

1. $\frac{1}{9}$.

2. $\frac{|\sin x|^3}{3}$, $x \in \mathbb{R}$.
3. $\frac{8}{15}$.
4. $\frac{a^{2(a+1)}}{2(a+1)}$.
5. $\frac{1}{(a+1)(a+2)}$.
6. $\frac{2}{15}$.
7. $\frac{\pi}{28} - \frac{16}{2205}$.

Problem 2.2. Answer:

1. $\frac{1}{\sqrt{2}}$.
2. $\frac{255}{4}$.
3. 0.
4. $\frac{\log 2}{6}$.
5. $\frac{\pi^2}{32}$.
6. $\frac{e-1}{2}$.
7. $\frac{\cos 1-1}{3}$.
8. $\log(\cos(\frac{1}{4}))$.
9. $\frac{e-1}{2}$.
10. $\frac{\pi}{6}$.
11. $\frac{8(b^4-a^4)}{3}$.
12. 20.
13. 7π.
14. $\frac{b^4-a^4}{2}$.
15. $\frac{10+3\pi}{6}$.
16. $\frac{a^3}{3}$.
17. 8.
18. $\frac{a^2}{2}$.
19. $\frac{4\pi}{3} + 4\log(2+\sqrt{3})$.
20. $\frac{\pi ab}{8}$.
21. 0.
22. $\frac{5\pi a^3}{2}$.
23. $\frac{\pi a^2 b}{4}$.
24. $-\frac{243}{70}$.
25. 0.
26. $\frac{16}{45}$.

Problem 2.5. Answer:

1. $\frac{16}{3}$.
2. $\frac{8\log 2-5}{16}$.
3. $\frac{5}{12}$.
4. $-\frac{4}{15}$.
5. $\frac{1}{8}$.

6. $\frac{1}{48}$.
7. $\frac{\pi}{6}$.
8. $\frac{1}{364}$.
9. $\frac{1}{96}$.

Problem 2.9. Answer:

1. $\frac{\sqrt{2}}{3}(b-a)R^3$.
2. $\frac{2(1-k)}{3\sqrt{1+k^2}}R^3$.
3. $\arctan|k|$.
4. $\frac{\pi a^2}{16}$.
5. $-\frac{1}{6}$.
6. $\frac{8-3\sqrt{3}}{3}$.
7. $\frac{2\pi-3\sqrt{3}}{2}$.
8. $\frac{7\pi a^3}{16}$.
9. $\frac{2a^3}{9}$.
10. $\frac{45\pi+20}{3}$.

Problem 2.10. Answer:

1. $\frac{\pi(\log 4-1)}{2}$.
2. $\frac{\pi}{3}$.
3. $\frac{3\pi}{8}$.
4. $\frac{16\sqrt{2}}{15}$.
5. $\frac{3\sqrt{3}-\pi}{108}$.
6. $\frac{1}{24}$.
7. $\frac{\pi}{32}$.
8. $\frac{1}{5}$.
9. $\frac{(2\sqrt{3}-9)a^2}{6}$.
10. $\frac{(15\pi-4)a^3}{9}$.

Problem 2.11. Answer:

1. 20.
2. $\frac{2}{9}$.
3. 0.
4. $\frac{c^2(e^a-1)(1-\cos b)}{2}$.
5. $c(ae^a+be^{-b}-(a+b)e^{a-b})$.
6. 0.
7. $\frac{3}{4}$.
8. -8.
9. 28.
10. 1.126.
11. $\frac{\pi\log 2}{4}$.

12. $\frac{31\pi}{15}$.
13. $\frac{R^5}{15}$.
14. $\frac{\pi a^2 R^3}{3(a^2+h^2)}$.
15. $\frac{\pi}{10}$.
16. $\frac{\pi R^2 H(3R^2+2H^2)}{12}$.
17. $\frac{7\pi}{4}$.
18. $\frac{16\pi}{3}$.
19. $\frac{\pi a^4}{12}$.
20. $\frac{4\pi abc}{5}$.
21. $\frac{4\pi abc(a^2+b^2)}{15}$.
22. $\frac{\pi^2 abc}{4}$.
23. $\frac{\pi R^2 h^2}{4}$.
24. $\frac{4\pi R^4}{3}$.
25. $\frac{\log 3-1}{16}$.
26. $\frac{a^4}{10}$.
27. $\frac{59\pi R^5}{480}$.
28. $\frac{3\sqrt{2}-4}{3}$.
29. $(\log 3)\log 5$.
30. $\frac{55}{72}$.
31. $\frac{27}{32}$.
32. $\frac{2}{27}(\frac{1}{\sqrt{a}} - \frac{1}{\sqrt{b}})(\frac{1}{\alpha^3} - \frac{1}{\beta^3})h^{\frac{9}{2}}$.
33. $\frac{1}{32}(\frac{1}{m^2} - \frac{1}{n^2})(b^8 - a^8)(\frac{1}{\alpha^2} - \frac{1}{\beta^2} + \beta^2 - \alpha^2 + 2\log\frac{\beta}{\alpha})$.

Problem 2.12. Answer:

$$f(a_2,b_2,c_2) - f(a_1,b_2,c_2) - f(a_2,b_1,c_2) - f(a_2,b_2,c_1) \\ + f(a_1,b_1,c_2) + f(a_1,b_2,c_1) + f(a_2,b_1,c_1) - f(a_1,b_1,c_1).$$

Problem 2.14. Answer: $4\pi t^2 f(t)$, $t \geq 0$.

Problem 2.15. Answer:

$$\int\limits_{x^2+y^2\leq t^2} f(x,y,t)\,dxdy, \quad t \geq 0.$$

Problem 2.16. Answer:

$$\int\limits_{Z(t)} f(x,y,t-x-y)\,dxdy, \quad t \geq 0,$$

where

$$Z(t)=\{(x,y)\in\mathbb{R}^2 : x\geq 0,\ y\geq 0,\ x+y\leq t\},\quad t\geq 0.$$

Problem 2.17. Answer:

$$f(x,y,z),\quad (x,y,z)\in X.$$

Problem 2.18. Answer.
1. $\frac{a^4}{4!}$.
2. $\frac{a^{\alpha+4}}{6(\alpha+4)}$.

Problem 2.19. Answer:
1. $\frac{a^8}{384}$.
2. $\frac{a^6}{40}$.

Problem 2.20. Answer:
1. $\frac{4\pi HR^3}{3}$.
2. $\frac{4\pi HR^3(9R^2+5H^2)}{45}$.

Problem 2.21. Answer:
1. $\frac{\pi a^3H^4}{3}$.
2. $\frac{2\pi a^5H^6}{15}$.

Problem 2.22. Answer:
1. $\pi^2a^2b^2$.
2. $\frac{\pi^2a^2b^2(a^2+b^2)}{2}$.

Problem 2.24. Answer:
1. $\frac{a^{n+p}}{p+1}$.
2. $\frac{na^{n+1}}{2}$.
3. $\frac{na^{n+p}}{p+1}$.
4. $$\sum_{k=0}^{n-1}(-1)^k\binom{n}{k}(n-k)^{p+n}\frac{a^{p+n}}{(p+1)(p+2)\cdots(p+n)}.$$
5. $$\prod_{k=1}^{n}\frac{e^{ac_k}-1}{c_k}.$$
6. $\frac{a^n}{2}$.

Problem 2.25. Answer:

1. $a^n n!$.
2. $\frac{a^{2n}}{(2n)!!}$.
3. $\frac{a^{n+1}}{2(n-1)!}$.

Problem 2.26. Answer:

1. $\frac{a^n}{n!}$.
2. $\frac{na^{n+1}}{(n+1)!}$.
3. $\frac{2na^{n+2}}{(n+2)!}$.
4. $$\frac{2a^{n+\frac{1}{2}}}{(n-1)!(2n+1)}.$$
5. $$\frac{a^{n+p}}{(n-1)!(n+p)}.$$

Problem 2.27. Answer:

1. $$2^n h_1 h_2 \cdots h_n \left|\det(a_{ij})\right|^{-1}.$$

Problem 2.28. Answer:

$$\frac{a_1 \cdots a_n}{n!}.$$

Problem 2.29. Answer:

$$r^{n-1}(\cos\psi_{n-1})^{n-2}(\cos\psi_{n-2})^{n-3}\cdots(\cos\psi_3)^2\cos\psi_2.$$

Problem 2.30. Answer: $HV_{n-1}(R)$, where $V_{n-1}(R)$ is the volume of the n-dimensional ball with radius R.

Problem 2.31. Answer: $\frac{1}{n}HV_{n-1}(aH)$, where $V_{n-1}(R)$ is the volume of the n-dimensional ball with radius R.

Problem 2.32. Answer:

$$\frac{1}{3}H^3 V_{n-1}(R),$$

where $V_{n-1}(R)$ is the volume of the $(n-1)$-dimensional ball with radius R.

Problem 2.33. Answer:

$$a_1 \cdots a_n V_n(1),$$

where $V_n(1)$ is the volume of the n-dimensional ball with radius 1.

Problem 2.34. Answer:

$$\frac{1}{2} V_{n+1}(R),$$

where $V_{n+1}(R)$ is the volume of the $(n+1)$-dimensional ball with radius R.

Problem 2.35. Answer:

$$nV_n(1) \int_0^R r^{n-1} f(r) dr,$$

where $V_n(1)$ is the volume of the one-dimensional ball with radius 1.

Problem 2.36. Answer: $f(x_1, \dots, x_n)$, $(x_1, \dots, x_n) \in X$.

Problem 2.37. Answer: $\frac{m}{n} a^n$.

Chapter 3.2

Problem 3.4. Answer:
1. Convergent for $\alpha > 1$ and $\beta > 1$. Divergent in the remaining cases.
2. Convergent for $\frac{1}{\alpha} + \frac{1}{\beta} < 1$. Divergent for $\frac{1}{\alpha} + \frac{1}{\beta} \geq 1$.
3. Convergent for $p > 0$ and $\frac{1}{\alpha} + \frac{1}{\beta} < p$. Divergent for $p \leq 0$ and $\frac{1}{\alpha} + \frac{1}{\beta} \geq p > 0$.
4. Convergent for $p > \frac{3}{2}$. Divergent for $p \leq \frac{3}{2}$.
5. Convergent for $p > \frac{3}{2}$. Divergent for $p \leq \frac{3}{2}$.
6. Convergent for $p > 2$. Divergent for $p \leq 2$.

Problem 3.5. Answer:
1. Convergent for $\alpha > 1$. Divergent for $\alpha \leq 1$.
2. Convergent for $\alpha > \frac{1}{4}$. Divergent for $\alpha \leq \frac{1}{4}$.
3. Convergent for $\alpha > \frac{1}{2}$. Divergent for $\alpha \leq \frac{1}{2}$.

Problem 3.6. Answer:
1. Convergent for $p < 1$. Divergent for $p \geq 1$.
2. Convergent for $p < 1$. Divergent for $p \geq 1$.
3. Convergent for $p < 1$. Divergent for $p \geq 1$.

Problem 3.8. Answer:
1. Convergent for $\alpha < \frac{3}{2}$. Divergent for $\alpha \geq \frac{3}{2}$.

2. Convergent for $\alpha < 2$. Divergent for $\alpha \geq 2$.
3. Convergent for $\alpha < 3$. Divergent for $\alpha \geq 3$.
4. Divergent.
5. Divergent.
6. Convergent.
7. Convergent for all $\alpha \in \mathbb{R}$.

Problem 3.11. Answer:

1. Convergent for $p > \frac{3}{2}$. Divergent for $p \leq \frac{3}{2}$.
2. Convergent for $\frac{1}{p} + \frac{1}{q} + \frac{1}{r} < 1$. Divergent for $\frac{1}{p} + \frac{1}{q} + \frac{1}{r} \geq 1$.
3. Convergent for $p < \frac{3}{2}$. Divergent for $p \geq \frac{3}{2}$.
4. Convergent for $p < 1$. Divergent for $p \geq 1$.
5. Convergent for $p < 1$. Divergent for $p \geq 1$.
6. Convergent for $p < 1$. Divergent for $p \geq 1$.

Chapter 4

Problem 4.1. Answer:

1. $1 + \sqrt{2}$.
2. $-\sqrt{5}\log 2$.
3. $\frac{ab(a^2+ab+b^2)}{3(a+b)}$.
4. 24.
5. $\frac{\pi a^3}{2}$.
6. $2\pi a^{2n+1}$.
7. $\frac{\pi a^2}{2}$.
8. $2a^2$.
9. $a^2\sqrt{2}$.
10. $\frac{2a^3\sqrt{2}}{3}$.
11. $2a^2(2 - \sqrt{2})$.
12. $4a^{\frac{7}{3}}$.
13. $\frac{32a^2}{3}$.
14. $\frac{256a^3}{15}$.
15. $2\pi^2 a^3(1 + 2\pi^2)$.
16. $\frac{((1+4\pi^2)^{\frac{3}{2}}-1)a^2}{3}$.

Problem 4.2. Answer:

1.

$$\frac{\sqrt{a^2 + b^2}}{ab}\arctan\left(\frac{2\pi b}{a}\right).$$

2.

$$\frac{2\pi\sqrt{a^2+b^2}(3a^2+4\pi^2b^2)}{3}.$$

3. $2\pi a^2$.
4. $\frac{a^4}{6}$.
5. $a^2\sqrt{2}$.
6. $\frac{2\pi a^3}{3}$.

Problem 4.3. Answer:

1. 0.
2. $\frac{2}{3}$.
3. 2.
4. $\frac{8}{15}$.
5. -11.
6. $\frac{5-\log 8}{3}$.
7. $\frac{\pi a^2}{2}$.
8. $\frac{7}{12}$.
9. 56.
10. 8.
11. 6.
12. $12+\log 5$.
13. 4.
14. $-\frac{1}{4}$.
15. 0.
16. $-2\pi ab$.
17. $-\frac{4ab^2}{3}$.
18. πa^2.
19. $\frac{3\pi a^{\frac{4}{3}}}{16}$.
20. -48.
21. 4.
22. $-\frac{1}{2}$.
23. 0.
24. $\frac{4}{3}$.
25. 0.
26. -2π.
27. 0.
28. 0.
29. 0.
30. $-\pi a^2$.
31. $-\pi a^2(\cos\alpha)^2$.

32. 13.
33. $3\sqrt{3}$.
34. a^3.

Problem 4.4. Answer:
1. $-\frac{\pi a^3}{8}$.
2. $\frac{1-e^{\pi}}{5}$.
3. 0.
4. $\frac{\pi a^2}{8}$.
5. −4.
6. $\frac{\pi R^4}{4}$.
7. −2.

Problem 4.5. Answer:
1. $\frac{(7\pi+3)ab}{12}$.
2. π.
3. $\frac{a^2}{6}$.
4. $\frac{4}{3}$.
5. $\frac{8\pi}{3}$.
6. a^2.
7. $\frac{5\pi a^2}{8}$.
8. $\frac{(3\sqrt{3}+4\pi)}{9\sqrt{3}}$.
9. $\frac{3}{2}$.
10. $\frac{4a^2}{3}$.
11. $\frac{1}{30}$.

Problem 4.6. Answer:
1. 1.
2. −4.
3. $-\frac{1148}{5}$.
4. 0.
5. 30.

Chapter 5

Problem 5.1. Answer:
1. $2\sqrt{3}a^4$.
2. $\pi r(r^3 + 2r^2H + rH^2 + \frac{2H^3}{3})$.
3. $\frac{125\sqrt{5}-1}{420}$.
4. $\frac{\pi}{\sqrt{2}}$.
5. $\frac{2\pi\sqrt{2}}{3}$.

6. $\frac{64\sqrt{2}}{15}$.
7. $\frac{29\pi\sqrt{2}}{8}$.
8. $\frac{4}{3}\pi abc(\frac{1}{a^2} + \frac{1}{b^2} + \frac{1}{c^2})$.
9. $4\pi abc$.
10. 4π.
11. If $n \neq 2$, then

$$\frac{2\pi R}{a(n-2)}(|a-R|^{2-n} - |a+R|^{2-n}).$$

If $n = 2$, then

$$\frac{2\pi R}{a}\log\left|\frac{a+R}{a-R}\right|.$$

12.

$$\frac{\pi \sin\alpha(\cos\alpha)^2}{2}.$$

13. $\pi^2(\sqrt{2} + \log(1+\sqrt{2}))$.
14. If $a \leq \sqrt{3}$, then

$$\frac{\pi(a^2-3)^2}{18}.$$

If $a > \sqrt{3}$, then 0.
15. $\frac{\pi(8-5\sqrt{2})R^4}{6}$.

Problem 5.2. Answer:
1. 0.
2. $-\pi R^4$.
3. $\frac{8\pi(a+b+c)R^3}{3}$.
4. $-\frac{\pi R^4}{2}$.
5. $-\frac{2\pi}{5}$.
6. 0.
7. $\frac{4\pi abc}{3}$.
8. 0.
9. $\frac{4\pi ab}{c}$.
10. $\frac{\pi abc^2}{4}$.
11. $\frac{2\pi(a^2+b^2)abc}{5}$.
12. $-\frac{3\pi H^4}{2}$.
13. 0.
14. $-\frac{\pi r^5}{6}$.

15. $(\frac{\pi H}{8} - \frac{r}{3})r^2 H$.
16. $-\frac{\pi}{3}$.
17. $-\frac{a^4}{3}$.

Problem 5.3. Answer:
1. $\frac{\pi abc^2}{2}$.
2. $\frac{3a^5}{20}$.
3. $\frac{12\pi R^5}{5}$.
4. 0.
5. $\frac{\pi R^6}{3}$.

A Elements of theory of curves

A.1 Frenet curves in $\mathbb{R}^n$

Let $a, b \in \mathbb{R}$, $a < b$.

Definition A.1. A curve in $\mathbb{R}^n$ is a continuous function $f : [a, b] \to \mathbb{R}^n$. The function f that defines the curve is called a parameterization of the curve, and the curve is a parameterized curve.

Example A.1. Suppose that $f : [1, 5] \to \mathbb{R}^3$ is given by

$$f(t) = (t - 1, t, t^2 - 5), \quad t \in [1, 5].$$

Then $f : [1, 5] \to \mathbb{R}^3$ is a curve.

Definition A.2. A regular parameterized curve is a function $f : [a, b] \to \mathbb{R}^n$ such that $f \in \mathscr{C}^1([a, b])$ and

$$f'(t) \neq (0, 0, \dots, 0), \quad t \in [a, b].$$

Definition A.3. Let $f : [a, b] \to \mathbb{R}^n$ be a parameterized curve such that $f \in \mathscr{C}^1([a, b])$. Then the vector $f'(t_0)$ is called the tangent vector to f at t_0, and the line spanned by this vector through $f(t_0)$ is called the tangent to f at this point.

Example A.2. Let $f : [2, 4] \to \mathbb{R}^2$ be given by

$$f(t) = (t, t^3 + t^2), \quad t \in [2, 4].$$

Here

$$\begin{aligned} f_1(t) &= t, \\ f_2(t) &= t^3 + t^2, \quad t \in [2, 4]. \end{aligned}$$

Hence

$$\begin{aligned} f_1'(t) &= 1, \\ f_2'(t) &= 3t^2 + 2t, \quad t \in [2, 4]. \end{aligned}$$

Thus

$$\begin{aligned} f'(t) &= (f_1'(t), f_2'(t)) \\ &= (1, 3t^2 + 2t) \\ &\neq (0, 0), \quad t \in [2, 4]. \end{aligned}$$

Thus the considered curve is a regular curve.

https://doi.org/10.1515/9783112219607-008

Example A.3. Let $f : [0, \pi] \to \mathbb{R}$ be given by

$$f(t) = \frac{1}{2}(\cos(2t), \sin(2t)), \quad t \in [0, \pi].$$

This curve is a circle of radius $\frac{1}{2}$. Here

$$\begin{aligned} f_1(t) &= \frac{1}{2}\cos(2t), \\ f_2(t) &= \frac{1}{2}\sin(2t), \quad t \in [0, \pi]. \end{aligned}$$

Then

$$\begin{aligned} f_1'(t) &= -\sin(2t), \\ f_2'(t) &= \cos(2t), \quad t \in [0, \pi]. \end{aligned}$$

Hence

$$\begin{aligned} f'(t) &= (f_1'(t), f_2'(t)) \\ &= (-\sin(2t), \cos(2t)) \\ &\neq (0, 0), \quad t \in [0, \pi]. \end{aligned}$$

Thus the considered curve is a regular curve.

Example A.4. Let $f : [0, 2\pi] \to \mathbb{R}$ be given by

$$f(t) = (\cos t, \sin t, t), \quad t \in [0, 2\pi].$$

This curve is called a (circular) helix. Here

$$\begin{aligned} f_1(t) &= \cos t, \\ f_2(t) &= \sin t, \\ f_3(t) &= t, \quad t \in [0, 2\pi]. \end{aligned}$$

Then

$$\begin{aligned} f_1'(t) &= -\sin t, \\ f_2'(t) &= \cos t, \\ f_3'(t) &= 1, \quad t \in [0, 2\pi]. \end{aligned}$$

Hence

$$\begin{aligned} f'(t) &= (f_1'(t), f_2'(t), f_3'(t)) \\ &= (-\sin t, \cos t, 1) \\ &\neq (0,0,0), \quad t \in [0, 2\pi]. \end{aligned}$$

Thus the considered curve is a regular curve.

Exercise A.1. Prove that the following curves are regular:

1. $n = 2$,

$$f(t) = (2t, -4t), \quad t \in [-10, 20].$$

2. $n = 3$,

$$f(t) = (2\cos(3t), 3\sin(4t), 5t), \quad t \in [0, 2\pi].$$

3. $n = 2$,

$$f(t) = (t^2, t^3), \quad t \in [1, 10].$$

Definition A.4. Let $[\alpha, \beta] \subset \mathbb{R}$, and let $\phi : [\alpha, \beta] \to [a, b]$ be such that $\phi \in \mathscr{C}^1([\alpha, \beta])$ and $\phi'(t) > 0$, $t \in [\alpha, \beta]$. Let also, $f : [a, b] \to \mathbb{R}$ be a regular curve. Then the curves f and $f \circ \phi$ are said to be equivalent.

Example A.5. Let $f : [0, \frac{2}{3}] \to \mathbb{R}^2$ be given by

$$f(t) = (1 + t, t^2), \quad t \in \left[0, \frac{2}{3}\right].$$

Here

$$\begin{aligned} f_1(t) &= 1 + t, \\ f_2(t) &= t^2, \quad t \in \left[0, \frac{2}{3}\right]. \end{aligned}$$

Then

$$\begin{aligned} f_1'(t) &= 1, \\ f_2'(t) &= 2t, \quad t \in \left[0, \frac{2}{3}\right], \end{aligned}$$

and

$$\begin{aligned} f'(t) &= (f_1'(t), f_2'(t)) \\ &= (1, 2t), \quad t \in \left[0, \frac{2}{3}\right], \end{aligned}$$

i. e., $f : [0, \frac{2}{3}] \to \mathbb{R}^2$ is a regular curve. Let also, $\phi : [0,2] \to [0, \frac{2}{3}]$ be given by

$$\phi(t) = \frac{t}{1+t}, \quad t \in [0,2].$$

Then $\phi : [0,2] \to [0, \frac{2}{3}]$, and

$$\begin{aligned}\phi'(t) &= \frac{1+t-t}{(1+t)^2}\\ &= \frac{1}{(1+t)^2}, \quad t \in [0,2].\end{aligned}$$

Next,

$$\begin{aligned}f \circ \phi(t) &= \left(1 + \frac{t}{1+t}, \frac{t^2}{(1+t)^2}\right)\\ &= \left(\frac{2t+1}{1+t}, \frac{t^2}{(1+t)^2}\right), \quad t \in [0,2].\end{aligned}$$

Therefore f and $f \circ \phi$ are equivalent.

Example A.6. Consider the Diocles cissoid

$$f(t) = \left(\frac{a}{t^2+1}, \frac{a}{t(t^2+1)}\right), \quad t \in [2,5],$$

with parameter $a > 0$. Here

$$\begin{aligned}f_1(t) &= \frac{a}{t^2+1},\\ f_2(t) &= \frac{a}{t(t^2+1)}, \quad t \in [2,5].\end{aligned}$$

Then

$$\begin{aligned}f_1'(t) &= -\frac{2at}{(t^2+1)^2},\\ f_2'(t) &= -\frac{a(3t^2+1)}{t^2(t^2+1)^2}, \quad t \in [2,5].\end{aligned}$$

Hence

$$\begin{aligned}f'(t) &= (f_1'(t), f_2'(t))\\ &\neq (0,0), \quad t \in [2,5],\end{aligned}$$

i. e., $f : [2,5] \to \mathbb{R}^2$ is a regular curve. Let

$$\phi(t) = 1 + t^2, \quad t \in [1,2].$$

Then $\phi : [1,2] \to [2,5]$,

$$\phi'(t) = 2t \\ > 0, \quad t \in [1,2],$$

and

$$\begin{aligned} f \circ \phi(t) &= f(\phi(t)) \\ &= \left(\frac{a}{1+(1+t^2)^2}, \frac{a}{(1+t^2)((1+t^2)^2+1)}\right) \\ &= \left(\frac{a}{2+2t^2+t^4}, \frac{a}{(1+t^2)(2+2t^2+t^4)}\right), \quad t \in [1,2]. \end{aligned}$$

Thus f and $f \circ \phi$ are equivalent curves.

Example A.7. Consider the witch of Maria Agnesi

$$f(t) = (a\cos t, a(\sin t)^2), \quad t \in \left[\frac{\pi}{4}, \frac{\pi}{2}\right],$$

with parameter $a > 0$. Here

$$\begin{aligned} f_1(t) &= a\cos t, \\ f_2(t) &= a(\sin t)^2, \quad t \in \left[\frac{\pi}{4}, \frac{\pi}{2}\right]. \end{aligned}$$

Then

$$\begin{aligned} f_1'(t) &= -a\sin t, \\ f_2'(t) &= a\sin(2t), \quad t \in \left[\frac{\pi}{4}, \frac{\pi}{2}\right], \end{aligned}$$

and

$$\begin{aligned} f'(t) &= (f_1'(t), f_2'(t)) \\ &= (-a\sin t, a\sin(2t)) \\ &\neq (0,0), \quad t \in \left[\frac{\pi}{4}, \frac{\pi}{2}\right], \end{aligned}$$

i. e., $f : [\frac{\pi}{4}, \frac{\pi}{2}] \to \mathbb{R}^2$ is a regular curve. Let

$$\phi(t) = \arcsin t, \quad t\left[\frac{\sqrt{2}}{2}, 1\right].$$

Then $\phi : [\frac{\sqrt{2}}{2}, 1] \to [\frac{\pi}{4}, \frac{\pi}{2}]$, and

$$\phi'(t) = \frac{1}{\sqrt{1-t^2}}$$
$$> 0, \quad t \in \left[\frac{\pi}{4}, \frac{\pi}{2}\right],$$

and

$$\begin{aligned} f \circ \phi(t) &= f(\phi(t)) \\ &= (a\cos(\arcsin t), a(\sin(\arcsin t))^2) \\ &= (a\sqrt{1-t^2}, at^2), \quad t \in \left[\frac{\sqrt{2}}{2}, 1\right]. \end{aligned}$$

We get that f and $f \circ \phi$ are equivalent curves.

Exercise A.2. Prove that the following curves are equivalent:

1. (strophoid)

$$f(t) = \left(\frac{2at^2}{1+t^2}, \frac{a+(t^2-1)}{1+t^2}\right), \quad t \in \left[\frac{1}{16}, \frac{1}{4}\right],$$

   and

$$g(t) = \left(\frac{2at^4}{1+t^4}, \frac{at^2(t^4-1)}{1+t^4}\right), \quad t \in \left[\frac{1}{4}, \frac{1}{2}\right].$$

2. (astroid)

$$f(t) = (a(\cos t)^3, a(\sin t)^3), \quad t \in \left[\frac{\pi}{4}, \frac{\pi}{2}\right],$$

   and

$$g(t) = (a(1-t^2)^{\frac{3}{2}}, at^3), \quad t \in \left[\frac{\sqrt{2}}{2}, 1\right].$$

3. (cycloid)

$$f(t) = (a(t-\sin t), a(1-\cos t)), \quad t \in \left[\frac{\pi}{4}, \frac{\pi}{2}\right],$$

   and

$$g(t) = (a(\arcsin t - t), a(1-\sqrt{1-t^2})), \quad t \in \left[\frac{\sqrt{2}}{2}, 1\right].$$

Hint A.1. Use the following functions:

1. $\phi(t) = t^2, t \in [\frac{1}{4}, \frac{1}{2}]$.
2. $\phi(t) = \arcsin t, t \in [\frac{\sqrt{2}}{2}, 1]$.
3. $\phi(t) = \arcsin t, t \in [\frac{\sqrt{2}}{2}, 1]$.

Definition A.5. Let $f : [a,b] \to \mathbb{R}^n$ be a regular parameterized curve,

$$f(t) = (f_1(t), f_2(t), \dots, f_n(t)), \quad t \in [a,b].$$

The arc length parameter $L_f(t,a)$, $t \in [a,b]$, is defined as follows:

$$L_f(t,a) = \int_a^t |f'(s)| ds.$$

Example A.8. Consider the circle given in Example A.3. By the computations in Example A.3 we get

$$\begin{aligned} |f'(t)| &= \sqrt{(f_1'(t))^2 + (f_2'(t))^2} \\ &= \sqrt{(-\sin(2t))^2 + (\cos(2t))^2} \\ &= 1, \quad t \in [0,\pi]. \end{aligned}$$

Hence

$$\begin{aligned} L_f(t,0) &= \int_0^t |f'(s)| ds \\ &= \int_0^t ds \\ &= t, \quad t \in [0,\pi]. \end{aligned}$$

Example A.9. Consider the circular helix given in Example A.4. By the computations in Example A.4 we find

$$\begin{aligned} |f'(t)| &= \sqrt{(f_1'(t))^2 + (f_2'(t))^2 + (f_3'(t))^2} \\ &= \sqrt{(-\sin t)^2 + (\cos t)^2 + 1} \\ &= \sqrt{2}, \quad t \in [0,2\pi]. \end{aligned}$$

Then

$$
\begin{aligned}
L_f(t,0) &= \int_0^t |f'(s)|ds \\
&= \int_0^t \sqrt{2}ds \\
&= \sqrt{2}t, \quad t \in [0, 2\pi].
\end{aligned}
$$

Example A.10. Consider the witch of Maria Agnesi given in Example A.7. By the computations in Example A.7 we find

$$
\begin{aligned}
|f'(t)| &= \sqrt{(f_1'(t))^2 + (f_2'(t))^2} \\
&= \sqrt{(-a\sin t)^2 + (a\sin(2t))^2} \\
&= a\sqrt{(\sin t)^2 + 4(\sin t)^2(\cos t)^2} \\
&= a\sin t\sqrt{1 + 4(\cos t)^2}, \quad t \in \left[\frac{\pi}{4}, \frac{\pi}{2}\right].
\end{aligned}
$$

Hence

$$
\begin{aligned}
L_f\left(t, \frac{\pi}{4}\right) &= \int_{\frac{\pi}{4}}^t |f'(s)|ds \\
&= a\int_{\frac{\pi}{4}}^t \sin s\sqrt{1 + 4(\cos s)^2}ds \\
&= -a\int_{\frac{\pi}{4}}^t \sqrt{1 + 4(\cos s)^2}d(\cos s), \quad t \in \left[\frac{\pi}{4}, \frac{\pi}{2}\right].
\end{aligned}
$$

Note that

$$
\int \sqrt{1 + 4y^2}dy = \frac{1}{2}y\sqrt{1 + 4y^2} + \frac{1}{4}\log(2y + \sqrt{1 + 4y^2}) + c, \quad y \in \mathbb{R},
$$

where c is a constant. Hence

$$
\begin{aligned}
L_f\left(t, \frac{\pi}{4}\right) &= -a\left(\frac{1}{2}y\sqrt{1 + 4y^2}\Big|_{y=\frac{\pi}{4}}^{y=t} + \frac{1}{4}\log(2y + \sqrt{1 + 4y^2})\Big|_{y=\frac{\pi}{4}}^{y=t}\right) \\
&= -a\left(\frac{1}{2}t\sqrt{1 + 4t^2} + \frac{1}{4}\log(2t + \sqrt{1 + 4t^2})\right) \\
&\quad + a\left(\frac{\pi}{8}\sqrt{1 + \frac{\pi^2}{4}} + \frac{1}{4}\log\left(\frac{\pi}{2} + \sqrt{1 + \frac{\pi^2}{4}}\right)\right), \quad t \in \left[\frac{\pi}{4}, \frac{\pi}{2}\right].
\end{aligned}
$$

Exercise A.3. Find the arc lengths of the following curves:

1.
$$f(t) = (t, t^{\frac{3}{2}}), \quad t \in [a,b].$$

2.
$$f(t) = (t, t^2), \quad t \in [a,b].$$

3.
$$f(t) = (t, \log t), \quad t \in [a,b].$$

4.
$$f(t) = \left(t, c\cosh\left(\frac{t}{c}\right)\right), \quad t \in [a,b],$$

   where $c > 0$ is a given constant.

5.
$$f(t) = (t, e^t), \quad t \in [a,b].$$

Answer A.1.

1.
$$L_f(t,a) = \frac{1}{27}((4+9t)^{\frac{3}{2}} - (4+9a)^{\frac{3}{2}}), \quad t \in [a,b].$$

2.
$$L_f(t,a) = \sqrt{1+t^2} - \sqrt{1+a^2} + \log\left|\frac{t}{a}\right| + \log\frac{1+\sqrt{1+a^2}}{1+\sqrt{1+t^2}}, \quad t \in [a,b].$$

3.
$$L_f(t,a) = c\left(\sinh\left(\frac{t}{c}\right) - \sinh\left(\frac{a}{c}\right)\right), \quad t \in [a,b].$$

4.
$$L_f(t,a) = t - a + \sqrt{1+e^{2t}} - \sqrt{1+e^{2a}} + \log\frac{1+\sqrt{1+e^{2a}}}{1+\sqrt{1+e^{2t}}}, \quad t \in [a,b].$$

Definition A.6. We will say that a regular parameterized curve is naturally parameterized if

$$|f'(s)| = 1, \quad s \in [a, b].$$

Usually, the natural parameter is denoted by s.

Exercise A.4. Prove that the arc lengths of any two equivalent curves are equal.

Solution. Let $f : [a, b] \to \mathbb{R}^n$ be a regular curve, $\phi : [\alpha, \beta] \to [a, b]$, $\phi(\alpha) = a$, $\phi(\beta) = b$, $\phi \in \mathscr{C}^1([\alpha, \beta])$, $\phi' > 0$ on $[\alpha, \beta]$. Then f and $f \circ \phi$ are equivalent. Set

$$g = f \circ \phi.$$

Then

$$\begin{aligned} L_g(\beta, \alpha) &= \int_\alpha^\beta |g'(\tau)| d\tau \\ &= \int_\alpha^\beta |f'(\phi(\tau))\phi'(\tau)| d\tau \\ &= \int_\alpha^\beta |f'(\phi(\tau))| \phi'(\tau) d\tau \\ &= \int_\alpha^\beta |f'(\phi(\tau))| d\phi(\tau) \\ &= \int_a^b |f'(t)| dt \\ &= L_f(b, a). \end{aligned}$$

This completes the solution.

Exercise A.5. Prove that for any regular parameterized curve, there is a naturally parameterized curve that is equivalent to it.

Solution. Let $f : [a, b] \to \mathbb{R}^n$ be a regular parameterized curve. Define

$$\phi(t) = \int_a^t |f'(\tau)| d\tau, \quad t \in [a, b].$$

We have that $\phi \in \mathscr{C}^1([a, b])$, $\phi > 0$ on $[a, b]$, and

$$\begin{aligned}\phi'(t) &= |f'(t)| \\ &> 0, \quad t \in [a,b].\end{aligned}$$

Hence $\phi' > 0$ on $[a,b]$, and $\phi : [a,b] \to [\alpha,\beta]$ is a diffeomorphism for some interval $[\alpha,\beta] \subset \mathbb{R}$. Note that

$$\begin{aligned}(\phi^{-1})'(s) &= \frac{1}{\phi'(\phi^{-1}(s))} \\ &= \frac{1}{|f'(\phi^{-1}(s))|}, \quad s \in [\alpha,\beta].\end{aligned}$$

Let

$$g(t) = f(\phi^{-1}(t)), \quad t \in [a,b].$$

Then

$$\begin{aligned}g'(t) &= f'(\phi^{-1}(t))(\phi^{-1})'(t) \\ &= \frac{f'(\phi^{-1}(t))}{|f'(\phi^{-1}(t))|}, \quad t \in [a,b],\end{aligned}$$

and

$$|g'(t)| = 1, \quad t \in [a,b].$$

This completes the solution.

Remark A.1. Note that

$$f(t) = g(s(t)), \quad t \in [a,b],$$

and

$$\begin{aligned}f'(t) &= g'(s(t))s'(t) \\ &= g'(s(t))|f'(t)|, \quad t \in [a,b].\end{aligned}$$

Example A.11. Consider the curve

$$f(t) = (at^3, bt^3), \quad t \in [2,10],$$

where $a, b \in \mathbb{R}$ are such that $a^2 + b^2 \neq 0$. Here

$$\begin{aligned}f_1(t) &= at^3, \\ f_2(t) &= bt^3, \quad t \in [2,10].\end{aligned}$$

Then

$$f_1'(t) = 3at^2,$$
$$f_2'(t) = 3bt^2, \quad t \in [2, 10],$$

and

$$\begin{aligned}|f'(t)| &= \sqrt{(f_1'(t))^2 + (f_2'(t))^2} \\ &= \sqrt{(3at^2) + (3bt^2)^2} \\ &= \sqrt{9a^2t^4 + 9b^2t^4} \\ &= 3t^2\sqrt{a^2 + b^2}, \quad t \in [2, 10].\end{aligned}$$

Thus $f : [2, 10] \to \mathbb{R}$ is a regular curve that is not naturally parameterized. For its arc length, we have

$$\begin{aligned}L_f(t, 2) &= -\int_2^t |f'(s)|ds \\ &= 3\sqrt{a^2 + b^2}\int_2^t s^2 ds \\ &= 3\sqrt{a^2 + b^2}\frac{s^3}{3}\Big|_{s=2}^{s=t} \\ &= \sqrt{a^2 + b^2}(t^3 - 8), \quad t \in [2, 10].\end{aligned}$$

Hence

$$\frac{1}{\sqrt{a^2 + b^2}}L_f(t, 2) = t^3 - 8, \quad t \in [2, 10],$$
$$t^3 = \frac{1}{\sqrt{a^2 + b^2}}L_f(t, 2) + 8, \quad t \in [2, 10],$$

and

$$t = \sqrt[3]{\frac{1}{\sqrt{a^2 + b^2}}L_f(t, 2) + 8}, \quad t \in [2, 10].$$

Therefore

$$L_f^{-1}(t, 2) = \sqrt[3]{\frac{1}{\sqrt{a^2 + b^2}}t + 8}, \quad t \in \left[0, 992\sqrt{a^2 + b^2}\right].$$

Consider the curve

$$g(t) = f \circ L_f^{-1}(t,2), \quad t \in [0, 992\sqrt{a^2+b^2}].$$

Then

$$\begin{aligned} g(t) &= \left(a(L_f^{-1}(t,2))^3, b(L_f^{-1}(t,2))^3\right) \\ &= \left(a\left(\frac{1}{\sqrt{a^2+b^2}} + 8\right), b\left(\frac{1}{\sqrt{a^2+b^2}}t + 8\right)\right), \quad t \in [0, 992\sqrt{a^2+b^2}]. \end{aligned}$$

Note that

$$\begin{aligned} g_1(t) &= a\left(\frac{1}{\sqrt{a^2+b^2}}t + 8\right), \\ g_2(t) &= b\left(\frac{1}{\sqrt{a^2+b^2}}t + 8\right), \quad t \in [0, 992\sqrt{a^2+b^2}], \end{aligned}$$

and

$$\begin{aligned} g_1'(t) &= \frac{a}{\sqrt{a^2+b^2}}, \\ g_2'(t) &= \frac{b}{\sqrt{a^2+b^2}}, \quad t \in [0, 992\sqrt{a^2+b^2}]. \end{aligned}$$

Then

$$\begin{aligned} |g'(t)| &= \sqrt{(g_1'(t))^2 + (g_2'(t))^2} \\ &= \sqrt{\frac{a^2}{a^2+b^2} + \frac{b^2}{a^2+b^2}} \\ &= 1, \quad t \in [0, 992\sqrt{a^2+b^2}]. \end{aligned}$$

Thus $g : [0, 992\sqrt{a^2+b^2}] \to \mathbb{R}^2$ is a naturally parameterized curve that is equivalent to the curve f.

Example A.12. Consider the circle

$$f(t) = \left(\frac{1}{a}\cos(at), \frac{1}{a}\sin(at)\right), \quad t \in [0, 2\pi].$$

Here

$$\begin{aligned} f_1(t) &= \frac{1}{a}\cos(at), \\ f_2(t) &= \frac{1}{a}\sin(at), \quad t \in [0, 2\pi]. \end{aligned}$$

Then

$$f_1'(t) = -\sin(at),$$
$$f_2'(t) = \cos(at), \quad t \in [0, 2\pi],$$

and

$$|f'(t)| = \sqrt{(f_1'(t))^2 + (f_2'(t))^2}$$
$$= \sqrt{(\sin(at))^2 + (\cos(at))^2}$$
$$= 1, \quad t \in [0, 2\pi].$$

Thus $f : [0, 2\pi] \to \mathbb{R}$ is a naturally parameterized curve.

Example A.13. Consider the circular helix

$$f(t) = (c\cos(at), c\sin(at), \beta t), \quad t \in [a, b],$$

where $\alpha, \beta, c \in \mathbb{R}$ are such that $\alpha^2 c^2 + \beta^2 \neq 0$. Here

$$f_1(t) = c\cos(at),$$
$$f_2(t) = c\sin(at),$$
$$f_3(t) = \beta t, \quad t \in [a, b].$$

Then

$$f_1'(t) = -ca\sin(at),$$
$$f_2'(t) = ca\sin(at),$$
$$f_3'(t) = \beta, \quad t \in [a, b],$$

and

$$|f'(t)| = \sqrt{(f_1'(t))^2 + (f_2'(t))^2 + (f_3'(t))^2}$$
$$= \sqrt{(-ca\sin(at))^2 + (ca\cos(at))^2 + \beta^2}$$
$$= \sqrt{c^2a^2(\sin(at))^2 + c^2a^2(\cos(at))^2 + \beta^2}$$
$$= \sqrt{c^2a^2 + \beta^2}, \quad t \in [a, b].$$

Thus $f : [a, b] \to \mathbb{R}^3$ is a regular curve that is not naturally parameterized. For its arc length, we have

$$
\begin{aligned}
L_f(t,a) &= \int_a^t |f'(s)|ds \\
&= \int_a^t \sqrt{c^2\alpha^2+\beta^2}ds \\
&= \sqrt{c^2\alpha^2+\beta^2}(t-a), \quad t \in [a,b],
\end{aligned}
$$

whereupon

$$
t-a = \frac{1}{\sqrt{c^2\alpha^2+\beta^2}}L_f(t,a), \quad t \in [a,b],
$$

and

$$
t = a + \frac{1}{\sqrt{c^2\alpha^2+\beta^2}}L_f(t,a), \quad t \in [a,b].
$$

Therefore

$$
L_f^{-1}(t,a) = a + \frac{1}{\sqrt{c^2\alpha^2+\beta^2}}t, \quad t \in [0,(b-a)\sqrt{c^2\alpha^2+\beta^2}].
$$

Consider the curve

$$
g(t) = f \circ L_f^{-1}(t,a), \quad t \in [a,b].
$$

We have

$$
\begin{aligned}
g(t) = \Bigg(& c\cos\left(\alpha\left(a + \frac{1}{\sqrt{c^2\alpha^2+\beta^2}}t\right)\right), c\sin\left(\alpha\left(a + \frac{1}{\sqrt{c^2\alpha^2+\beta^2}}t\right)\right), \\
& \beta\left(a + \frac{1}{\sqrt{c^2\alpha^2+\beta^2}}t\right)\Bigg), \quad t \in [0,(b-a)\sqrt{c^2\alpha^2+\beta^2}].
\end{aligned}
$$

Here

$$
\begin{aligned}
g_1(t) &= c\cos\left(\alpha\left(a + \frac{1}{\sqrt{c^2\alpha^2+\beta^2}}\right)\right), \\
g_2(t) &= c\sin\left(\alpha\left(a + \frac{1}{\sqrt{c^2\alpha^2+\beta^2}}t\right)\right), \\
g_3(t) &= \beta\left(a + \frac{1}{\sqrt{c^2\alpha^2+\beta^2}}t\right), \quad t \in [0,(b-a)\sqrt{c^2\alpha^2+\beta^2}].
\end{aligned}
$$

Then

$$
\begin{aligned}
g_1'(t) &= -\frac{c\alpha}{\sqrt{c^2\alpha^2+\beta^2}}\sin\left(\alpha\left(a+\frac{1}{\sqrt{c^2\alpha^2+\beta^2}}t\right)\right),\\
g_2'(t) &= \frac{c\alpha}{\sqrt{c^2\alpha^2+\beta^2}}\cos\left(\alpha\left(a+\frac{1}{\sqrt{c^2\alpha^2+\beta^2}}\right)\right),\\
g_3(t) &= \frac{\beta}{\sqrt{c^2\alpha^2+\beta^2}}, \qquad t\in[0,(b-a)\sqrt{c^2\alpha^2+\beta^2}],
\end{aligned}
$$

and

$$
\begin{aligned}
|g'(t)| &= \left(\left(-\frac{c\alpha}{\sqrt{c^2\alpha^2+\beta^2}}\cos\left(\alpha\left(a+\frac{1}{\sqrt{c^2\alpha^2+\beta^2}}t\right)\right)\right)^2\right.\\
&\quad \left.+\left(\frac{c\alpha}{\sqrt{c^2\alpha^2+\beta^2}}\sin\left(\alpha\left(a+\frac{1}{\sqrt{c^2\alpha^2+\beta^2}}t\right)\right)\right)^2+\left(\frac{\beta}{\sqrt{c^2\alpha^2+\beta^2}}\right)^2\right)^{\frac{1}{2}}\\
&= \sqrt{\frac{c^2\alpha^2}{c^2\alpha^2+\beta^2}+\frac{\beta^2}{c^2\alpha^2+\beta^2}}\\
&= 1, \quad t\in[0,(b-a)\sqrt{c^2\alpha^2+\beta^2}].
\end{aligned}
$$

Thus $g : [0,(b-a)\sqrt{c^2\alpha^2+\beta^2}] \to \mathbb{R}^3$ is a naturally parameterized curve, and f and g are equivalent.

Exercise A.6. Prove that the curves in Exercise A.3 are not naturally parameterized.

A.2 Analytical representations of curves

Let $I \subset \mathbb{R}$.

Definition A.7. A subset $M \subset \mathbb{R}^n$ is called a regular curve or a one-dimensional smooth manifold of $\mathbb{R}^n$ if for each point $t_0 \in M$, there is a regular parameterized curve $f : I \to \mathbb{R}^n$ whose support $f(I)$ is an open neighborhood in M of the point t_0, i. e., is a set of the form $M \cap U$, where U is an neighborhood of t_0 in $\mathbb{R}^n$, and the map $f : I \to f(I)$ is a homeomorphism with respect to the topology of the subspace of $f(I)$. A parameterized curve with these properties is called a local parameterization of the curve M around the point t_0. If for a curve M, there is a local parameterization that is global, i. e., $f(I) = M$, then the curve is called a simple curve.

A.2.1 Plane curves

Definition A.8. A regular curve $M \subset \mathbb{R}^3$ is called a plane curve if it is contained in a plane π. We will usually assume that the plane π coincides with the coordinate plane xOy with $O = (0, 0)$.

A.2.1.1 Parametric representation

We choose an arbitrary local parameterization $(I, f(t)) = (f_1(t), f_2(t), f_3(t))$ of the curve. Then the support $f(t)$ of this local parameterization is an open subset of the curve. For a global parametrization of a simple curve, $f(I)$ is the entire curve. Thus each point t_0 of the curve has an open neighborhood that is the support of the parameterized curve

$$\begin{aligned} x &= f_1(t), \\ y &= f_2(t). \end{aligned} \tag{A.1}$$

Definition A.9. Equations (A.1) are called the parametric equations of the curve in a neighborhood of the point t_0. Usually, unless the curve is simple, we cannot use the same set of equations to describe the points of the entire curve.

Example A.14. Let $I = [0, 20]$. Then

$$\begin{aligned} f_1(t) &= t^2 + t + 1, \\ f_2(t) &= 10 + \sin t + \cos t + \frac{1 - t + t^2}{1 + t^2 + t^4}, \qquad t \in I, \end{aligned}$$

is a parametric representation of a plane curve.

Example A.15. Let $I = [0, 2\pi]$. Then

$$\begin{aligned} f_1(t) &= a(\cos t)^3, \\ f_2(t) &= a(\sin t)^3, \quad t \in [0, 2\pi], \end{aligned}$$

where a is a positive constant, is a parametric representation of the astroid.

Example A.16. Let $I = [1, 2\pi]$. Then

$$\begin{aligned} f_1(t) &= \frac{a}{2}\left(t + \frac{1}{t}\right), \\ f_2(t) &= \frac{b}{2}\left(t - \frac{1}{t}\right), \quad t \in I, \end{aligned}$$

where a and b are positive constants, is a parametric representation of the hyperbola.

A.2.1.2 Explicit representation

Suppose that I is an open interval in $\mathbb{R}$ and $f : I \to \mathbb{R}$ is a smooth function, i. e., $f \in \mathscr{C}^1(I)$. Then its graph

$$C = \{(t, f(t)) : t \in I\} \tag{A.2}$$

is a simple curve, which has the global representation

$$\begin{aligned} x &= t, \\ y &= f(t), \quad t \in I. \end{aligned}$$

Definition A.10. The equation

$$y = f(x)$$

is called the explicit equation of the curve (A.2). Sometimes, an explicit representation of a plane curve is also called a nonparametric form.

Example A.17. Let $I = (1, 128)$. Then

$$y = \frac{x+1}{1+x+x^2}, \quad x \in I,$$

is an explicit representation of a plane curve.

Example A.18. Let $I = (0, 15)$. Then

$$y = x^2, \quad x \in I,$$

is an explicit representation of a parabola.

Example A.19. Let $I = (1, 24)$. Then

$$y = \frac{1}{1+x}, \quad x \in I,$$

is an explicit equation of a hyperbola.

A.2.1.3 Implicit representation

Let $D \subset \mathbb{R}^2$. Let $F : D \to \mathbb{R}$ be a smooth function, and let

$$C = \{(x, y) \in D : F(x, y) = 0\}$$

be the 0-level set of the function F. In the general case, C is not a regular curve. Nevertheless, if at a point $(x_0, y_0) \in C$, the vector gradient

$$\operatorname{grad} F(x_0,y_0) = (F'_x(x_0,y_0), F'_y(x_0,y_0))$$

is not vanishing, then there exists an open neighborhood U of (x_0,y_0) and a smooth function $y = f(x)$ defined on an open neighborhood $I \subset \mathbb{R}$ of x_0 such that

$$C \cap U = \{(x,f(x)) : x \in I\}.$$

If $\operatorname{grad} F \neq 0$ at all points of C, then C is a regular curve.

Example A.20. The equation

$$(x^2+y^2)^2 - 2b^2(x^2-y^2) = a^4 - b^4,$$

where a and b are positive constants, is the equation of the Cassini oval.

Example A.21. The equation

$$(x^2+y^2)^3 - 4a^2x^2y^2 = 0$$

is the equation of the four-petaled rosette.

Example A.22. The equation

$$(x^2+y^2-2ax)^2 = 4b^2(x^2+y^2)$$

is the equation of the cardioid.

Remark A.2. Note that the condition for non-singularity of $\operatorname{grad} F$ is only a sufficient condition for the equation $F(x,y) = 0$ to represent a curve. If $\operatorname{grad} F(x_0,y_0) = 0$ for some $(x_0,y_0) \in D$, then we cannot claim that the equation represents a curve in a neighborhood of that point and the opposite.

A.2.2 Space curves

A.2.2.1 Parametric representation

As in the case of plane curve with local parameterization

$$\begin{aligned} x &= f_1(t), \\ y &= f_2(t), \\ z &= f_3(t), \quad t \in I, \end{aligned}$$

we can represent either the entire curve or only a neighborhood of one of its points.

Example A.23. Let $I = \mathbb{R}$. Then

$$
\begin{aligned}
x &= t+1,\\
y &= t^2+1,\\
z &= \frac{1}{1+t+t^2}, \quad t \in \mathbb{R},
\end{aligned}
$$

is a parametric representation of a space curve.

Example A.24. Let $I = \mathbb{R}$. The equations

$$
\begin{aligned}
x &= at\cos t,\\
y &= at\sin t,\\
z &= \frac{a^2t^2}{2p}, \quad t \in I,
\end{aligned}
$$

where a and p are positive constants, are a parametric representation of the Archimed spiral.

Example A.25. Let $I = \mathbb{R}$. The equations

$$
\begin{aligned}
x &= \cos(3t),\\
y &= \sin(3t),\\
z &= 2t, \quad t \in I,
\end{aligned}
$$

are a parametric representation of the circular helix.

A.2.2.2 Explicit representation

Let $f, g : I \to \mathbb{R}$ be two smooth functions on an open interval $I \subset \mathbb{R}$. Then the set

$$C = \{(x, f(x), g(x)) \subset \mathbb{R}^3 : x \in I\}$$

is a simple curve with a global representation given by

$$
\begin{aligned}
x &= t,\\
y &= f(t),\\
z &= g(t), \quad t \in I.
\end{aligned}
$$

Definition A.11. The equations

$$
\begin{aligned}
y &= f(x),\\
z &= g(x), \quad x \in I,
\end{aligned}
$$

are called explicit equations of the curve.

Example A.26. Let $I = [1, 20]$. Then

$$y = x + 1,$$
$$z = x^2 + x + 1, \quad x \in I,$$

is an explicit representation of a space curve.

Example A.27. Let $I = [0, 2\pi]$. Then

$$x = \cos(4z),$$
$$y = \sin(4z),$$
$$z = z, \quad z \in I,$$

is an explicit representation of the circular helix.

A.2.2.3 Implicit representation

Let $D \subset \mathbb{R}^3$, and let $F, G : D \to \mathbb{R}$ be smooth functions. Consider the set

$$C = \{(x, y, z) \in D : F(x, y, z) = 0,\ G(x, y, z) = 0\},$$

i. e., the set of solutions of the system

$$F(x, y, z) = 0,$$
$$G(x, y, z) = 0.$$

In the general case, the set C is not a regular curve. Nevertheless, if $a = (x_0, y_0, z_0) \in C$ and

$$\operatorname{rank}\begin{pmatrix} F'_x(a) & F'_y(a) & F'_z(a) \\ G'_x(a) & G'_y(a) & G'_z(a) \end{pmatrix} = 2,$$

then there is an open neighborhood $U \subset D$ of the point a such that $C \cap U$ is a curve. If the rank of the matrix

$$\begin{pmatrix} F'_x & F'_y & F'_z \\ G'_x & G'_y & G'_z \end{pmatrix}$$

is equal to two, then C is a curve.

Example A.28. The equations

$$x^2 + y^2 + z^2 = 4b^2,$$
$$(x - b)^2 + y^2 = b^2,$$

where b is a positive constant, are the equations of the temple of Viviani.

A.3 The tangent and the normal plane

Suppose that $I \subseteq \mathbb{R}$, $t_0 \in I$, and $f : I \to \mathbb{R}^n$ is a curve such that $f'(t_0) \neq 0$.

Definition A.12. The line passing through $f(t_0)$ and having direction of the vector $f'(t_0)$ is called the tangent of the curve at the point $f(t_0)$ (or at the point t_0).

The equations of the tangent line read as follows:

$$\begin{aligned} F(\tau) &= f(t_0) + \tau f'(t_0) \\ &= (f_1(t_0), f_2(t_0), \ldots, f_n(t_0)) + \tau(f_1'(t_0), f_2'(t_0), \ldots, f_n'(t_0)) \\ &= (f_1(t_0), f_2(t_0), \ldots, f_n(t_0)) + (\tau f_1'(t_0), \tau f_2'(t_0), \ldots, \tau f_n'(t_0)) \\ &= (f_1(t_0) + \tau f_1'(t_0), f_2(t_0) + \tau f_2'(t_0), \ldots, f_n(t_0) + \tau f_n'(t_0)), \quad \tau \in \mathbb{R}. \end{aligned}$$

Example A.29. Let

$$f(t) = (t^2, t^3 + t, t), \quad t \in \mathbb{R}.$$

Here

$$\begin{aligned} f_1(t) &= t^2, \\ f_2(t) &= t^3 + t, \\ f_3(t) &= t, \quad t \in \mathbb{R}. \end{aligned}$$

Hence

$$\begin{aligned} f_1'(t) &= 2t, \\ f_2'(t) &= 3t^2 + 1, \\ f_3'(t) &= 1, \quad t \in \mathbb{R}, \end{aligned}$$

and therefore the equation of the line tangent to f at an arbitrary point is given by

$$F(\tau) = (t^2 + 2\tau t, t^3 + t + \tau(3t^2 + 1), t + \tau), \quad \tau, t \in \mathbb{R}.$$

Example A.30. Consider the curve

$$f(t) = (a \cosh t, a \sinh t, ct), \quad t \in \mathbb{R},$$

where $a, c \in \mathbb{R}$, $c \neq 0$. Here

$$\begin{aligned} f_1(t) &= a \cosh t, \\ f_2(t) &= a \sinh t, \\ f_3(t) &= ct, \quad t \in I. \end{aligned}$$

Hence

$$f_1'(t) = a \sinh t,$$
$$f_2'(t) = a \cosh t,$$
$$f_3'(t) = c, \quad t \in \mathbb{R}.$$

Then the equation of the tangent line at an arbitrary point is as follows:

$$(F_1(\tau), F_2(\tau), F_3(\tau)) = (a \cosh t + \tau a \sinh t, a \sinh t + \tau a \cosh t, ct + c\tau), \quad t, \tau \in \mathbb{R},$$

or

$$\frac{F_1(\tau) - a \cosh t}{a \sinh t} = \frac{F_2(\tau) - a \sinh t}{a \cosh t} = \frac{F_3(\tau) - ct}{c}, \quad \tau, t \in \mathbb{R}.$$

Exercise A.7. Find the equation of the tangent at the corresponding points of the following curves:

1.

$$f(t) = (t, t^2 + 4t + 3), \quad t \in \mathbb{R}, \ (-1, 0), \ (0, 3), \ (1, 8).$$

2.

$$f(t) = (t, t^3), \quad t \in \mathbb{R}, \ (0, 0), \ (1, 1).$$

3.

$$f(t) = (t, \sin t), \quad t \in \mathbb{R}, \quad (0, 0), \ \left(\frac{\pi}{2}, 1\right), \ (\pi, 0).$$

4.

$$f(t) = (t, \tan t), \quad t \in \mathbb{R}, \ (0, 0), \ \left(\frac{\pi}{4}, 1\right).$$

5.

$$f(t) = (e^t, e^{-t}, t^2), \quad t \in \mathbb{R}, \ (e, e^{-1}, 1).$$

Answer A.2.

1.

$$2F_1(\tau) - F_2(\tau) + 2 = 0,$$
$$4F_1(\tau) - F_2(\tau) + 3 = 0,$$
$$6F_1(\tau) - F_2(\tau) + 2 = 0, \quad \tau \in \mathbb{R}.$$

2.

$$F_2(\tau) = 0,$$
$$3F_1(\tau) - F_2(\tau) - 2 = 0, \quad \tau \in \mathbb{R}.$$

3.

$$F_1(\tau) = F_2(\tau),$$
$$F_2(\tau) = 1,$$
$$F_1(\tau) + F_2(\tau) = \pi, \quad \tau \in \mathbb{R}.$$

4.

$$F_1(\tau) = F_2(\tau),$$
$$F_1(\tau) = -F_2(\tau), \quad \tau \in \mathbb{R}.$$

5.

$$\frac{F_1(\tau) - e}{e} = \frac{F_2(\tau) - e^{-1}}{-e^{-1}} = \frac{F_3(\tau) - 1}{2}, \quad \tau \in \mathbb{R}.$$

Exercise A.8. Prove that the tangent vectors of two equivalent curves are collinear at the corresponding points and the tangent lines coincide.

Solution. Let (I, f) and (J, g) be two equivalent curves. Let $s : I \to J$ be the parameter change. Then

$$f(t) = g(s(t)), \quad t \in I.$$

Hence

$$f'(t) = g'(s(t))s'(t), \quad t \in I.$$

This completes the solution.

Now suppose that $f : I \to \mathbb{R}^n$ is a regular curve. For h close enough to 0, or $h \to 0$, by the Taylor formula, we have

$$f(t_0 + h) = f(t_0) + hf'(t_0) + h\varepsilon,$$

where $\varepsilon \to 0$ as $h \to 0$. Let l be an arbitrary line passing through $f(t_0)$ and having the unit direction vector m. Set

$$d(h) = d(f(t_0 + h), l).$$

Exercise A.9. Prove that l is the tangent line to the regular parameterized curve f at the point t_0 if and only if

$$\lim_{h\to 0}\frac{d(h)}{|h|}=0.$$

Solution. We have

$$\begin{aligned} d(h) &= \left|(f(t_0+h)-f(t_0))\times m\right| \\ &= \left|(hf'(t_0)+h\varepsilon)\times m\right| \\ &= \left|h(f'(t_0)\times m)+h(\varepsilon\times m)\right| \\ &= |h|\left|(f'(t_0)\times m)+(\varepsilon\times m)\right| \end{aligned}$$

and

$$\frac{d(h)}{|h|}=\left|(f'(t_0)\times m)+(\varepsilon\times m)\right|.$$

Hence, using that $\varepsilon\to 0$ in $|(f'(t_0)\times m)+(\varepsilon\times m)|$ as $h\to 0$, we get

$$\begin{aligned} \lim_{h\to 0}\frac{d(h)}{|h|} &= \lim_{h\to 0}\left|(f'(t_0)\times m)+(\varepsilon\times m)\right| \\ &= \left|f'(t_0)\times m\right|. \end{aligned}$$

1. Let l be the tangent line to f at t_0. Then $f'(t_0)$ and m are collinear, and

$$\left|f'(t_0)\times m\right|=0.$$

   Hence

$$\lim_{h\to 0}\frac{d(h)}{|h|}=0. \tag{A.3}$$

2. Now suppose that (A.3) holds. Then

$$\left|f'(t_0)\times m\right|=0,$$

   and $f'(t_0)$ and m are collinear. Thus l is a tangent line to f at t_0. This completes the solution.

Definition A.13. Let (I,f) be a parametric curve, and let $t_0\in I$. The normal plane at $f(t_0)$ is the plane through $f(t_0)$ that is perpendicular to the tangent line to the curve at the point $f(t_0)$.

The equation for the normal plane is as follows:

$$(R(t) - f(t_0)) \cdot f'(t_0) = 0, \quad t \in I,$$

where $\cdot$ denotes the inner product. We have

$$\begin{aligned} 0 &= ((R_1(t), R_2(t), \dots, R_n(t)) - (f_1(t_0), f_2(t_0), \dots, f_n(t_0))) \cdot (f_1'(t_0), f_2'(t_0), \dots, f_n'(t_0)) \\ &= (R_1(t) - f_1(t_0), R_2(t) - f_2(t_0), \dots, R_n(t) - f_n(t_0)) \cdot (f_1'(t_0), f_2'(t_0), \dots, f_n'(t_0)) \\ &= (R_1(t) - f_1(t_0))f_1'(t_0) + \cdots + (R_n(t) - f_n(t_0))f_n'(t_0). \end{aligned}$$

Example A.31. Let $f : \mathbb{R} \to \mathbb{R}^3$ be defined by

$$f(t) = (t^2, e^{-t^2+1}, e^{-t+2}), \quad t \in \mathbb{R}.$$

Here

$$\begin{aligned} f_1(t) &= t^2, \\ f_2(t) &= e^{-t^2+1}, \\ f_3(t) &= e^{-t+2}, \quad t \in \mathbb{R}. \end{aligned}$$

Hence

$$\begin{aligned} f_1'(t) &= 2t, \\ f_2'(t) &= -2te^{-t^2+1}, \\ f_3'(t) &= -e^{-t+2}, \quad t \in \mathbb{R}. \end{aligned}$$

Then the equation of the normal plane at arbitrary point $t_0 \in \mathbb{R}$ is as follows:

$$2t_0(R_1(t) - t_0^2) - 2t_0e^{-t_0^2+1}(R_2(t) - e^{-t_0^2+1}) - e^{t_0+2}(R_3(t) - e^{-t_0+2}) = 0.$$

Example A.32. Let $f : \mathbb{R} \to \mathbb{R}^4$ be defined by

$$f(t) = (t, e^{-t}, e^{-t^2}, e^{-t^3}), \quad t \in \mathbb{R}.$$

Here

$$\begin{aligned} f_1(t) &= t, \\ f_2(t) &= e^{-t}, \\ f_3(t) &= e^{-t^2}, \\ f_4(t) &= e^{-t^3}, \quad t \in \mathbb{R}. \end{aligned}$$

Hence

$$f_1'(t) = 1,$$
$$f_2'(t) = -e^{-t},$$
$$f_3'(t) = -2te^{-t^2},$$
$$f_4'(t) = -3t^2e^{-t^3}, \quad t \in \mathbb{R},$$

and the equation of the normal plane at arbitrary point $t_0 \in \mathbb{R}$ is as follows:

$$(R_1(t) - t_0) - e^{-t_0}(R_2(t) - e^{-t_0}) - 2t_0e^{-t_0^2}(R_3(t) - e^{-t_0^2}) - 3t_0^2e^{-t_0^3}(R_4(t) - e^{-t_0^3}) = 0.$$

Exercise A.10. Find the equation of the normal plane at $t = 0$ for the curve

$$f(t) = (2\cos t, 2\sin t, 4t), \quad t \in \mathbb{R}.$$

Answer A.3.

$$R_2(\tau) + 2R_3(\tau) = 0, \quad \tau \in \mathbb{R}.$$

B Elements of theory of surfaces

B.1 Parameterized surfaces

Suppose that $U \subseteq \mathbb{R}^2$.

Definition B.1. A regular parameterized surface in $\mathbb{R}^3$ is a smooth map $f : U \to \mathbb{R}^3$, $(t_1, t_2) \to f(t_1, t_2)$, such that

$$f_{t_1} \times f_{t_2} \neq 0 \quad \text{on } U. \tag{B.1}$$

Definition B.2. Condition (B.1) is called the regularity condition.

Example B.1. Let $f : \mathbb{R}^2 \to \mathbb{R}^3$ be given by

$$f(t_1, t_2) = (t_1 + t_2, e^{t_1^2 + t_2}, e^{t_1 + t_2^2}), \quad (t_1, t_2) \in \mathbb{R}^2.$$

Here

$$\begin{aligned} f_1(t_1, t_2) &= t_1 + t_2, \\ f_2(t_1, t_2) &= e^{t_1^2 + t_2}, \\ f_3(t_1, t_2) &= e^{t_1 + t_2^2}, \quad (t_1, t_2) \in \mathbb{R}^2. \end{aligned}$$

Then

$$\begin{aligned} f_{1t_1}(t_1, t_2) &= 1, \\ f_{1t_2}(t_1, t_2) &= 1, \\ f_{2t_1}(t_1, t_2) &= 2t_1 e^{t_1^2 + t_2}, \\ f_{2t_2}(t_1, t_2) &= e^{t_1^2 + t_2}, \\ f_{3t_1}(t_1, t_2) &= e^{t_1 + t_2^2}, \\ f_{3t_2}(t_1, t_2) &= 2t_2 e^{t_1 + t_2^2}, \quad (t_1, t_2) \in \mathbb{R}^2. \end{aligned}$$

Hence

$$\begin{aligned} f_{t_1}(t_1, t_2) &= (f_{1t_1}(t_1, t_2), f_{2t_1}(t_1, t_2), f_{3t_1}(t_1, t_2) \\ &= (1, 2t_1 e^{t_1^2 + t_2}, e^{t_1 + t_2^2}), \\ f_{t_2}(t_1, t_2) &= (f_{1t_2}(t_1, t_2), f_{2t_2}(t_1, t_2), f_{3t_2}(t_1, t_2)) \\ &= (1, e^{t_1^2 + t_2}, 2t_2 e^{t_1 + t_2^2}), \quad (t_1, t_2) \in \mathbb{R}^2, \end{aligned}$$

and

https://doi.org/10.1515/9783112219607-009

$$f_{t_1}(t_1,t_2)\times f_{t_2}(t_1,t_2)=((4t_1t_2-1)e^{t_1+t_1^2+t_2+t_2^2},(1-2t_2)e^{t_1+t_2^2},(1-2t_1)e^{t_1^2+t_2}),\quad (t_1,t_2)\in\mathbb{R}^2.$$

Then

$$f_{t_1}(t_1,t_2)\times f_{t_2}(t_1,t_2)\neq 0$$

if and only if $t_1\neq\frac{1}{2}$ or $t_2\neq\frac{1}{2}$. Thus the considered surface is regular for those $(t_1,t_2)\in\mathbb{R}^2$ for which $t_1\neq\frac{1}{2}$ or $t_2\neq\frac{1}{2}$.

Example B.2. Consider the unit sphere

$$f(t_1,t_2)=(\cos t_1\cos t_2,\sin t_1\cos t_2,\sin t_2),\quad t_1\in(0,2\pi),\ t_2\in\left(-\frac{\pi}{2},\frac{\pi}{2}\right).$$

Here

$$\begin{aligned}
f_1(t_1,t_2)&=\cos t_1\cos t_2,\\
f_2(t_1,t_2)&=\sin t_1\cos t_2,\\
f_3(t_1,t_2)&=\sin t_2,\quad t_1\in(0,2\pi),\ t_2\in\left(-\frac{\pi}{2},\frac{\pi}{2}\right).
\end{aligned}$$

Then

$$\begin{aligned}
f_{1t_1}(t_1,t_2)&=-\sin t_1\cos t_2,\\
f_{1t_2}(t_1,t_2)&=-\cos t_1\sin t_2,\\
f_{2t_1}(t_1,t_2)&=\cos t_1\cos t_2,\\
f_{2t_2}(t_1,t_2)&=-\sin t_1\sin t_2,\\
f_{3t_1}(t_1,t_2)&=0,\\
f_{3t_2}(t_1,t_2)&=\cos t_2,\quad t_1\in(0,2\pi),\ t_2\in\left(-\frac{\pi}{2},\frac{\pi}{2}\right),
\end{aligned}$$

and

$$\begin{aligned}
f_{t_1}(t_1,t_2)&=(f_{1t_1}(t_1,t_2),f_{2t_1}(t_1,t_2),f_{3t_1}(t_1,t_2))\\
&=(-\sin t_1\cos t_2,\cos t_1\cos t_2,0),\\
f_{t_2}(t_1,t_2)&=(f_{1t_2}(t_1,t_2),f_{2t_2}(t_1,t_2),f_{3t_2}(t_1,t_2))\\
&=(-\cos t_1\sin t_2,-\sin t_1\sin t_2,\cos t_2),\quad t_1\in(0,2\pi),\ t_2\in\left(-\frac{\pi}{2},\frac{\pi}{2}\right).
\end{aligned}$$

Hence

$$\begin{aligned}
&f_{t_1}(t_1,t_2)\times f_{t_2}(t_1,t_2)\\
&\quad=(\cos t_1(\cos t_2)^2,\sin t_1(\cos t_2)^2,(\sin t_1)^2\sin t_2\cos t_2+(\cos t_1)^2\sin t_2\cos t_2)
\end{aligned}$$

$$= (\cos t_1(\cos t_2)^2, \sin t_1(\cos t_2)^2, \sin t_2 \cos t_2)$$
$$\neq 0, \quad t_1 \in (0, 2\pi), \; t_2 \in \left(-\frac{\pi}{2}, \frac{\pi}{2}\right).$$

Thus the unit sphere is a regular surface in $(0, 2\pi) \times (-\frac{\pi}{2}, \frac{\pi}{2})$.

Example B.3. Consider the torus

$$f(t_1, t_2) = ((a + b\cos t_1)\cos t_2, (a + b\cos t_1)\sin t_2, b\sin t_1), \quad (t_1, t_2) \in [0, 2\pi] \times [0, 2\pi],$$

where $a, b \in \mathbb{R}$, $a > 2$, $b \in (0, 1)$. Here

$$\begin{aligned} f_1(t_1, t_2) &= (a + b\cos t_1)\cos t_2, \\ f_2(t_1, t_2) &= (a + b\cos t_1)\sin t_2, \\ f_3(t_1, t_2) &= b\sin t_1, \quad (t_1, t_2) \in [0, 2\pi] \times [0, 2\pi]. \end{aligned}$$

Then

$$\begin{aligned} f_{1t_1}(t_1, t_2) &= -b\sin t_1 \cos t_2, \\ f_{1t_2}(t_1, t_2) &= -(a + b\cos t_1)\sin t_2, \\ f_{2t_1}(t_1, t_2) &= -b\sin t_1 \sin t_2, \\ f_{2t_2}(t_1, t_2) &= (a + b\cos t_1)\cos t_2, \\ f_{3t_1}(t_1, t_2) &= b\cos t_1, \\ f_{3t_2}(t_1, t_2) &= 0, \quad (t_1, t_2) \in [0, 2\pi] \times [0, 2\pi], \end{aligned}$$

and

$$\begin{aligned} f_{t_1}(t_1, t_2) &= (f_{1t_1}(t_1, t_2) < f_{2t_1}(t_1, t_2), f_{3t_1}(t_1, t_2)) \\ &= (-b\sin t_1 \cos t_2, -b\sin t_1 \sin t_2, b\cos t_1), \\ f_{t_2}(t_1, t_2) &= (f_{2t_1}(t_1, t_2), f_{2t_2}(t_1, t_2), f_{3t_2}(t_1, t_2)) \\ &= (-(a + b\cos t_1)\sin t_2, (a + b\cos t_1)\cos t_2, 0), \quad (t_1, t_2) \in [0, 2\pi] \times [0, 2\pi]. \end{aligned}$$

Hence

$$\begin{aligned} & f_{t_1}(t_1, t_2) \times f_{t_2}(t_1, t_2) \\ &\quad = (b(a + b\cos t_1)\cos t_2 \cos t_1, -b(a + b\cos t_1)\sin t_2 \cos t_1, \\ &\qquad - b(a + b\cos t_1)\sin t_1(\cos t_2)^2 - b(a + b\cos t_1)\sin t_1(\sin t_2)^2) \\ &\quad = (b(a + b\cos t_1)\cos t_1 \cos t_2, -b(a + b\cos t_1)\cos t_1 \sin t_2, -b(a + b\cos t_1)\sin t_1) \\ &\quad \neq (0, 0, 0), \quad (t_1, t_2) \in [0, 2\pi] \times [0, 2\pi]. \end{aligned}$$

Thus the considered torus is a regular surface.

Exercise B.1. Prove that the elliptic cylinder

$$f(t_1, t_2) = (a\cos t_2, b\sin t_2, t_1), \quad t_1 \in \mathbb{R},\ t_2 \in [0, 2\pi],$$

where $a, b \in \mathbb{R}$, $a, b > 0$, is a regular surface.

Definition B.3. The set $f(U) \subseteq \mathbb{R}^3$ is said to be the support of the parameterized surface (U, f).

Example B.4. Let $U = \mathbb{R}^2$ and

$$f(t_1, t_2) = (at_1 + a_0, bt_1 + b_0, ct_1 + c_0), \quad (t_1, t_2) \in \mathbb{R}^2,$$

where $a, a_0, b, b_0, c, c_0 \in \mathbb{R}$ are given parameters. We will find the support of f. Here

$$\begin{aligned} f_1(t_1, t_2) &= at_1 + a_0, \\ f_2(t_1, t_2) &= bt_1 + b_0, \\ f_3(t_1, t_2) &= ct_1 + c_0, \quad (t_1, t_2) \in \mathbb{R}^2. \end{aligned}$$

We have the following cases.

1. Let $a^2 + b^2 + c^2 \neq 0$. Then the support of f is the line

$$\frac{f_1 - a_0}{a} = \frac{f_2 - b_0}{b} = \frac{f_3 - c_0}{c}.$$

2. Let $a^2 + b^2 + c^2 = 0$. Then the support of f is the point (a_0, b_0, c_0).

Example B.5. Let $U = \mathbb{R}^2$ and

$$f(t_1, t_2) = (t_2 + 2, 3t_1 + 4t_2 + 5, 6t_1 + 7t_2 + 8), \quad (t_1, t_2) \in \mathbb{R}^2.$$

We will find the support of f. Here

$$\begin{aligned} f_1(t_1, t_2) &= t_2 + 2, \\ f_2(t_1, t_2) &= 3t_1 + 4t_2 + 5, \\ f_3(t_1, t_2) &= 6t_1 + 7t_2 + 8, \quad (t_1, t_2) \in \mathbb{R}^2. \end{aligned}$$

Then

$$\begin{aligned} t_2 &= f_1 - 2, \\ f_2 &= 3t_1 + 4(f_1 - 2) + 5, \\ f_3 &= 6t_1 + 7(f_1 - 2) + 8, \end{aligned}$$

or

$$\begin{aligned} t_2 &= f_1 - 2, \\ f_2 &= 3t_1 + 4f_1 - 3, \\ f_3 &= 6t_1 + 7f_1 - 6, \end{aligned}$$

whereupon

$$\begin{aligned} t_2 &= f_1 - 1, \\ t_1 &= \frac{f_2 - 4f_1 - 3}{3}, \\ t_2 &= \frac{f_3 - 7f_1 - 6}{6}. \end{aligned}$$

Hence

$$\frac{f_2 - 4f_1 - 3}{3} = \frac{f_3 - 7f_1 - 6}{6}$$

and

$$2(f_2 - 4f_1 - 3) = f_3 - 7f_1 - 6,$$

or

$$2f_2 - 8f_1 - 6 = f_3 - 7f_1 - 6,$$

or the support of the considered surface is the plane

$$f_1 - 2f_2 + f_3 = 0.$$

Example B.6. Let $U = \mathbb{R}^2$ and

$$f(t_1, t_2) = (\cos t_1 \cos t_2, \cos t_1 \sin t_2, \sin t_1), \quad (t_1, t_2) \in \mathbb{R}^2.$$

Here

$$\begin{aligned} f_1(t_1, t_2) &= \cos t_1 \cos t_2, \\ f_2(t_1, t_2) &= \cos t_1 \sin t_2, \\ f_3(t_1, t_2) &= \sin t_1, \quad (t_1, t_2) \in \mathbb{R}^2. \end{aligned}$$

Observe that

$$\begin{aligned} f_1^2 + f_2^2 + f_3^2 &= (\cos t_1)^2 (\cos t_2)^2 + (\cos t_1)^2 (\sin t_2)^2 + (\sin t_1)^2 \\ &= (\cos t_1)^2 + (\sin t_1)^2 \\ &= 1, \end{aligned}$$

i. e., the support of the considered surface is the sphere

$$f_1^2 + f_2^2 + f_3^2 = 1.$$

Exercise B.2. Find the support of the surface

$$f(t_1, t_2) = (t_1 \cos t_2, t_1 \sin t_2, t_2), \quad (t_1, t_2) \in \mathbb{R}^2.$$

Answer B.1. The cylinder

$$f_1^2 + f_2^2 = 1.$$

Definition B.4. Let $m, n \in \mathbb{N}$, $M \subseteq \mathbb{R}^m$, and $N \subseteq \mathbb{R}^n$. A differentiable map $f : M \to N$ is called a homeomorphism if it is a bijection, $f \in \mathscr{C}(M)$, and its inverse $f^{-1} : N \to M$ is continuous.

Definition B.5. Let $m, n \in \mathbb{N}$, $M \subseteq \mathbb{R}^m$, and $N \subseteq \mathbb{R}^n$. A differentiable map $f : M \to N$ is called a diffeomorphism if it is a bijection and its inverse $f^{-1} : N \to M$ is differentiable. If $f \in \mathscr{C}^r(M)$ and $f^{-1} \in \mathscr{C}^r(N)$, then f is said to be a $\mathscr{C}^r$- diffeomorphism.

Example B.7. Consider the map $f : \mathbb{R}^2 \to \mathbb{R}^2$ given by

$$f(t_1, t_2) = (t_1^2 + t_2^3, t_1^2 - t_2^3), \quad (t_1, t_2) \in \mathbb{R}^2.$$

Here

$$\begin{aligned} f_1(t_1, t_2) &= t_1^2 + t_2^3, \\ f_2(t_1, t_2) &= t_1^2 - t_2^3, \quad (t_1, t_2) \in \mathbb{R}^2. \end{aligned}$$

Then

$$\begin{aligned} f_{1t_1}(t_1, t_2) &= 2t_1, \\ f_{1t_2}(t_1, t_2) &= 3t_2^2, \\ f_{2t_1}(t_1, t_2) &= 2t_1, \\ f_{2t_2}(t_1, t_2) &= -3t_2^2, \quad (t_1, t_2) \in \mathbb{R}^2. \end{aligned}$$

In the considered case, the Jacobian matrix is

$$\begin{aligned} J_f(t_1, t_2) &= \begin{pmatrix} f_{1t_1}(t_1, t_2) & f_{1t_2}(t_1, t_2) \\ f_{2t_1}(t_1, t_2) & f_{2t_2}(t_1, t_2) \end{pmatrix} \\ &= \begin{pmatrix} 2t_1 & 3t_2^2 \\ 2t_1 & -3t_2^2 \end{pmatrix}, \quad (t_1, t_2) \in \mathbb{R}^2, \end{aligned}$$

and for its determinant, we have

$$\begin{aligned}\det J_f(t_1, t_2) &= -6t_1t_2^2 - 6t_1t_2^2 \\ &= -12t_1t_2^2, \quad (t_1, t_2) \in \mathbb{R}^2.\end{aligned}$$

From here we conclude that the considered map is a diffeomorphism away from the t_1-axis and t_2-axis.

Example B.8. Consider the map $f : \mathbb{R}^2 \to \mathbb{R}^2$ given by

$$f(t_1, t_2) = (\sin(t_1^2 + t_2^2), \cos(t_1^2 + t_2^2)), \quad (t_1, t_2) \in \mathbb{R}^2.$$

Here

$$\begin{aligned}f_1(t_1, t_2) &= \sin(t_1^2 + t_2^2), \\ f_2(t_1, t_2) &= \cos(t_1^2 + t_2^2), \quad (t_1, t_2) \in \mathbb{R}^2.\end{aligned}$$

Then

$$\begin{aligned}f_{1t_1}(t_1, t_2) &= 2t_1 \cos(t_1^2 + t_2^2), \\ f_{1t_2}(t_1, t_2) &= 2t_2 \cos(t_1^2 + t_2^2), \\ f_{2t_1}(t_1, t_2) &= -2t_1 \sin(t_1^2 + t_2^2), \\ f_{2t_2}(t_1, t_2) &= -2t_2 \sin(t_1^2 + t_2^2), \quad (t_1, t_2) \in \mathbb{R}^2.\end{aligned}$$

The Jacobian matrix is

$$\begin{aligned}J_f(t_1, t_2) &= \begin{pmatrix} f_{1t_1}(t_1, t_2) & f_{1t_2}(t_1, t_2) \\ f_{2t_1}(t_1, t_2) & f_{2t_2}(t_1, t_2) \end{pmatrix} \\ &= \begin{pmatrix} 2t_1 \cos(t_1^2 + t_2^2) & 2t_2 \cos(t_1^2 + t_2^2) \\ -2t_1 \sin(t_1^2 + t_2^2) & -2t_2 \sin(t_1^2 + t_2^2) \end{pmatrix}, \quad (t_1, t_2) \in \mathbb{R}^2,\end{aligned}$$

and its determinant is

$$\begin{aligned}\det J_f(t_1, t_2) &= -4t_1t_2 \sin(t_1^2 + t_2^2) \cos(t_1^2 + t_2^2) + 4t_1t_2 \sin(t_1^2 + t_2^2) \cos(t_1^2 + t_2^2) \\ &= 0, \quad (t_1, t_2) \in \mathbb{R}^2.\end{aligned}$$

Thus the considered map is not a diffeomorphism.

Exercise B.3. Prove that the map $f : [1, 2] \times [1, 2] \to \mathbb{R}^2$ given by

$$f(t_1, t_2) = (t_1 \cos t_2, t_1 \sin t_2), \quad (t_1, t_2) \in [1, 2] \times [1, 2],$$

is a diffeomorphism.

Definition B.6. Parameterized surfaces (U,f) and (V,g) are said to be equivalent if there is a diffeomorphism $\phi : U \to V$ such that

$$f = g(\phi).$$

Definition B.7. A subset S of $\mathbb{R}^3$ is called a regular surface if for each point $a \in S$, there are a neighborhood W in S and a homeomorphism $f : U \to W$ such that $f(U) = W$ and (U,f) is a parameterized surface. The pair (U,f) is said to be a local parameterization of the surface S. The support $f(U)$ is called the domain of the representation. If $f(U) = S$, then S is said to be a simple surface.

We have the following representations of the surfaces.

1. If (U,f) is a parameterized surface and

$$f(t_1,t_2) = (f_1(t_1,t_2), f_2(t_1,t_2), f_3(t_1,t_2)), \quad (t_1,t_2) \in U,$$

then the equations

$$\begin{aligned} f_1 &= f_1(t_1,t_2), \\ f_2 &= f_2(t_1,t_2), \\ f_3 &= f_3(t_1,t_2), \quad (t_1,t_2) \in U, \end{aligned}$$

are said to be parametric equations of the parameterized surface (U,f).

2. If $g : U \to \mathbb{R}$, $g \in \mathscr{C}^1(U)$, and

$$f(t_1,t_2) = (t_1, t_2, g(t_1,t_2)),$$

then the representation of the parameterized surface (U,f) is said to be an explicit representation.

3. Let $W \subseteq \mathbb{R}^3$ and $F \in \mathscr{C}^1(\mathbb{R}^3)$.

Definition B.8. The set

$$S = \{(t_1,t_2,t_3) \in \mathbb{R}^3 : F(t_1,t_2,t_3) = 0\}$$

is said to be the 0-level set of F.

Definition B.9. The vector

$$\operatorname{grad} F(t_1,t_2,t_3) = (F_{t_1}(t_1,t_2,t_3), F_{t_2}(t_1,t_2,t_3), F_{t_3}(t_1,t_2,t_3)), \quad (t_1,t_2,t_3) \in W,$$

is said to be the gradient vector of F.

Exercise B.4. Let

$$\operatorname{grad} F(t_1, t_2, t_3) \neq (0, 0, 0), \quad (t_1, t_2, t_3) \in W.$$

Prove that S is a regular surface.

Solution. Let $(t_1^0, t_2^0, t_3^0) \in W$. Without loss of generality, suppose that

$$F_{t_3}(t_1^0, t_2^0, t_3^0) \neq 0.$$

Then by the implicit theorem it follows that there are a neighborhood M of (t_1^0, t_2^0, t_3^0) and $f \in \mathscr{C}^1$ such that

$$t_3 = f(t_1, t_2), \quad (t_1, t_2, t_3) \in M.$$

This completes the solution.

Example B.9. We will find the parametric equations of the ellipsoid

$$\frac{x^2}{a^2} + \frac{y^2}{b^2} + \frac{z^2}{c^2} = 1,$$

where $a, b, c \in \mathbb{R}$, $(a, b, c) \neq (0, 0, 0)$. Let

$$\begin{aligned} x &= a \cos u \cos v, \\ y &= b \sin u \cos v, \\ z &= c \sin v, \quad u \in [0, 2\pi],\ v \in [0, \pi]. \end{aligned}$$

We have

$$\begin{aligned} \frac{x^2}{a^2} + \frac{y^2}{b^2} + \frac{z^2}{c^2} &= \frac{1}{a^2}(a \cos u \cos v)^2 + \frac{1}{b^2}(b \sin u \cos v)^2 + \frac{1}{c^2}(c \sin v)^2 \\ &= (\cos u)^2(\cos v)^2 + (\sin u)^2(\cos v)^2 + (\sin v)^2 \\ &= ((\cos u)^2 + (\sin u)^2)(\cos v)^2 + (\sin v)^2 \\ &= (\cos v)^2 + (\sin v)^2 \\ &= 1, \quad u \in [0, 2\pi],\ v \in [0, \pi]. \end{aligned}$$

Thus the parametric equation of the ellipsoid is

$$f(u, v) = (a \cos u \cos v, b \sin u \cos v, c \sin v), \quad u \in [0, 2\pi],\ v \in [0, \pi].$$

Example B.10. Consider the one-sheeted hyperboloid

$$\frac{x^2}{a^2} + \frac{y^2}{b^2} - \frac{z^2}{c^2} = 1,$$

where $a, b, c \in \mathbb{R}$, $(a, b, c) \neq (0, 0, 0)$. Let

$$x = a \cosh u \cos v,$$
$$y = b \cosh u \sin v,$$
$$z = c \sinh u, \quad u \in \mathbb{R},\ v \in [0, 2\pi].$$

Then

$$\begin{aligned}\frac{x^2}{a^2} + \frac{y^2}{b^2} - \frac{z^2}{c^2} &= \frac{1}{a^2}(a \cosh u \cos v)^2 + \frac{1}{b^2}(\cosh u \sin v)^2 - \frac{1}{c^2}(c \sinh u)^2 \\ &= (\cosh u)^2(\cos v)^2 + (\cosh u)^2(\sin v)^2 - (\sinh u)^2 \\ &= (\cosh u)^2((\cos v)^2 + (\sin v)^2) - (\sinh u)^2 \\ &= (\cosh u)^2 - (\sinh u)^2 \\ &= 1.\end{aligned}$$

Thus the parametric equation of the one-sheeted hyperboloid is

$$f(u, v) = (a \cosh u \cos v, b \cosh u \sin v, c \sinh u), \quad u \in \mathbb{R},\ v \in [0, 2\pi].$$

Example B.11. Consider the two-sheeted hyperboloid

$$\frac{x^2}{a^2} + \frac{y^2}{b^2} - \frac{z^2}{c^2} = -1,$$

where $a, b, c \in \mathbb{R}$, $(a, b, c) \neq (0, 0, 0)$. Let

$$x = a \sinh u \cos v,$$
$$y = b \sinh u \sin v,$$
$$z = c \cosh u, \quad u \in \mathbb{R},\ v \in [0, 2\pi].$$

Then

$$\begin{aligned}\frac{x^2}{a^2} + \frac{y^2}{b^2} - \frac{z^2}{c^2} &= \frac{1}{a^2}(a \sinh u \cos v)^2 + \frac{1}{b^2}(\sinh u \sin v)^2 - \frac{1}{c^2}(c \cosh u)^2 \\ &= (\sinh u)^2(\cos v)^2 + (\sinh u)^2(\sin v)^2 - (\cosh u)^2 \\ &= (\sinh u)^2((\cos v)^2 + (\sin v)^2) - (\cosh u)^2 \\ &= (\sinh u)^2 - (\cosh u)^2 \\ &= -1.\end{aligned}$$

Thus the parametric equation of the two-sheeted hyperboloid is

$$f(u, v) = (a \sinh u \cos v, b \sinh u \sin v, c \cosh u), \quad u \in \mathbb{R},\ v \in [0, 2\pi].$$

Example B.12. Consider the elliptic paraboloid

$$\frac{x^2}{a^2} + \frac{y^2}{b^2} = z,$$

where $a, b \in \mathbb{R}$, $(a, b) \neq (0, 0)$. Let

$$\begin{aligned} x &= au\cos v, \\ y &= bu\sin v, \\ z &= u^2, \quad u \in \mathbb{R},\ v \in [0, 2\pi]. \end{aligned}$$

Then

$$\begin{aligned} \frac{x^2}{a^2} + \frac{y^2}{b^2} &= \frac{1}{a^2}(au\cos v)^2 + \frac{1}{b^2}(bu\sin v)^2 \\ &= u^2(\cos v)^2 + u^2(\sin v)^2 \\ &= u^2((\cos v)^2 + (\sin v)^2) \\ &= u^2 \\ &= z, \quad u \in \mathbb{R},\ v \in [0, 2\pi]. \end{aligned}$$

Thus the parametric equation of the elliptic paraboloid is

$$f(u, v) = (au\cos v, bu\sin v, u^2), \quad u \in \mathbb{R},\ v \in [0, 2\pi].$$

Example B.13. Consider the elliptic cylinder

$$\frac{x^2}{a^2} + \frac{y^2}{b^2} = 1,$$

where $a, b \in \mathbb{R}$, $(a, b) \neq (0, 0)$. Let

$$\begin{aligned} x &= a\cos u, \\ y &= a\sin u, \\ z &= z, \quad u \in [0, 2\pi[,\ u \in \mathbb{R}. \end{aligned}$$

Then

$$\begin{aligned} \frac{x^2}{a^2} + \frac{y^2}{b^2} &= \frac{1}{a^2}(a\cos u)^2 + \frac{1}{b^2}(b\sin u)^2 \\ &= (\cos u)^2 + (\sin u)^2 \\ &= 1, \quad u \in [0, 2\pi],\ v \in \mathbb{R}. \end{aligned}$$

Thus the parametric equation of the elliptic cylinder is

$$f(u,v) = (a\cos u, b\sin u, v), \quad u \in [0,2\pi],\ v \in \mathbb{R}.$$

Example B.14. Consider the cone

$$\frac{x^2}{a^2} + \frac{y^2}{b^2} = \frac{z^2}{c^2},$$

where $a, b, c \in \mathbb{R}$, $(a,b,c) \neq (0,0,0)$. Let

$$\begin{aligned} x &= au\cos v, \\ y &= bu\sin v, \\ z &= cu, \quad u \in \mathbb{R},\ v \in [0,2\pi]. \end{aligned}$$

Then

$$\begin{aligned} \frac{x^2}{a^2} + \frac{y^2}{b^2} &= \frac{1}{a^2}(au\cos v)^2 + \frac{1}{b^2}(bu\sin v)^2 \\ &= u^2(\cos v)^2 + u^2(\sin v)^2 \\ &= u^2((\cos v)^2 + (\sin v)^2) \\ &= u^2 \\ &= \frac{z^2}{c^2}. \end{aligned}$$

Thus the parametric equation of the cone is

$$f(u,v) = (au\cos v, bu\sin v, cu), \quad u \in \mathbb{R},\ v \in [0,2\pi].$$

Example B.15. Consider the hyperbolic paraboloid

$$\frac{x^2}{a^2} - \frac{y^2}{b^2} = 2z,$$

where $a, b \in \mathbb{R}$, $(a,b) \neq (0,0)$. Let

$$\begin{aligned} x &= a(u+v), \\ y &= b(v-u), \\ z &= 2uv, \quad (u,v) \in \mathbb{R}^2. \end{aligned}$$

We have

$$\begin{aligned} \frac{x^2}{a^2} - \frac{y^2}{b^2} &= \frac{1}{a^2}(a(u+v))^2 - \frac{1}{b^2}(b(v-u))^2 \\ &= (u+v)^2 - (v-u)^2 \\ &= u^2 + 2uv + v^2 - (v^2 - 2uv + u^2) \end{aligned}$$

$$= 4uv$$
$$= 2z, \quad (u, v) \in \mathbb{R}^2.$$

Thus the parametric equation of the hyperbolic paraboloid is

$$f(u, v) = (a(u + v), b(v - u), 2uv), \quad (u, v) \in \mathbb{R}^2.$$

Exercise B.5. Prove that the equations

$$f(u, v) = \left(\frac{u}{u^2 + v^2}, \frac{v}{u^2 + v^2}, \frac{1}{u^2 + v^2} \right), \quad (u, v) \in \mathbb{R}^2,$$

and

$$g(t_1, t_2) = (t_1 \cos t_2, t_1 \sin t_2, t_1^2), \quad t_1 \in \mathbb{R},\ t_2 \in [0, 2\pi],$$

determine the same surface.

B.2 The equivalence of local representations

Definition B.10. Let S be a surface, let (U, f) be its local parameterization, and let $W = f(U)$. Then the map $f^{-1} : W \to U$ is a bijection and will be called a line coordinate system on S or a chart on S.

Exercise B.6. Let (U, f) be a local parameterization of the surface S, $f(U) = W$, and $f^{-1} : W \to U$. Then for each point $a \in W$, there are an open set B in the topology of $\mathbb{R}^3$ and a smooth map $G : B \to V$ such that $a \in B$ and

$$f^{-1}\Big|_{W\cap B} = G\Big|_{W\cap B}.$$

Solution. Let

$$f(t_1, t_2) = (f_1(t_1, t_2), f_2(t_1, t_2), f_3(t_1, t_2)), \quad (t_1, t_2) \in U,$$

and

$$a = f(t_1^0, t_2^0).$$

Note that

$$\operatorname{rank} \begin{pmatrix} f_{1t_1} & f_{1t_2} \\ f_{2t_1} & f_{2t_2} \\ f_{3t_1} & f_{3t_2} \end{pmatrix} = 2.$$

Without loss of generality, suppose that

$$\begin{vmatrix} f_{1t_1} & f_{1t_2} \\ f_{2t_1} & f_{2t_2} \end{vmatrix} \neq 0.$$

Then by the inverse function theorem it follows that there are an open neighborhood V of the point (t_1^0, t_2^0) in U and an open neighborhood V_1 of the point $(f_1(t_1^0, t_2^0), f_2(t_1^0, t_2^0))$ such that $f : V \to V_1$ is a diffeomorphism. Because $f : U \to W$ is a homeomorphism, we have that $f(V)$ is an open neighborhood in S of the point a. Therefore there is an open neighborhood of the point a such that

$$f(V) = B \cap S = B \cap W.$$

Now define the map $\phi : \mathbb{R}^3 \to \mathbb{R}^2$ as follows:

$$\phi(t_1, t_2, t_3) = (t_1, t_2), \quad (t_1, t_2, t_3) \in \mathbb{R}^3.$$

Let

$$G = (f^{-1}(\phi))\big|_B : B \to U.$$

Note that G is a smooth map. Next, to each point $(s_1, s_2, s_3) \in B \cap W$, there corresponds a single point

$$(t_1, t_2) = f^{-1}(s_1, s_2, s_3) \in V.$$

Also, to each point $(s_1, s_2) \in V_1$, there corresponds the point

$$(t_1, t_2) = f^{-1}(s_1, s_2) \in V.$$

Thus, if $(s_1, s_2, s_3) \in B \cap W$, then

$$\begin{aligned} f^{-1}(s_1, s_2, s_3) &= (t_1, t_2) \\ &= f^{-1}(s_1, s_2) \\ &= f^{-1}(\phi(s_1, s_2, s_3)) \\ &= G(s_1, s_2, s_3). \end{aligned}$$

This completes the proof.

Exercise B.7. Let (U, f) and $(U - 1, f_1)$ be two local parameterizations of a surface S, and let $f(U) = f_1(U_1)$. Then there is a homeomorphism $\phi : U \to U_1$ such that

$$f = f_1(\phi).$$

Solution. Let $W = f(U) = f_1(U_1)$ and

$$\phi = f_1^{-1} \cdot f.$$

Then $\phi : U \to U_1$. Since $f_1 : U_1 \to W$ is a homeomorphism, we have that $f_1^{-1} : W \to U_1$ is a homeomorphism. Therefore $\phi : U \to U_1$ is a homeomorphism. Now we will prove that each point $(t_1^0, t_2^0) \in U$ has a open neighborhood $V \subset U$ such that $\phi_{|_V}$ is a smooth map. Let

$$a = f_1(\phi(t_1^0, t_2^0)).$$

It follows that there is an open set B of $\mathbb{R}^3$ such that $G : B \to U_1$ is a smooth map,

$$f_1^{-1}\Big|_{B\cap W} = G\Big|_{B\cap W},$$

and

$$V = f^{-1}(B \cap W).$$

Then

$$\begin{aligned} \phi\Big|_V &= f_1^{-1} \cdot f\Big|_V \\ &= G \cdot f\Big|_V, \end{aligned}$$

and $\phi_{|_V}$ is a smooth map. As above, ϕ^{-1} is a smooth map. This completes the proof.

Exercise B.8. Let (U,f) be a regular parameterized surface. Then each point $(t_1^0, t_2^0) \in U$ has an open neighborhood $V' \subset U$ such that $f(V)$ is a simple surface in $\mathbb{R}^3$ for which $(V, f_{|_V})$ is a global representation.

Solution. Let

$$f(t_1, t_2) = (f_1(t_1, t_2), f_2(t_1, t_2), f_3(t_1, t_2)), \quad (t_1, t_2) \in U,$$

and

$$\begin{vmatrix} f_{1t_1}(t_1^0, t_2^0) & f_{1t_2}^{v}(t_1^0, t_2^0) \\ f_{2t_1}(t_1^0, t_2^0) & f_{2t_2}(t_1^0, t_2^0) \end{vmatrix} \neq 0.$$

Then by the inverse function theorem there are an open neighborhood $V \subset U$ of the point (t_1^0, t_2^0) and an open neighborhood V_1 of the point $(s_{10}, s_{20}, s_{30}) = f(t_1^0, t_2^0)$ such that $f : V \to V_1$ is a diffeomorphism. Let now $(t_1, t_2), (y_1, y_2) \in V$ be such that

$$f_1(t_1, t_2) = f_1(y_1, y_2),$$

$$f_2(t_1,t_2)=f_2(y_1,y_2).$$

Since $f:V\to V_1$ is a diffeomorphism, it is an injective map, and thus

$$(t_1,t_2)=(y_1,y_2).$$

Since $f:U\to\mathbb{R}^3$ is continuous, we have that $f_{|_V}:V\to f(V)$ is continuous. Note that

$$(x,y,z)\in f(V)\to(u,v,w)\in V_1\to(t_1,t_2)=f^{-1}(u,v,w).$$

Thus the inverse map of $f_{|_V}:V\to f(V)$ is continuous. This completes the proof.

B.3 Curves on surfaces

Exercise B.9. Let (U,f) be a parameterization of the surface S, and let $(I,g=g(t))$ be a smooth parameterized curve whose support is included in $f(U)$. Then there is a unique smooth parameterized curve (I,g_1) on U such that

$$g(t)=f(g_1(t)),\quad t\in I. \tag{B.2}$$

Conversely, any smooth parameterized curve g_1 on U defines, by (B.2), a smooth curve on $f(U)$. The regularity of g at t is equivalent to the regularity of g_1 at t.

Solution. Since $f:U\to f(U)$ is a homeomorphism and $g(I)\subset f(U)$, we get

$$g_1=f^{-1}\circ g.$$

We have that g_1 is continuous because it is a composition of two continuous maps. Let $t\in I$. Then $g(t)\in f(U)$. It follows that there are an open neighborhood B of the point $g(t)$ and a smooth map $G:B\to U$ such that

$$f^{-1}_{|_{B\cap f(U)}}=G_{|_{B\cap f(U)}}.$$

Therefore the map g_1 can be represented in the neighborhood of the point t as a composition of Rg and G. Since G and g are smooth, we have that g_1 is multiplicative smooth. The converse statement follows because g is smooth by (B.2) and the smoothness of f and g_1. Let

$$g_1(t)=(u(t),v(t)),\quad t\in I.$$

Then by (B.2) we have

$$g(t)=f(u(t),v(t)),\quad t\in I.$$

Now differentiating the last equation with respect to t, we get

$$g(t) = f_{t_1}(u(t), v(t))u(t) + f_{t_2}(u(t), v(t))v(t), \quad t \in I.$$

Since

$$f_{t_1} \times f_{t_2} \neq 0 \quad \text{on } U,$$

we have that f_{t_1} and f_{t_2} are not collinear. Therefore

$$g(t) = 0, \quad t \in I,$$

if and only if

$$g_1(t) = 0, \quad t \in I.$$

This completes the proof.

Definition B.11. Let U, f, g, g_1, and I be as in Theorem B.9. Then the parameterized curve g_1 on U is called a local parameterization of g in the local parameterization (U, f). The equations

$$u = u(t),$$
$$v = v(t), \quad t \in I,$$

are called local equations of g.

B.4 The tangent vector space, the tangent plane, the multiplicative normal to a surface

Let $a \in \mathbb{R}^3$. By $\mathbb{R}^3_a$ we denote the space of all vectors with the origin a.

Definition B.12. A vector $p \in \mathbb{R}^3$ is called a tangent vector to the surface S if there are a parameterized curve $(I, g = g(t))$ on S and $t_0 \in I$ such that $g(t_0) = a$ and

$$p = g^{\Delta}(t_0).$$

Thus a tangent vector to a surface S is a tangent vector to a parameterized curve on S.

By $T_a S$ we denote the set of all tangent vectors at a to S. If

$$g(t) = f(u(t), v(t)), \quad t \in I,$$

then

$$g'(t) = f_u(u(t), v(t))v(t) + f_v(u(t), v(t))v(t), \quad t \in I.$$

Exercise B.10. The set T_aS is a two-dimensional vector subspace of $\mathbb{R}^3$. If (U, f) is a local parameterization of S, $a = f(u_0, v_0)$, then $f_u(u_0, v_0)$ and $f_v(u_0, v_0)$ make up a basis to T_aS.

Solution. Let $(I, g = g(t))$ be a parameterized curve on S, and let $g(t_0) = a$ for some $t_0 \in I$. Assume that $g(I) \subset f(U)$ and

$$g(t) = f(u(t), v(t)), \quad t \in I.$$

Then

$$g'(t) = f_u(u(t), v(t))v(t) + f_v(u(t), v(t))v(t), \quad t \in I.$$

Note that any vector of the form

$$h = \alpha f_u(u_0, v_0) + \beta f_v(u_0, v_0)$$

for some $\alpha, \beta \in \mathbb{R}$ is a tangent vector to the curve with equations

$$u = u_0 + \alpha t,$$
$$v = v_0 + \beta t,$$

which is a curve on S passing through the point a for $t = t_0$. Thus $h \in T_{\star a}S$. This completes the solution.

Definition B.13. The vector space T_aS is called the tangent space to S at a. The plane passing through a and having T_aS as a directing plane is called the tangent plane of S at a.

If

$$a = (f_1(t_1^0, t_2^0), f_2(t_1^0, t_2^0), f_3(t_1^0, t_2^0)),$$

then the equation of the tangent plane is given by

$$0 = \begin{vmatrix} X - f_1(t_1^0, t_2^0) & Y - f_2(t_1^0, t_2^0) & Z - f_3(t_1^0, t_2^0) \\ f_{1t_1}(t_1^0, t_2^0) & f_{2t_1}(t_1^0, t_2^0) & f_{3t_1}(t_1^0, t_2^0) \\ f_{1t_2}(t_1^0, t_2^0) & f_{2t_2}(t_1^0, t_2^0) & f_{3t_2}(t_1^0, t_2^0) \end{vmatrix}.$$

If the surface S is given by

$$z = f(x, y), \quad (x, y) \in U,$$

where $U \subset \mathbb{R}^2$ and $f \in \mathscr{C}^1(U)$, then the equation of the tangent plane at (x_0, y_0, z_0), $(x_0, y_0) \in U$, $z_0 = f(x_0, y_0)$, is given by

$$z - z_0 = f_x(x_0, y_0)(x - x_0) + f_y(x_0, y_0)(y - y_0).$$

If the surface S is given by

$$F(x, y, z) = 0, \quad (x, y, z) \in V,$$

where $V \subset \mathbb{R}^3$ and $F \in \mathscr{C}^1(V)$, then the equation of the tangent plane at $(x_0, y_0, z_0) \in V$ is

$$F_x(x_0, y_0, z_0)(x - x_0) + F_y(x_0, y_0, z_0)(y - y_0) + F_z(x_0, y_0, z_0)(z - z_0) = 0.$$

Example B.16. We will find the equation of the tangent plane to the surface

$$z = xy, \quad (x, y) \in \mathbb{R}^2,$$

at the point $(2, 1, 2)$. We have

$$\begin{aligned} z_x(x, y) &= y, \\ z_y(x, y) &= x, \quad (x, y) \in \mathbb{R}^2, \\ z_x(2, 1) &= 1, \\ z_y(2, 1) &= 2. \end{aligned}$$

Then the equation of the tangent plane is

$$z - 2 = (x - 2) + 2(y - 1),$$

or

$$z - 2 = x - 2 + 2y - 2,$$

or

$$x + 2y - z = 2.$$

Example B.17. We will find the equation of the tangent plane to the surface

$$x^2 + y^2 + z^2 = 169, \quad (x, y, z) \in \mathbb{R}^3,$$

at the point $(3, 4, -12)$. Let

$$F(x, y, z) = x^2 + y^2 + z^2 - 169, \quad (x, y, z) \in \mathbb{R}^3.$$

Then

$$F_x(x,y,z) = 2x,$$
$$F_y(x,y,z) = 2y,$$
$$F_z(x,y,z) = 2z, \quad (x,y,z) \in \mathbb{R}^3,$$

and

$$F_x(3,4,-12) = 6,$$
$$F_y(3,4,-12) = 8,$$
$$F_z(3,4,-12) = -24.$$

Thus the equation of the tangent plane is

$$6(x-3) + 8(y-4) - 24(z+12) = 0,$$

or

$$6x - 18 + 8y - 32 - 24z - 288 = 0,$$

or

$$6x + 8y - 24z = -338,$$

or

$$3x + 4y - 12z = 0.$$

Example B.18. We will find the equation of the tangent plane to the surface

$$f(u,v) = (u+v, u^2+v^2, u^3+v^3), \quad (u,v) \in \mathbb{R}^2,$$

at the point $(3,5,9)$. Here

$$f_1(u,v) = u+v,$$
$$f_2(u,v) = u^2+v^2,$$
$$f_3(u,v) = u^3+v^3, \quad (u,v) \in \mathbb{R}^2.$$

Then

$$f_{1u}(u,v) = 1,$$
$$f_{1v}(u,v) = 1,$$
$$f_{2u}(u,v) = 2u,$$
$$f_{2v}(u,v) = 2v,$$

$$f_{3u}(u,v) = 3u^2,$$
$$f_{3v}(u,v) = 3v^2, \quad (u,v) \in \mathbb{R}^2.$$

Next, we have

$$u + v = 3,$$
$$u^2 + v^2 = 5,$$
$$u^3 + v^3 = 9$$

or

$$v = 3 - u,$$
$$u^2 + (3-u)^2 = 5,$$
$$u^3 + (3-u)^3 = 9.$$

By the second equation of the last system we find

$$5 = u^2 + (3-u)^2$$
$$= u^2 + 9 - 6u + u^2$$
$$= 2u^2 - 6u + 9,$$

whereupon

$$2u^2 - 6u + 4 = 0,$$

or

$$u^2 - 3u + 2 = 0.$$

Thus

$$(u_1, v_1) = (1, 2), \quad (u_2, v_2) = (2, 1).$$

From this we get

$$f_{1u}(u_1, v_1) = 1,$$
$$f_{1v}(u_1, v_1) = 1,$$
$$f_{2u}(u_1, v_1) = 2,$$
$$f_{2v}(u_1, v_1) = 4,$$
$$f_{3u}(u_1, v_1) = 3,$$
$$f_{3v}(u_1, v_1) = 12$$

and

$$\begin{aligned}
f_{1u}(u_2, v_2) &= 1,\\
f_{1v}(u_2, v_2) &= 1,\\
f_{2u}(u_2, v_2) &= 4,\\
f_{2v}(u_2, v_2) &= 2,\\
f_{3u}(u_2, v_2) &= 12,\\
f_{3v}(u_2, v_2) &= 3.
\end{aligned}$$

The equation of the tangent plane is

$$\begin{aligned}
0 &= \begin{vmatrix} x-3 & y-5 & z-9 \\ 1 & 2 & 3 \\ 1 & 4 & 12 \end{vmatrix}\\
&= 24(x-3) + 3(y-5) + 4(z-9) - 2(z-9) - 12(x-3) - 12(y-5)\\
&= 12(x-3) - 9(y-5) + 2(z-9)\\
&= 12x - 36 - 9y + 45 + 2z - 18\\
&= 12x - 9y + 2z - 9,
\end{aligned}$$

or

$$12x - 9y + 2z = 9.$$

Exercise B.11. Find the equation of the tangent plane to the following surfaces at the given points:

1.

$$z = x^2 + y^2, \quad (1, 1, 2).$$

2.

$$z = 2x^2 - 4y^2, \quad (-2, 1, 4).$$

3.

$$z = (x - y)^2 - x + 2y, \quad (1, 1, 1).$$

4.

$$z = x^3 - 3xy + y^3, \quad (1, 1, -1).$$

5.

$$z = \sqrt{x^2 + y^2} - xy, \quad (-3, 4, 17).$$

6.

$$xy^2 + z^3 = 12, \quad (1, 2, 2).$$

7.

$$x^3 + y^3 + z^3 + xyz = 6, \quad (1, 2, -1).$$

8.

$$xyz(z^2 - x^2) = 6 + y^5, \quad (1, 2, -1).$$

9.

$$\sqrt{x^2 + y^2 + z^2} = x + y + z - 4, \quad (2, 3, 6).$$

10.

$$e^z - z + xy = 3, \quad (2, 1, 0).$$

11.

$$(u, u^2 - 2uv, u^3 - 3u^2v), \quad (1, 3, 4).$$

Answer B.2.

1.

$$2x + 2y - z = 2.$$

2.

$$8x + 8y + z + 4 = 0.$$

3.

$$x - 2y + z = 0.$$

4.

$$z = -1.$$

5.

$$23x - 19y + 5z + 60 = 0.$$

6.

$$x + y + 3z = 9.$$

7.

$$x + 11y + 3z = 18.$$

8.

$$2x + y + 11z = 25.$$

9.

$$5x + 4y + z = 28.$$

10.

$$x + 2y = 4.$$

11.

$$6x + 3y - 2z = 7.$$

Let now $(U, f = f(u, v))$ be a parameterized surface S, and let $(u_0, v_0) \in U$. Then

$$f(u_0 + \alpha h, v_0 + \beta h) = f(u_0, v_0) + h\big(\alpha f_{t_1}(u_0, v_0) + \beta f_{t_2}(u_0, v_0)\big) + h\varepsilon,$$

where

$$\lim_{h \to 0} \varepsilon = 0.$$

Let Π be the plane in $\mathbb{R}^3$ passing through the point $a_0 = f(u_0, v_0)$, let d be the distance from the point

$$a = f(u_0 + \alpha h, v_0 + \beta h)$$

to the plane Π, and let δ be the distance between a_0 and a.

Exercise B.12. Prove that the plane Π is the tangent plane to S at a_0 if and only if for all $\alpha, \beta \in \mathbb{R}$, $\alpha^2 + \beta^2 \neq 0$, we have

$$\lim_{h\to 0} d/\delta = 0. \tag{B.3}$$

Solution. Let n be the vector of the normal to Π. Then

$$\begin{aligned} d &= \langle f(u_0 + \alpha h, v_0 + \beta h) - f(u_0, v_0), n\rangle, \\ \delta &= \left|f(u_0 + \alpha h, v_0 + \beta h) - f(u_0, v_0)\right|. \end{aligned}$$

Then

$$\begin{aligned} \lim_{h\to 0} d/\delta &= \lim_{h\to 0}(\langle f(u_0 + \alpha h, v_0 + \beta h) - f(u_0, v_0), n\rangle \\ &\quad /\left|f(u_0 + \alpha h, v_0 + \beta h) - f(u_0, v_0)\right|) \\ &= \lim_{h\to 0}(\langle h(\alpha f_{t_1}(u_0, v_0) + \beta f_{t_2}(u_0, v_0)) + h\varepsilon, n\rangle \\ &\quad /\left|h(\alpha f_{t_1}(u_0, v_0) + \beta f_{t_2}(u_0, v_0)) + h\varepsilon\right|) \\ &= \pm(\langle \alpha f_{t_1}(u_0, v_0) + \beta f_{t_2}(u_0, v_0), n\rangle \\ &\quad /\left|\alpha f_{t_1}(u_0, v_0) + \beta f_{t_2}(u_0, v_0)\right|. \end{aligned}$$

Thus (B.3) holds if and only if

$$\langle \alpha f_{t_1}(u_0, v_0) + \beta f_{t_2}(u_0, v_0), n\rangle = 0. \tag{B.4}$$

1. Let Π be the tangent plane to S at $f(u_0, v_0)$. Then

$$n \perp f_{t_1}(u_0, v_0), \quad n \perp f_{t_2}(u_0, v_0),$$

   and (B.4) holds. Hence (B.3) holds.
2. Let (B.3) hold. Then (B.4) holds. For $\alpha = 1$ and $\beta = 0$, we get

$$\langle f_{t_1}(u_0, v_0), n\rangle = 0.$$

   For $\alpha = 0$ and $\beta = 1$, we get

$$\langle f_{t_2}(u_0, v_0), n\rangle = 0.$$

   Thus

$$n \perp f_{t_1}(u_0, v_0), \quad n \perp f_{t_2}(u_0, v_0),$$

   and Π is the tangent plane to S at $f(u_0, v_0)$. This completes the proof.

Definition B.14. The straight line passing through a point of the surface S that is perpendicular to the tangent plane of the surface at that point is called the normal to the surface S at the considered point.

Let $(U, f = f(u,v))$ be a surface S, $a = f(u_0, v_0) \in S$, and

$$f(u,v) = (f_1(u,v), f_2(u,v), f_3(u,v)), \quad (u,v) \in U.$$

Then the equations of the normal to S at a are given by the following equations:

$$\frac{X - f_1(u_0,v_0)}{\begin{vmatrix} f_{2u}(u_0,v_0) & f_{3u}(u_0,v_0) \\ f_{2v}(u_0,v_0) & f_{3v}(u_0,v_0) \end{vmatrix}} = \frac{Y - f_2(u_0,v_0)}{\begin{vmatrix} f_{3u}(u_0,v_0) & f_{1u}(u_0,v_0) \\ f_{3v}(u_0,v_0) & f_{1v}(u_0,v_0) \end{vmatrix}} = \frac{Z - f_3(u_0,v_0)}{\begin{vmatrix} f_{1u}(u_0,v_0) & f_{2u}(u_0,v_0) \\ f_{1v}(u_0,v_0) & f_{2v}(u_0,v_0) \end{vmatrix}}.$$

If the surface S is given by

$$z = f(x,y), \quad (x,y) \in U,$$

where $U \subset \mathbb{R}^2$, then the equation of the normal to S at (x_0, y_0, z_0), $(x_0, y_0) \in U$, $z_0 = f(x_0, y_0)$, is given by

$$\frac{x - x_0}{f_x(x_0,y_0)} = \frac{y - y_0}{f_y(x_0,y_0)} = \frac{z - z_0}{-1}.$$

If the surface S is given by

$$F(x,y,z) = 0, \quad (x,y,z) \in V,$$

where $V \subset \mathbb{R}^3$, then the equation of the normal at $(x_0, y_0, z_0) \in V$ is given by

$$\frac{x - x_0}{F_x(x_0,y_0,z_0)} = \frac{y - y_0}{F_y(x_0,y_0,z_0)} = \frac{z - z_0}{F_z(x_0,y_0,z_0)}.$$

Example B.19. Consider the surface in Example B.16. Then the equation of the normal at the point $(1, 1, 2)$ is

$$\frac{x-1}{1} = \frac{y-1}{2} = \frac{z-2}{-1}.$$

Example B.20. Consider the surface in Example B.17. Then the equation of the normal at the point $(3, 4, -12)$ is

$$\frac{x-3}{6} = \frac{y-4}{8} = \frac{z+12}{-24},$$

or

$$\frac{x-3}{3} = \frac{y-4}{4} = \frac{z+12}{-12}.$$

Example B.21. Consider the surface in Example B.18. Then the equation of the normal at the point (3, 5, 9) is

$$\frac{x-3}{\begin{vmatrix} 2 & 3 \\ 4 & 12 \end{vmatrix}} = \frac{y-5}{\begin{vmatrix} 3 & 1 \\ 12 & 1 \end{vmatrix}} = \frac{z-9}{\begin{vmatrix} 1 & 2 \\ 1 & 4 \end{vmatrix}},$$

or

$$\frac{x-3}{12} = \frac{y-5}{-9} = \frac{z-9}{2}.$$

Exercise B.13. Find the equation of the normal to the surfaces in Exercise 3.4 at the given point.

Answer B.3.

1.

$$\frac{x-1}{2} = \frac{y-1}{2} = \frac{z-2}{-1}.$$

2.

$$\frac{x+2}{8} = \frac{y-1}{8} = z-4.$$

3.

$$x-1 = \frac{y-1}{-2} = z-1.$$

4.

$$\begin{cases} x = 1, \\ y + 1. \end{cases}$$

5.

$$\frac{x+3}{23} = \frac{y-4}{-19} = \frac{z-17}{5}.$$

6.

$$x-1 = y-2 = \frac{z-2}{3}.$$

7.

$$x-1 = \frac{y-2}{11} = \frac{z+1}{5}.$$

8.

$$\frac{x-1}{2} = y-1 = \frac{z-2}{11}.$$

9.

$$\frac{x-2}{5} = \frac{y-3}{4} = z-6.$$

10.

$$x-2 = \frac{y-1}{2} = \frac{z}{0}.$$

11.

$$\frac{x-1}{6} = \frac{y-3}{3} = \frac{z-4}{-2}.$$

Exercise B.14. Let (x_0, y_0, z_0) be a point of the surface given by the equation

$$F(x,y,z) = 0.$$

Prove that the gradient vector

$$\operatorname{grad} F(x_0, y_0, z_0) = (F_x(x_0, y_0, z_0), F_y(x_0, y_0, z_0), F_z(x_0, y_0, z_0))$$

is perpendicular to the tangent plane of the surface at this point.

Solution. Let

$$f = (x(u,v), y(u,v), z(u,v))$$

be a local parameterization of the considered surface.Then

$$\begin{aligned} f_u &= (x_u, y_u, z_u), \\ f_v &= (x_v, y_v, z_v), \\ 0 &= F_x \cdot x_u + F_y \cdot y_u + F_z z_u \\ &= \langle \operatorname{grad} F, f_u \rangle, \end{aligned}$$

and

$$\begin{aligned} 0 &= F_x \cdot x_v + F_y \cdot y_v + F_z z_v \\ &= \langle \operatorname{grad} F, f_v \rangle. \end{aligned}$$

Thus,

$$\operatorname{grad} F \perp f_u, \quad \operatorname{grad} F \perp f_v.$$

This completes the solution.

Definition B.15. An orientation of a surface S is a choice of a normal vector n to $T_{\star a}S$.

Definition B.16. Let S be an oriented surface with orientation $n(a)$. A local parameterization (U,f) of S is said to be compatible if

$$n = (f_u \times f_v)/|f_u \times f_v|.$$

Index

https://doi.org/10.1515/9783112219607-010

www.ingramcontent.com/pod-product-compliance
Lightning Source LLC
Chambersburg PA
CBHW080710220326
41598CB00033B/5363